DFG

Essential
Biomonitoring Methods

The MAK-Collection

Current Volumes

Part I: MAK Value Documentations, Volume 21
2005. 3-527-31134-3

Part II: BAT Value Documentations, Volume 4
Drexler, H. (ed.)
2005. ISBN 3-527-27049-3

Part III: Air Monitoring Methods, Volume 9
Parlar, H. (ed.)
2005. ISBN 3-527-31138-6

Part IV: Biomonitoring Methods, Volume 10
Angerer, J. (ed.)
2006. ISBN 3-527-31137-8

The MAK-Collection online
www.mak-collection.com

A special selection to start with...

Essential MAK Value Documentations
Greim, H. (ed.)
2006. ISBN 3-527-31394-X

Essential BAT Value Documentations
Drexler, H./Greim H. (eds.)
2006. ISBN 3-527-31477-6

Essential Air Monitoring Methods
Parlar, H. / Greim, H. (eds.)
2006. ISBN 3-527-31476-8

DFG Deutsche Forschungsgemeinschaft

Essential Biomonitoring Methods

from the MAK-Collection for Occupational Health and Safety

Edited by Jürgen Angerer
Working Group Analytical Chemistry

Commission for the Investigation of Health Hazards of Chemical
Compounds in the Work Area
(Chairman: Helmut Greim)

WILEY-VCH

WILEY-VCH Verlag GmbH & Co. KGaA

Prof. Dr. Helmut Greim
DFG-Senatskommission zur Prüfung
gesundheitsschädlicher Arbeitsstoffe
Hohenbachernstr. 15–17
85354 Freising
Germany

Prof. Dr. Jürgen Angerer
Institut für Arbeits-, Sozial-
und Umweltmedizin
Universität Erlangen-Nürnberg
Schillerstr. 25/29
91054 Erlangen
Germany

Important Notice

This volume presents a selection of documentations or methods taken from published volumes of the MAK-Collection. These include occupational exposure values (MAK and BAT values, EKA, BLW) or classifications which may have changed since the publication of the respective volume. Readers are advised to consult the latest edition of the "List of MAK and BAT Values", which is published annually both in print and online (www.mak-collection.com), for current occupational exposure values and classifications.

Library of Congress Card No.: applied for

British Library Cataloging-in-Publication Data: A catalogue record for this book is available from the British Library.

Bibliographic information published by Die Deutsche Bibliothek
Die Deutsche Bibliothek lists this publication in the Deutsche Nationalbibliografie; detailed bibliographic data is available in the Internet at http://dnb.ddb.de.

Typesetting K+V Fotosatz GmbH, Beerfelden
Printing betz-druck GmbH, Darmstadt
Binding Litges & Dopf Buchbinderei GmbH, Heppenheim

Printed in the Federal Republic of Germany
Printed on acid-free paper

ISBN-13: 978-3-527-31478-2
ISBN-10: 3-527-31478-4

Preface

The year 2005 has marked both the 50th anniversary of the *Commission for the Investigation of Health Hazards of Chemical Compounds in the Work Area* of the Deutsche Forschungsgemeinschaft (DFG, German Research Foundation) as well as the launch of the new international reference series, the *MAK-Collection for Occupational Health and Safety*. Together with the annual *List of MAK and BAT Values*, this series comprises the Commission's English language publications, which are available both in print and online. The series has four parts, each part continuing an established book series:

- MAK Value Documentations (prior title: *Occupational Toxicants*)
- BAT Value Documentations (prior title: *Biological Exposure Values for Occupational Toxicants*)
- Air Monitoring Methods (prior title: *Analyses of Hazardous Substances in Air*)
- Biomonitoring Methods (prior title: *Analyses of Hazardous Substances in Biological Materials*)

For those unfamiliar with these publications and the wealth and quality of information offered through the *MAK-Collection*, we have an additional new design feature on offer: special editions, named *Essentials* with 20 highlights from each of the four parts of the Collection. These *Essentials* are particularly relevant for professionals and researchers in the fields of occupational health and safety. This *Essentials* features a special selection of *Biomonitoring Methods* taken from previous volumes of the *MAK-Collection*. The selected methods are exemplary for the standards maintained by the Commission, which works strictly according to scientific criteria. Also included is a general chapter on the use of gas chromatography – mass spectrometry in biological monitoring. We hope that this selection will prove useful for fostering occupational health and safety as well as toxicological research and analytical methodology around the globe.

We would like to thank the members and guests of the Commission's working group "Analytical Chemistry" and its subgroup "Analyses of Hazardous Substances in Biological Materials", who have contributed with all their expertise to the respective documentations. Thanks go also to the Commission's scientific secretariat who have thoroughly reviewed the documentations and for the assistance of the translators, all of whom have made them ready for publication with outstanding accuracy. We would also like to thank the publisher Wiley-VCH for editorial guidance, swift production and world-wide distribution. Last, but not least, we are thankful for the continuing support of the Deutsche Forschungsgemeinschaft (DFG).

December 2005

J. Angerer
Chairman of the Working Group
"Analytical Chemistry"

H. Greim
Chairman of the Commission for the
Investigation of Health Hazards of
Chemical Compounds in the Work Area

Essential Biomonitoring Methods. DFG, Deutsche Forschungsgemeinschaft
Copyright © 2006 WILEY-VCH Verlag GmbH & Co. KGaA, Weinheim
ISBN: 3-527-31478-4

Contents

General Aspects

Substances

Essential Biomonitoring Methods. DFG, Deutsche Forschungsgemeinschaft
Copyright © 2006 WILEY-VCH Verlag GmbH & Co. KGaA, Weinheim
ISBN: 3-527-31478-4

General Aspects

Essential Biomonitoring Methods. DFG, Deutsche Forschungsgemeinschaft
Copyright © 2006 WILEY-VCH Verlag GmbH & Co. KGaA, Weinheim
ISBN: 3-527-31478-4

Preliminary remarks

Introduction

It is the duty of occupational and environmental medicine to evaluate the health risk posed by hazardous chemicals in order to guard against impairment of health. In the last 20 years biological monitoring has proven extremely valuable for this purpose.

The following definition of biological monitoring, which has been adopted in a slightly modified form by the "Commission for the Investigation of Health Hazards of Chemical Compounds in the Work Area" of the Deutsche Forschungsgemeinschaft, was formulated by the international seminar "Assessment of Toxic Agents at the Work Place", organized in 1980 by *CEC, OSHA*, and *NIOSH*. This definition reflects how biological monitoring is viewed by the countries of the European Community, where it has been practiced for a number of years:

"Biological monitoring is the directed systematic continuous or repetitive health-related activity for collection of biological samples for the measurement and assessment of hazardous chemical compounds, their metabolites of their specific biochemical effect-parameters. The objective is to evaluate the exposure and health risk of exposed persons by comparing the obtained data with appropriate reference values, leading to corrective actions if necessary."

The difficulties of biological monitoring are primarily analytical because occupational and environmental medicine must measure
 − a minute quantity of substance (down to the picogram level)
 − in a small volume (a few milliliters)
 − of complex specimen material (blood, urine, etc.)
This demands a high-performance analytical method. Often only the most up-to-date instrumental techniques are capable of achieving the required level of specificity and sensitivity. The lower the amount of analyte present and the higher the instrumental complexity, the more important is a continuous check on analytical reliability. For this reason both internal and external quality control are essential for investigations within the realm of occupational and environmental toxicology. The so-called preanalytical phase (specimen collection, transport, storage, sample aliquotation etc.) is also important within the framework of trace analyses and deserves more attention than it has received up to now.

An evaluation of the results is necessary from an analytical as well as a medical viewpoint and should be based on the collaborative efforts of a body of experts.

Analysis in biological materials

Biological specimen

The biological material used for the analysis must be representative of the exposure to the particular hazardous substance. The concentration of the substance in the critical organ would provide the optimum measure for an individual exposure, but this is usually not

Essential Biomonitoring Methods. DFG, Deutsche Forschungsgemeinschaft
Copyright © 2006 WILEY-VCH Verlag GmbH & Co. KGaA, Weinheim
ISBN: 3-527-31478-4

possible. Furthermore, within the framework of biological monitoring, the collection of the biological specimen should not place any noticeable strain on the donor.

The preferred biological materials for occupational health are therefore

– blood
– urine
– exhaled air.

The determination of a hazardous substance in the blood is one of the diagnostically most reliable ways to quantify an exposure. Since the organ concentrations of the hazardous substance are maintained in equilibrium with that of the circulating blood, its concentration in the critical organ can be estimated from its concentration in the blood. Moreover, the risk of exogenous contamination or manipulation of the specimen is considerably less for blood than for other biological test materials.

Urine has the advantage of being easier to collect and more available than blood. However, urinary analyses also have drawbacks stemming from, for example, the different functions of the kidneys, the variation in volumes drunk and excreted, and the fact that the metabolic products of some industrial substances are also excreted physiologically. Under working conditions, spontaneous urine has to be used because unfortunately the collection of urine over a 24 h period is usually impracticable. This leads to difficulties in the interpretation of analytical results from spontaneous urine specimens. Parallel analyses of a reference para-meter such as creatinine, osmolality, or density are in practice still unable to resolve this problem satisfactorily. This is especially true for very dilute or highly concentrated urine specimens. Parallel analyses of the above-mentioned parameters could, however, make possible the preselection of urine specimens suitable for occupational health studies.

From an analytical and medical viewpoint the analysis of exhaled air offers several advan-tages. However, because of difficulties in collection, transport, and storage the analysis of exhaled air is still of limited practical importance.

In structures like hair, fingernails, and teeth an exposure to a hazardous chemical can be determined only after a longer period of time. Hence these materials are far better suited for environmental than occupational health studies.

Toxicological investigation in occupational health

In occupational medicine, as for the field of medicine in general, the goal of every investiga-tion is a diagnosis related to the individual patient. In the case of an exposure to hazardous (chemical) industrial materials this diagnosis is usually based on a chemical analysis of the substance, its metabolites, or other parameters of its intermediary metabolism in the biologi-cal material and on the analytical-medical evaluation of the obtained value.

Thus, in occupational medicine a diagnosis is reached essentially in three steps:

– the preanalytical phase
– the analytical determination and assessment
– the medical evaluation.

Preanalytical phase

There is a growing recognition of the importance of the preanalytical phase for the accuracy of the results. Any factor that can alter the analytical result between the time of specimen collection up to its analysis in the laboratory is termed an *interference factor*. Such a change may occur for example as a result of evaporation or chemical transformation of the analyte. Interference factors are independent of the reliability of the method used and are, therefore, not covered by the statistical quality control. They can be minimized only by the appropriate standardization of:

– specimen collection (syringes, needles, containers, etc.)
– transport and storage (duration, temperature, etc.)
– aliquotation of the biological specimen (homogenization, etc.).

Which preventive measures are to be taken during the preanalytical phase depends on the kind of biological test material as well as the substances to be determined. To satisfy these requirements and the practical demands of monitoring, the preanalytical phase should be designed to be applicable to as many analytes as possible.

The Working Group "Analytical Chemistry" has worked out some recommendations along these lines. These suggestions for occupational health and toxicological monitoring investigations on urine and blood specimens can be summarized as follows:

Time of specimen collection. Whole blood or spontaneous urine specimens should be collected at the end of the work shift, if possible after 3 workdays. For special working materials different conditions for specimen collection have to be accepted. Information pertaining to this can be found in the annual report of the Commission of the Deutsche Forschungsgemeinschaft for the Investigation of Health Hazards of Chemical Compounds in the Work Area and in the most recent Arbeitsmedizinisch-toxikologische Begründung von BAT-Werten (Occupational Hygiene and Toxicological Justification of BAT Values).

Collection of urine specimens. Plastic containers (about 200 mL, wide-mouthed) that have been rinsed with acid are used for urine collection. The urine is voided directly into the container. Care must be taken that the hands have been washed and the work clothes exchanged for street clothes before the specimen is collected. Contamination by dust, gas, or vapor from the work area must be strictly avoided.
Specimen volume: If possible the volume of urine should be at least 50 mL.

Collection of whole blood specimens. Venous blood to which anticoagulant has been added is used for the analytical investigations. *Coagulation must be scrupulously avoided by thorough rotation of the specimen containers!*
Determination of inorganic substances (metals):
Disposable needles, syringes, and containers (e.g. Monovettes and Vacutainers), can be used for collection. Monovettes and Vacutainers already contain anticoagulants (e.g., K-EDTA) in the required amounts. They can also serve as containers for transport and storage.
Specimen volume: For most analyses 5 mL whole blood is sufficient, but the volume of blood should be at least this amount.
Determination of highly volatile organic substances:
Disposable syringes (5 mL) and needles are used for collection. Usually disinfection of the puncture site with alcohol must be waived.
Specimen volume: A 4 mL sample of venous blood from the arm is divided equally between two 20 mL "head-space sample flasks" containing an anticoagulant, e.g., 50 mg ammonium oxalate, and the flasks are sealed airtight with PTFE-coated butyl rubber stoppers. The flasks serve also as storage and transport containers.

In a single case it may be necessary to use a procedure that deviates from the one described here. It is then presented in detail in the method descriptions.
Storage and shipping of the specimens. Arrangements should be made to ship the blood and urine specimens as soon as possible after collection. If it is not possible to ship them right after collection, the specimens can be stored at 4 °C for up to 5 days.

Sample aliquotation. A major source of error in the analysis of blood and urine is the inhomogeneity of the biological material due to coagulation or sedimentation. To avoid gross analytical errors appropriate precautions must be observed in the treatment of the blood and urine specimens to ensure that the analyzed portion (sample) is representative of the specimen.

Blood

Blood specimens to be analyzed for nonvolatile organic and inorganic constituents are homogenized by careful mixing. The so-called roller mixer has proven especially well suited for this purpose. The specimen containers are placed on their side in the roller mixer and are turned continuously around their long axis. At the same time they are tipped slightly back and forth so that the test material is swished around in the container. After the blood specimens have been mixed on this apparatus for 1 h, a sample is taken for analysis.

To measure volatile substances in the blood the head-space sample flasks containing the specimens are incubated at a constant temperature of, for example, 60 °C. By 60 min the concentration of the substance in the blood and in the head space of the flask have reached an equilibrium.

Urine

Before aliquotation the urine specimens are agitated on a shaking machine for 30 min. Directly before the samples are pipetted, each urine specimen must be shaken thoroughly again by hand.

Analytical determination and assessment

Laboratories studying problems concerning occupational health and toxicology rely mainly on those techniques of analytical chemistry that distinguish themselves by especially high specificity and selectivity.

The quality of the investigative method can be characterized by reliability criteria. Practical application of a method must be accompanied by statistical quality control. This is the only way to ensure that the analytical results are consistently reliable and is also a prerequisite for the comparison of results from different laboratories.

Analytical reliability criteria

The analytical reliability of a method can be satisfactorily described by the criteria
– sensitivity
– imprecision
– inaccuracy
– detection limit
– specificity.

Sensitivity

In chemical analysis the sensitivity H is the differential quotient of the calibration function. A graph of the calibration function is obtained by plotting the observed value of the measure x (ordinate) as a function of the analyte concentration c (abscissa). The sensitivity H is then the slope of this calibration function:

$$H: = \frac{\mathrm{d}x}{\mathrm{d}c}$$

Imprecision

Imprecision is a measure for the reproducibility of the results from a given experimental design. The 1980 IFCC definition for imprecision is:
"Imprecision: Standard deviation or coefficient of variation of the results in a set of replicate measurements. The mean value and number of replicates must be stated, and the design used must be described in such a way that other workers can repeat it. This is particularly important whenever a specific term is used to denote a particular type of imprecision, such as within-day, between-day, or between-laboratory."

To establish the **within-series imprecision** one person performs several successive analytical determinations on a ready-made sample pool using the same reagents, instruments, and technical aids. Within-series imprecision is a measure for the reproducibility of an individual determination under identical conditions. The length of the series should be set so that the possibility of time-dependent effects can be excluded, but the number of analyses (n) should be at least 10. Ideally this check on imprecision should be carried out at three different concentrations, chosen to encompass as much of the linearity range of the method as possible. Preferably the specimens should be from donors who have been exposed to the hazardous substance in question.

To establish the **between-day imprecision** determinations on the same material are carried out on different days. The preparation of the sample pool should be described in detail, and information should be provided as to whether the same reagents, standards, and instruments were used and whether all determinations were performed by the same person.

The **between-laboratory imprecision** is given for some of the methods. In these cases samples from the same pool were analyzed by workers in different laboratories.

Imprecision is defined by the relative standard deviation s and the prognostic range u. The **relative standard deviation** s is the standard deviation relative to the mean and is given in percent.

The **prognostic range** u defines an interval that contains the analytical result for an identical sample with a probability $P = 95\%$. Due to practical considerations the prognostic range u is given in relative units (percent) with respect to the analytical results. The prognostic range u is determined by the Student's t_p factor for $P = 95\%$ and the standard deviation s of the complete method, which can be calculated in two ways:

1. By replicate analyses at a given analyte concentration (standard deviation s_w)

$$s_w = \sqrt{\frac{\sum\limits_{i=1}^{n}(c_i - \bar{c})^2}{n-1}} \cdot \frac{1}{\bar{c}}$$

where

n = number of analyses
c_i = analytical result of the ith analysis
$\bar{c}$ = mean of n analytical results

2. By double analyses at different concentrations within a defined concentration range (standard deviation s_d)

$$s_d = \sqrt{\frac{\sum\limits_{i=1}^{n} (c_{i1} - c_{i2})^2}{2z}} \cdot \frac{1}{\bar{c}}$$

where

z = number of double analyses
c_{i1}, c_{i2} = results of the ith double analysis
$\bar{c}$ = mean of the $2z$ analytical results

For the first case (replicate analyses) the prognostic range is given by:

$u = t_p \cdot s_w$

For the second case (double analyses) the prognostic range is given by:

$u = t_p \cdot s_d$

where t_p = Student's t factor for $P = 95\%$.
The effect of the number of analyses, n or z, on the Student's t factor for $P = 95\%$ is shown in the following table.

$n-1$ or z	5	10	15	20	25	30	35	50	60
t_p	2.57	2.23	2.13	2.09	2.06	2.04	2.03	2.01	2.00

Since imprecision can be effected not only by the chemical preparation of the biological material and the physical technique used but also by individual variations in the composition of the sample matrix, the second method for calculating imprecision is preferable. The biological samples from exposed donors should cover as much as possible of the range of concentrations relevant to occupational medicine.

Inaccuracy

The inaccuracy of a method is defined by the deviation of the analytical results from the true value. It is a measure for the agreement between the amount of substance actually present and the amount determined by the analysis. The IFCC gives the following definition for inaccuracy:
"Inaccuracy: Numerical difference between the mean of a set of replicate measurements and the true value. This difference (positive or negative) may be expressed in the units in which the quantity is measured, or as a percentage of the true value."
The following ways of determining the inaccuracy of a method are arranged in descending order of validity as applied to this criterion:

- Comparison of results from the method in question with results from a definitive method.
- On the basis of reference materials having a certified value for the concentration of analyte in the matrix to be analyzed.
- Comparison of results from the method in question with results from a second method that differs as much as possible from the first in all procedural steps.

- By recovery experiments. In this case the analyte must be added to the biological material in at least three concentrations, covering the concentration range of interest. The analysis must be based on aqueous calibration standards, and the experimental conditions should be described in detail.

Detection limit

A generally applicable definition of the limit of detection must be based on mathematical statistics, involving a "signal-to-noise ratio" (in a generalized sense).

At very low concentrations it is uncertain whether an observed value is due to the presence of the substance to be determined or to uncontrolled chance perturbations. Perturbations can be caused by umpurities in reagents, losses from sputtering or absorption, weighing or titration errors, temperature fluctuations in light sources, secondary reactions, thermal electronic noise in amplifiers, observational error, and contamination. **All** such uncontrolled fluctuations are subsumed under the concept of **"analytical noise"**. This "analytical noise" can be grasped only in statistical terms. For example, it is essentially the scatter of the **"blank measures"** from "blank analyses" and can be numerically described by the standard deviation $s_{bl,abs}$ of the blank measures.

Because the blank scatter of an analytical procedure is a very complex phenomenon, it cannot be calculated by theory. However, in practice it can be found **experimentally** by making a sufficiently larger number of blank analyses (at least 20) and then treating the measured x_{bl} statistically, i.e., the mean x_{bl} and the absolute standard deviation $s_{bl,abs}$ are calculated.

Whether an observed value of the measure x can be accepted as genuine or whether it must be rejected because it is suspected to be only an accidentally high blank measure is decided according to the following inequality:

$$x \geq \overline{x}_{bl} + 3\,s_{bl,abs}$$

The detection limit can be read from the calibration curve as the concentration, corresponding to the lowest observed measure, that still satisfies this inequality. This is, therefore the **lowest concentration that is acceptable with the given analytical procedure.**

This definition of detection limit may appear questionable when the observed reagent blank values are of the same order of magnitude as the concentrations to be measured in the biological material. The Working Group "Analytical Chemistry" of the Commission of the Deutsche Forschungsgemeinschaft for the Investigation of Health Hazards of Chemical Compounds in the Work Area is considering this problem and discussing a redefinition of detection limit as the lowest still detectable concentration. This would be accompanied by the introduction of the concept *"determination limit"* to designate the lowest concentration that can still be determined with the degree of reliability that characterizes the method.

Specificity

Specificity refers to the power of a method to differentiate one chemical compound from other potentially detectable compounds. The specificity of an analytical method depends heavily on the type of chemical procedure used for sample preparation and on the physicochemical principles of the method.

In 1980 the IFCC gave the following definition for specificity:
"Specificity: The ability of an analytical method to determine solely the component(s) it purports to measure. It has no numerical value. It is assessed on the evidence available on the components that contribute to the result and on the extent to which they do."

Statistical quality control

Statistical quality control in the Federal Republic of Germany follows the guidelines established by the Bundesärztekammer (German Medical Association) in 1971. Similar recommendations have also been formulated by other countries.
Detailed procedural directives can be found in, e.g., Regulation TRgA 410 of the Arb-StoffV; Richtlinien der Bundesärztekammer (1971); and *Angerer* and *Schaller* (1976). Quality control provides for:

internal quality control: the continual monitoring of
– imprecision
– inaccuracy

external quality control: analysis of the same specimens by different laboratories to check
– inaccuracy

Internal quality control. A control sample, whose analyte concentration is known to the analyst, is processed with each analytical series. These control values are recorded in a control chart with their mean and the two- and threefold standard deviations of the mean (Fig. 1). A control sample with an analyte concentration known only to the director of the laboratory is analyzed after every fourth series.

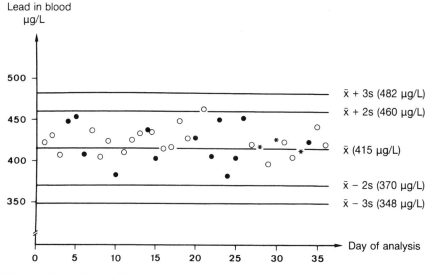

Figure 1. Control chart (Determination of lead in blood, see p. 161, internal quality control).

External quality control. The inaccuracy of the results is checked by round-robin experiments on the national and international level in which the same specimens are analyzed by different laboratories. The analyte concentration of these circulating specimens is known only to the organizer of these experiments.

In the Federal Republic of Germany statistical quality control falls under the jurisdiction of the Arbeitsstoffverordnung (Code on Hazardous Working Materials), TRgA 410 (Regulation 410). Control materials are commercially available for these purposes in concentration ranges relevant to environmental and occupational medicine (Behring, Marburg).

When no control material is commercially available for the determination of a particular parameter, control samples must be prepared in the laboratory from pooled biological material. A year's supply of such samples can be made up at one time and stored in the deep-freeze until needed.

Evaluation and interpretation of analytical results in biological material

In the evaluation and interpretation of analytical results in occupational medicine and toxicology certain particulars specific to these fields must be taken into consideration. An example of this is the variation from individual to individual of sensitivity and metabolizing ability. In addition to such individual-specific parameters, in occupational medicine an evaluation is based on biological limit values. These are set on the basis of the relationships between external exposure to internal exposure or between internal exposure to effects. In the Federal Republic of Germany this has been codified as the BAT value (Biological Tolerance Value for Working Materials).

The BAT values are based on the Arbeitsmedizinisch-toxikologische Begründung von BAT-Werten (Occupational Hygiene and Toxicological Justifications of BAT Values) published by the Commission of the Deutsche Forschungsgemeinschaft for the Investigation of Health Hazards of Chemical Compounds in the Work Area. A list of updated BAT values is put out yearly.

Supplementary literature:

General:
- *A. Berlin, R. E. Yodaiken,* and *D. C. Logan:* Summary report of the international seminar on the assessment of toxic agents at the workplace. Roles of ambient and biological monitoring, Luxembourg, 8 – 12 December, 1980. Int. Arch. Occup. Environ. Health *50, 197 – 207* (1982)
- *R. Roi, W. G. Town, W. G. Hunter,* and *L. Alessio:* Occupational health guidelines for chemical risk. Commission of the European Communities. EUR 8513, Brüssel/Luxemburg, 1983
- *R. R. Lauwerys* (ed.): Industrial chemical exposure: Guidelines for biological monitoring. Biomedical Publications, Davis, California, 1983

Preanalytical phase:

- *J. Angerer, K. H. Schaller,* and *H. Seiler:* The pre-analytical phase of toxicological monitoring examinations in occupational medicine. TrAC (Trend in Analytical Chemistry) *2,*
257–261 (1983)

Reliability criteria:

- *J. Büttner, R. Borth, J. H. Boutwell, P. M. G. Broughton,* and *R. C. Bowyer* (International Federation of Clinical Chemistry, Committee on Standards, Expert Panel on
Nomenclature and Principles of Quality Control in Clinical Chemistry): Approved recom-
mendation (1978) on quality control in clinical chemistry. Part 1. General principles and
terminology. J. Clin. Chem. Clin. Biochem. *18,* 69–77 (1980)
- *International Union of Pure and Applied Chemistry,* Analytical Chemistry Division, Commission on Spectrochemical and Other Optical Procedures for Analysis: Nomenclature,
symbols, units and their usage in spectrochemical analysis. II. Terms and symbols related
to analytical functions and their figures of merit. Information Bulletin, Appendices on
Tentative Nomenclature, Symbols, Units and Standards. Number 26, Oxford 1972

Statistical quality control:

- TRgA 410: Statistische Qualitätssicherung. Ausgabe April 1979. In: *W. Weinmann,* and
H.-P. Thomas: Arbeitsstoffverordnung Teil 2: Technische Regeln (TRgA) und ergän-
zende Bestimmungen zur Verordnung über gefährliche Arbeitsstoffe. Carl Heymanns
Verlag KG, Köln, 8th supplement 1982

- *J. Angerer* and *K. H. Schaller:* Erfahrungen mit der statistischen Qualitätskontrolle im
arbeitsmedizinisch-toxikologischen Laboratorium. Arbeitsmed. Sozialmed. Präventivmed. *11,* 311–312 (1976) and *12,* 33–35 (1977)
- Richtlinien der Bundesärztekammer zur Durchführung der statistischen Qualitätskontrolle und von Ringversuchen im Bereich der Heilkunde. Deutsches Ärzteblatt, Heft 31,
2228–2231 (1971)
- *D. Stamm:* A new concept for quality control of clinical laboratory investigations in the
light of clinical requirements and based on reference method values. J. Clin. Chem. Clin.
Biochem. *20,* 817–824 (1982)

Evaluation and interpretation:

- *Deutsche Forschungsgemeinschaft:* Maximale Arbeitsplatzkonzentrationen und Biologische Arbeitsstofftoleranzwerte. Mitteilung XIX der Senatskommission zur Prüfung
gesundheitsschädlicher Arbeitsstoffe. Verlag Chemie, Weinheim 1983
- *D. Henschler* and *G. Lehnert (ed.):* Biologische Arbeitsstoff-Toleranz-Werte (BAT-
Werte). Arbeitsmedizinisch-toxikologische Begründungen. Verlag Chemie, Weinheim,
1st edition 1983
- *American Conference of Governmental Industrial Hygienists:* Threshold limit values for
chemical substances in the work environment, 1983–84. Cincinnati, U.S.A. 1983

The use of gas chromatography-mass spectrometry in biological monitoring

Contents

Abbreviations

ADBI	4-Acetyl-1,1-dimethyl-6-*tert.*-butyldihydroindene
AHDI	6-Acetyl-1,1,2,3,3,5-hexamethyldihydroindene
AHTN	7-Acetyl-1,1,3,4,4,6-hexamethyltetrahydronaphthalene
CI	Chemical ionisation
CAD	Collision-activated decomposition
CID	Collision-induced dissociation
DCP	Dichlorophenol

Essential Biomonitoring Methods. DFG, Deutsche Forschungsgemeinschaft
Copyright © 2006 WILEY-VCH Verlag GmbH & Co. KGaA, Weinheim
ISBN: 3-527-31478-4

DDE	Dichlorodiphenyldichloroethene
DDT	Dichlorodiphenyltrichloroethane
ECD	Electron capture detector
EI	Electron impact ionisation
eV	Electron volt
FID	Flame ionisation detector
FPD	Flame photometric detector
GC	Gas chromatography
GC/MS	Gas chromatography/mass spectrometry
HCB	Hexachlorobenzene
HCH	Hexachlorocyclohexane
HHCB	1,3,4,6,7,8-Hexahydro-4,6,6,7,8,8-hexamethylcyclopenta-[g]-2-benzopyrane
HMEPMU	1-(4-(1-Hydroxy-1-methylethyl)phenyl)-3-methylurea
HPLC	High performance liquid chromatography
IP	Ionisation potential
m/z	Mass/charge ratio
MALDI	Matrix-assisted laser desorption ionisation
MEV	N-Methylvaline
MRM	Multiple reaction monitoring
MS	Mass spectrometry
MSD	Mass selective detector
MS/MS	Tandem mass spectrometer
NCI	Negative chemical ionisation
PAH	Polycyclic aromatic hydrocarbon
PCB	Polychlorinated biphenyls
PCI	Positive chemical ionisation
PFPTH	Pentafluorophenylthiohydantoin
PND	Phosphorus nitrogen detector
rpm	Revolutions per minute
SIM	Selected ion monitoring
SRM	Selected reaction monitoring
TIC	Total ion chromatogram
TCD	Thermal conductivity detector
TCP	Trichlorophenol
TeCP	Tetrachlorophenol

1 Introduction

In recent years gas chromatography coupled with mass spectrometric detection (GC/MS) has become a standard technique for the assay of organic compounds and their metabolites in biological material. Whereas capillary gas chromatography has already been in use for some time, e.g. in combination with flame ionisation detectors (FID), electron capture detectors (ECD) or substance-specific detectors (such as the phosphorus nitrogen detector, PND), mass spectrometry has become established as a routine technique only as a consequence of price reductions in the last ten years. This is especially true for the widely used "quadrupole" instruments with a performance that is often sufficient to meet the requirements of investigations in the fields of occupational and environmental medicine. However, in special cases more powerful systems, which are technically more complex and more expensive are required, and such systems have been employed only in relatively few laboratories to date.

With the establishment of GC/MS occupational and environmental medicine now have a technique that permits specific, sensitive and reproducible measurement in the trace range, i.e. in the ppm, ppt and sub-ppt range. Thus mass spectrometry is the only detection method combined with gas chromatography to meet all the key requirements for the qualitative and quantitative trace analysis of organic compounds. However, the most important limitation of GC/MS analysis arises from the coupling of mass spectrometry with gas chromatography, i.e. the range of substances that can be analysed is restricted to those substances that can be vaporised without being decomposed.

Gas chromatography has already been comprehensively described as a separation method in combination with various detectors in the "Analyses of Hazardous Substances in Biological Materials" series [1]. Therefore this review of GC/MS analysis focuses on mass spectrometry as a detection method. In addition to the presentation of the underlying physico-chemical principles and current instrumental technology, special aspects of the coupling and method development are dealt with here.

2 Mass spectrometry in occupational and environmental chemical analysis

The origins of mass spectrometric analysis can be traced back to the studies of Thomson on the separation and detection of the neon isotopes 20 and 22 in 1910 [2]. Four years earlier Tswett had first successfully performed the chromatographic separation of leaf pigments on calcium carbonate [3]. This principle in the form of liquid chromatography (HPLC) and gas chromatography (GC) is one of the most efficient and versatile separation methods, and advances continue to be made in its development. Mass spectrometry was predominantly used for investigations in nuclear physics, e.g. for the separation of isotopes, until after the Second World War. Electron impact ionisation, ion focusing by means of electromagnetic lenses, ion separation in an electromagnetic field and detection by means of photoplates were developed in this context [4].

The publications of McLafferty et al. in the early 1960s triggered systematic research on fragmentation reactions and their assignment to intramolecular and intermolecular reaction mechanisms. Instrumental techniques were also developed during that period and they have been continually improved since then, e.g. electrical and magnetic sector field separation for high resolution, quadrupole analysers and time-of-flight mass spectrometers. Today access to these developments is no longer restricted to highly specialised research laboratories. Towards the end of the 1980s simple GC/MS systems with packed separation columns were devised and made affordable for a wide range of users [4].

With the introduction of fused silica capillary columns and the technical advances in mass spectrometry for routine investigations, GC/MS analysis has made great progress in every aspect (e.g. reduced space requirements, powerful vacuum systems, ionisation techniques), and it is now in widespread use. However, mass spectrometry has prevailed as a standard procedure in occupational and environmental medicine only in recent years. This tendency is clearly demonstrated in the series "Analysis of Hazardous Substances in Biological Materials" by the Working Group of the Deutsche Forschungsgemeinschaft's Commission for the Investigation of Health Hazards of Chemical Compounds in the Work Area (Fig. 1).

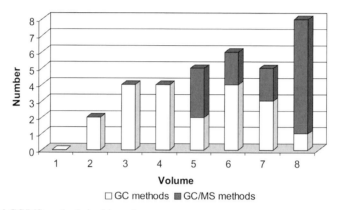

Fig. 1. GC and GC/MS methods in this series

Gas chromatography, generally in combination with flame ionisation or electron capture detectors, was described as a separation method for biological monitoring from the very beginning of the series. From 1975 to 1981 an average of five GC methods were published in each volume. The first GC/MS method to be included in the German loose-leaf collection of methods was the assay of thiodiacetic acid, a metabolite of vinyl chloride, in 1982 [5]. In this case gas chromatography was still performed using a packed column.

In general, collective methods for GC analysis, e.g. to determine aromatic, aliphatic and chlorinated hydrocarbons in blood by means of headspace gas chromatography, were published until the 2nd volume in this series in 1988. Since then the number of GC methods has risen with every volume, but further GC/MS methods were only included in the 5th volume of this collection in 1996. Seven of the eight GC methods published in the 8th volume in 2003 were based on the use of a mass spectrometer as a detector.

Mass spectrometry is of special significance for biological monitoring on account of its high substance specificity and its great power of detection. As limit values have been reduced at the workplace and on account of the generally low levels of expo-

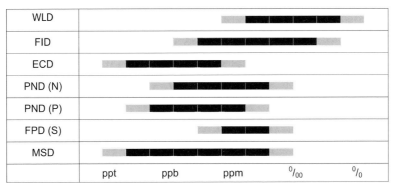

Fig. 2. Working ranges of various detectors used with gas chromatography (TCD thermal conductivity detector, FID flame ionisation detector, ECD electron capture detector, PND phosphorus nitrogen detector, FPD flame photometric detector, MSD mass selective detector) (according to [6])

sure to xenobiotics from environmental sources, demand has continued to rise for methods to enable sure and reliable determination of the correspondingly lower substance concentrations in the body. Although other detectors achieve detection limits of magnitudes similar to those of a mass spectrometer (Fig. 2), GC/MS exhibits a relatively wide dynamic working range in which the signals show a linear relationship with respect to the substance concentration [2].

However, the instrumental parameters of mass spectrometry can be optimised to ensure compatibility with the target molecule to a large extent, and therein lies the decisive practical advantage of this technique. Molecules or molecular fragments can even be selectively enriched in the mass filter by means of the "ion trap" technique before detection. This technical option enables the measured signal to be intensified, while the analytical background noise is minimised. The combination of gas chromatographic separation on capillary columns with mass spectrometry permits trace analysis in complex matrices, e.g. in blood and urine, which are the samples of choice for investigations in the fields of occupational and environmental medicine.

The methods that have been published in this series to date were therefore developed for the assay of such compounds of relevance at the workplace and in the environment, which are present in very low concentrations and for which the risk of interference from concomitant substances in the matrix is very high (Table 1). This applies particularly to the analysis of dioxins and furans, organochlorine compounds and "protein adducts" (N-terminal alkylated globins, aromatic amines).

A literature search on the use of GC/MS for biological monitoring since 2000 (keywords: GC-MS, biological monitoring, urine, blood) showed a similar focus on this application (Table 2). In addition to the aromatic amines, which are detected in their acetylated form in urine or as haemoglobin adducts from a complex matrix (e.g. Kaaria et al. 2001 [21]; Sabbioni et al. 2001 [22]; Weiss and Angerer 2002 [23]; Sennbro et al. 2003 [24]), various procedures for the assay of polychlorinated biphenyls, dioxins, furans and other organochlorine compounds have been described (e.g. Dmitrovic et al. 2002 [25]; Turci et al. 2003 [27]; Focant et al. 2004 [28]).

Table 1. GC/MS methods in this series (including the 8th volume) (DCP: dichlorophenol, TCP: trichlorophenol, TeCP: tetrachlorophenol, DDT: dichlorodiphenyltrichloroethane, DDE: dichlorodiphenyldichloroethene, HCB: hexachlorobenzene, HCH: hexachlorocyclohexane, HMEPMU: 1-(4-(1-hydroxy-1-methylethyl)-phenyl)-3-methylurea, PCB: polychlorinated biphenyls)

Parameter	Working material	Author
Thiodiacetic acid	Vinyl chloride	Müller (1982) [5]
S-Phenylmercapturic acid	Benzene	Müller and Jeske (1996) [7]
Haemoglobin adducts of alkylating compounds	Ethylene oxide, acrylnitrile, methylating substances	van Sittert et al. (1996) [8]
Chlorophenoxycarboxylic acids	Chlorophenoxycarboxylic acids	Krämer et al. (1996) [9]
Pyrethroid metabolites	Pyrethroids	Angerer et al. (1996) [10]
Pentachlorophenol	Pentachlorophenol	Hoppe (1999) [11]
2,4-, 2,5-, 2,6-DCP 2,3,4-, 2,4,5-, 2,4,6-TCP 2,3,4,6-TeCP	Chlorophenols	Angerer (2001) [12]
Haemoglobin adducts of aromatic amines	Aromatic amines	Lewalter and Gries (2001) [13]
N-Benzylvaline	Benzylchloride	Lewalter et al. (2003) [14]
Cotinine	Nicotine	Müller (2003) [15]
Hexamethylene diisocyanate and hexamethylenediamine	Hexamethylene diisocyanate, hexamethylenediamine	Lewalter et al. (2003) [16]
HMEPMU	Isoproturon	Bader et al. (2003) [17]
DDT, DDE, HCB, α-, β-, γ-HCH, PCB 28, 52, 101, 138, 153, 180	Organochlorine compounds	Hoppe and Weiss (2003) [18]
Dioxins, furans, PCB	Dioxins, furans, PCB	Ball (2003) [19]
Cyclophosphamide, Ifosfamide	Oxazaphosphorines	Hauff and Schierl (2003) [20]

The GC/MS technique has proved particularly successful for determining environmental exposure, for example to pesticides and their metabolites, on account of its great power of detection combined with its high substance specificity (e.g. Hardt and Angerer 2000 [29]; Perry et al. 2001 [30]; Wittke et al. 2001 [31]; Heudorf and Angerer 2001 [32]; Leng et al. 2003 [33]).

In certain cases GC/MS has also been used to improve analytical specificity, for instance in the case of the metabolites of the BTX aromatic compounds (e.g. Laurens et al. 2002 [34]; Barbieri et al. 2002 [35]; Jacobson and McLean 2003 [36]) or inhalation anaesthetics (e.g. Accorsi et al. 2001 [37]; Gentili et al. 2004 [38]).

Table 2 clearly shows that the "quadrupole technique" is by far the most frequently applied analytical variation (see Section 4.4) of mass spectrometry used to assay hazardous substances and their metabolites. At present quadrupole analytical instruments offer an optimum compromise between the high performance of a GC/MS system

Table 2. Publications on the use of gas chromatography-mass spectrometry in the field of biological monitoring since 2000 (Searches based on: PubMed service of the U.S. National Library of Medicine, Bethesda, MD, USA. URL: *http://www.ncbi.nlm.nih.gov/entrez/query.fcgi*, keywords: biological monitoring, GC-MS, urine, blood)

Analyte(s)	Matrix	Detection	Author(s)
Aniline	Haemoglobin	Quadrupole	Zwirner-Baier et al. 2003 [40]
Aromatic amines from diisocyanates	Urine	Quadrupole	Kaaria et al. 2001 [21] Rosenberg et al. 2002 [41] Sennbro et al. 2003 [24]
Aromatic amines from toluene diisocyanate	Haemoglobin	Quadrupole	Sabbioni et al. 2001 [22]
Aromatic amines and metabolites of the nitroaromatic compounds	Urine	Quadrupole	Weiss & Angerer 2002 [23]
Organic arsenic compounds	Haemoglobin	Quadrupole	Fidder et al. 2000 [42]
Atrazine	Urine	Quadrupole	Perry et al. 2001 [30]
Benzene	Urine	Quadrupole	Prado et al. 2004 [43]
Bromoxynil	Urine	Quadrupole	Semchuk et al. 2004 [44]
Metabolites of the BTX aromatic compounds	Urine	Quadrupole	Laurens et al. 2002 [34] Szucs et al. 2002 [45] Jacobson & McLean 2003 [36]
1,3-Butadiene	Haemoglobin	Quadrupole	van Sittert et al. 2000 [46] Begemann et al. 2001 [47]
Butoxyethoxyacetic acid	Urine	Quadrupole	Göen et al. 2002 [48]
Dialkylphosphates	Urine	Quadrupole	Hardt & Angerer 2000 [29]
Dichloroanilines from pesticides	Urine	Quadrupole MS/MS	Wittke et al. 2001 [31]
2,4-Dichlorophenoxy-acetic acid	Urine	Quadrupole	Hughes et al. 2001 [49]
Dimethylphenyl-mercapturic acids	Urine	Quadrupole	Gonzalez-Reche et al. 2003 [50]
Dioxins, furans and poly-chlorinated biphenyls	Hair	High-resolution	Nakao et al. 2002 [39]
Hydroxyterpenes	Urine	Quadrupole	Sandner et al. 2002 [51]
Inhalation anaesthetics	Urine	Quadrupole	Accorsi et al. 2001 [37] Accorsi et al. 2003 [52] Gentili et al. 2004 [38]
Ethylenethiourea	Urine	Quadrupole	Fustinoni et al. 2005 [53]
5-Hydroxy-N-methyl-2-pyrrolidone 2-Hydroxy-N-methylsuccinimide	Urine	Quadrupole	Åkesson & Jönsson 2000 [54] Akrill et al. 2002a [55] Jönsson & Åkesson 2003 [56]
Nitroaromatic compounds	Urine	Quadrupole	Letzel et al. 2003 [57]
Nitroglycerin	Urine	Quadrupole	Akrill et al. 2002b [58]

Table 2 (continued)

Analyte(s)	Matrix	Detection	Author(s)
Opioids	Urine	Quadrupole	van Nimmen et al. 2004 [59]
Organochlorine compounds	Serum	Ion trap	Moreno Frias et al. 2004 [60]
Metabolites of the organophosphates	Urine	Quadrupole	Heudorf and Angerer 2001 [32]
Phenylenediamine	Blood, urine	Ion trap	Stambouli et al. 2004 [61]
S-Phenylmercapturic acid	Urine	Quadrupole	Aston et al. 2002 [62]
Polybrominated diphenyl ethers	Plasma, serum, milk	Quadrupole, high-resolution	Thomsen et al. 2002 [26]
Polychlorinated biphenyls	Serum	Quadrupole	Turci et al. 2003 [27]
Polychlorinated biphenyls	Serum	Time-of-flight	Focant et al. 2004 [28]
Polychlorinated biphenyls and various pesticides	Serum	Quadrupole	Dmitrovic et al. 2002 [25]
Polychlorinated biphenyls and organophosphates	Blood	Quadrupole	Liu & Pleil 2002 [63]
Polychlorinated biphenyls and endocrine-affecting pesticides	Serum	Ion trap	Martinez Vidal et al. 2002 [64]
Polycyclic aromatic hydrocarbons (PAH)	Urine	Quadrupole	Waidyanatha et al. 2003 [65]
PAH metabolites	Urine	High-resolution	Smith et al. 2002 [66]
Pyrethroid metabolites	Urine	Quadrupole	Elflein et al. 2003 [67] Hardt & Angerer 2003 [68] Leng et al. 2003 [33]
m-Toluidine	Urine	Quadrupole	Schettgen et al. 2001 [69]
Toxaphenes	Serum	High-resolution	Barr et al. 2004 [70]
trans,trans-Muconic acid	Urine	Quadrupole	Barbieri et al. 2002 [35]
Trichloroethylene	Urine	Quadrupole	Imbriani et al. 2001 [71]

and the necessary investment and maintenance costs. Other mass filters, such as ion traps and MS/MS, are rarely used for biological monitoring, while there is no practical alternative to high-resolution sector field mass spectrometry for differentiation of complex mixtures of structural isomers (e.g. dioxins, furans, PCB and polybrominated diphenyl ethers) (e.g. Nakao et al. 2002 [39]; Turci et al. 2003 [27]; Ball 2003 [19]; Focant et al. 2004 [28]; Thomsen et al. [26]).

One advantage of mass spectrometry over conventional detection methods is the possibility of including an isotope-labelled reference substance to clearly establish the identity of the analyte and at the same time to serve as an optimum internal standard to compensate for losses due to processing.

However, it must be emphasised in this context that despite the option of "suppressing" the analytical background interference effectively using the GC/MS technique, sample preparation (extraction, work-up, derivatisation, etc.) still plays a key role in an analytical method. The reduced effort required for sample treatment was long regarded as an advantage of GC/MS. In fact, some manufacturers promote their instruments with explicit claims of the efficient elimination of undesirable matrix constituents. But concomitant substances from the work-up, by-products of derivatisation reactions or contaminants in solvents can cause rapid contamination of the GC/MS system (e.g. injector, capillary column, interface, ion source, analyser). In the case of matrix constituents being co-eluted with the analyte a "quenching" effect may also occur, e.g. due to competitive ionisation reactions [4].

In the most unfavourable case the affected component groups have to be laboriously cleaned after only a few injections. Moreover, increasing contamination lowers both the reproducibility of the system and its useful operational life. Therefore GC/MS analysis does not differ significantly from other detection methods with regard to general method development. The specificity and sensitivity of GC/MS can be utilised to achieve low detection limits and substance-characteristic measurement signals only after optimum sample preparation.

3 Principle of gas chromatography/mass spectrometry

Gas chromatography/mass spectrometry (GC/MS) is a so-called "coupled technique", in which two fundamentally independent and separately functioning systems are linked together (Fig. 3). In a simplified view, the gas chromatograph represents only a "sample inlet system" for a mass spectrometer compared with other detectors (e.g. FID, ECD, PND, TCD) that can be operated only in combination with a gas chromatograph. The greatest technical problem of coupling gas chromatography and mass spectrometry is the pressure difference between the two systems: the components to be separated are transported by a stream of carrier gas at 0.5 to 2 mL/min in the GC and this leads to a column pressure of approx. 1 to 2 bar at the end of the column.

In contrast, typical mass spectrometers require a stable high vacuum of 10^{-5} to 10^{-6} mbar. Therefore pressure adjustment is necessary at the interface between the gas

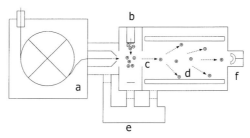

Fig. 3. Schematic structure of a GC/MS (a=gas chromatograph, b=ion source, c=ion focusing, d=mass filter, e=vacuum system, f=detector)

chromatograph and the mass spectrometer. However, this adjustment should have little effect on the gas chromatographic separation of components and, at the same time, should admit as many sample molecules as possible into the mass spectrometer. This is primarily achieved by devices for distributing the volume flow rate or by efficient carrier gas extraction using vacuum pumps.

After gas chromatographic separation (Fig. 3 a), ionisation and excitation of the molecules to be analysed are caused e.g. by electron bombardment or intermolecular charge exchange (Fig. 3 b). Then an acceleration voltage focuses the positively or negatively charged ions (depending on the ionisation method) emitted from the ion source in the direction of an electromagnetic field in the analyser of the mass spectrometer (Fig. 3 c). There the ions are deflected from their flight path to a greater or lesser degree, depending on their mass/charge ratio (m/z) (Fig. 3 d).

This deflection is utilised for ion separation, as only molecules with certain mass/charge ratios are allowed to pass to the detector. The high vacuum of at least 10^{-5} mbar (Fig. 3 e) minimises undesirable subsidiary reactions, e.g. charge exchanges as a result of intermolecular impact and reactions as a result of collision with the walls. The ions finally reach a detector that converts the intensity of the ionic beam into an electrically processable signal (Fig. 3 f). In addition to the charged molecule ions, substance-characteristic fragments are formed as a result of rapid chemical decomposition reactions. The total information obtained (molecule ion + fragment ions) provide indications of the chemical structure of the ionised compound. Thus the GC/MS system yields important substance-specific data for the identification of target components in addition to the gas chromatographic retention time.

GC/MS systems can be described on the basis of six characteristic component groups:

1. Gas chromatograph (for the separation of substance mixtures)
2. Interface (between the gas chromatograph and the mass spectrometer)
3. Ion source (to generate electrically charged and excited molecules)
4. Ion focusing and mass filters (for ion separation)
5. Detector (to record the ions)
6. Vacuum pumps (to generate a high vacuum in the mass spectrometer).

The technology of these components (with the exception of the gas chromatograph) is presented, and the analytically relevant aspects are discussed in the following sections. For a more detailed consideration of individual topics, especially of fragmentation reactions and spectral interpretation, the reader is referred to relevant review articles and textbooks [2, 4, 72–74].

4 Component groups of the mass spectrometer

4.1 Coupling (interface)

Coupling a gas chromatograph to the high vacuum of the mass spectrometer poses special technical problems on account of the pressure difference between the two systems. Depending on the dimensions of the installed separation column and the pressure of the carrier gas at the injector, the volume flow rate is between approximately 0.5 mL/min and 4 mL/min with a pressure difference of several orders of magnitude between the capillary outlet of the gas chromatographic separation column and the high vacuum of the mass spectrometer. Most of the vacuum pumps normally used in GC/MS systems are set to operate at extraction rates of about 2 mL/min. High-performance vacuum pumps must be used for instruments with the option of chemical ionisation or for high flow rates, in some cases the ion source and the analyser are equipped with separate vacuum pumps.

In older GC/MS systems with packed columns and less powerful vacuum technology the carrier gas stream had to be diminished before entering the ion source by "open" coupling of two columns or by the use of gas flow separators (jet and membrane separators). However, "direct" coupling prevailed after the introduction of capillary columns with their lower gas flow rates. In this technique the end of the capillary column is positioned immediately before the ion source or is inserted directly into the ion source. Direct coupling is advantageous for trace analysis, as practically all the analyte in a sample reaches the ion source with the carrier gas, and thus high sensitivity is achieved. The sample volume is primarily determined by the injection technique of the gas chromatograph (e.g. unpulsed and pulsed split/splitless injection, on-column injection, thermal desorption).

Direct coupling also has disadvantages, the most serious of which, from a practical point of view, is the inevitable ventilation of the entire GC/MS system when replacing the column. Each time it is opened, the mass spectrometer can be contaminated with dust, fibres or gases from the ambient air in the laboratory, which may lead to mechanical damage to the vacuum pumps. After the instrument is put into operation, undiscovered leaks may cause corrosion of the heated parts and may impair ionisation efficiency.

Two aspects must be considered, regardless of the type of the interface:

1. No "cold" or "active" sites may be present in the interface. Therefore the temperature must be permanently at least 10 °C higher than the final temperature of the analytical separation column. This can cause problems, if columns with a low operational temperature and notable "column bleeding" are used. Therefore most manufacturers offer "low-bleed" or MS columns, which are specially designed for GC/MS analysis.

2. The vacuum system must be adjusted to suit the carrier gas stream. This is practically always the case in modern instruments, and when capillary columns are used. When an instrument is purchased, it is advisable to select the most powerful vacuum pump available (most manufacturers offer two or three alternatives), as

the quality and constancy of the high vacuum has a direct influence on the sensitivity of the mass spectrometric analysis.

4.2 Vacuum system

Most GC/MS techniques require a high vacuum in the ion source, in the analyser and at the detector. The following factors make this necessary:

(a) *Intermolecular reactions and collisions with the walls must be minimised by ensuring that the mean free path of the molecules is as long as possible.* Each impact of an ion with another molecule or with the surfaces in the mass spectrometer leads to a deflection in the flight path and to a reduction in the ion beam reaching the detector. The mean free path should be approximately equivalent to the distance between the ion source and the detector. Therefore the greater the distance between the ion source and the detector the lower the pressure that is required. This distance is about 50 cm in modern quadrupole instruments, the operational vacuum is approximately 10^{-6} mbar.

(b) *The background flow of molecules from the air (e.g. due to leaks) must be kept to a minimum.* Atmospheric gases, such as nitrogen, oxygen, argon and water vapour may be ionised in the mass spectrometer and be accelerated towards the detector. The resulting background causes a higher baseline and lowers the sensitivity of the detector. In addition, the signals cause interference to analysis in the lower mass range (18 to 40 m/z).

(c) *It is essential to avoid electrical discharges.* In some cases potentials of several kilovolts are present in the ion source, at the focusing apertures and in the analyser of the mass spectrometer. High gas pressure (e.g. due to leaks) or surface soiling can lead to discharges, short circuits and considerable damage to the instrument.

(d) *The incandescent filament in the ion source reacts sensitively to overheating.* On account of the enhanced thermal conductivity at a higher pressure in the ion source the current at the electron-generating incandescent filament is automatically regulated in order to maintain a constant emission of thermal electrons. The elevated current leads to more rapid wear and tear of the filament, and in extreme cases the filament burns out. In addition to leaks, the solvent peak after a sample injection leads to higher pressure in the ion source. For this reason all GC/MS instruments allow a "solvent delay", i.e. the ion source is switched on only when the solvent peak has passed through the instrument and the gas pressure has fallen to about 10^{-6} mbar again. The "solvent delay" to protect the filament is an important parameter of a GC/MS method.

As a rule, the vacuum system of a GC/MS is composed of two components: a mechanical "pre-vacuum pump" initially generates an underpressure of the magnitude of 10^{-2} to 10^{-3} mbar. Then a high-performance pump switches on automatically to attain the final vacuum of 10^{-5} to 10^{-6} mbar. The pre-vacuum pump is normally a rotary-vane pump, which pumps the gas from the space to be evacuated by means of

Table 3. Advantages and disadvantages of typical high-vacuum pumps

	Oil diffusion pump	Turbomolecular pump
Advantages	• Inexpensive • Low maintenance (only oil change) • Pump capacity increases as the molecule size decreases (favourable for carrier gases and solvents)	• Higher pump performance • Vacuum is achieved more rapidly • Short residual running time after switching off e.g. to replace the column
Disadvantages	• Poorly volatile solvents can condense in the oil and reduce the performance • Risk of contamination of the MS due to rising oil vapours • Long cooling phase	• Complex mechanical parts, expensive to replace the bearings • Sensitive to mechanical stress and dust • Pump capacity increases as the molecule size increases (unfavourable for carrier gas and solvents)

two eccentric rotors in series. Either oil diffusion pumps or turbomolecular pumps are used to create the high vacuum of 10^{-5} to 10^{-6} mbar (Table 3). Both pumps require a pre-vacuum of about 10^{-2} mbar in order to function.

The oil diffusion pump is directly coupled via a high-vacuum flange to the mass spectrometer and it has a connection to the pre-pump on its lower end. The pre-pump creates the initial vacuum, while a poorly volatile synthetic oil is being heated in a bottom sump of the diffusion pump. The hot oil vapours ascend in a central tube and subsequently diffuse at high speed from downwards-facing nozzles with a small diameter into the pre-vacuum. This supersonic stream of gas carries over residual gas molecules in the direction of the connected pre-vacuum pump. While the oil vapours condense on the cooled walls of the pump and flow back into the bottom sump, the residual gas molecules are sucked through the pre-pump. Residual pressures of up to 10^{-8} mbar can be achieved using oil diffusion pumps.

Turbomolecular pumps function on the principle of the suction turbine: gas molecules at the inlet of the pump are accelerated by impact with very rapidly rotating inclined lamellae (approx. 60,000 rpm) in the direction of the pre-pump and there they are sucked away. Fixed blades inclined in the opposite direction are installed between two adjacent lamellar rotors in each case, thus forming the "stator", and these blades prevent the molecules diffusing back. Turbomolecular pumps can attain lower final pressures than oil diffusion pumps (10^{-10} mbar).

A basic rule of thumb is that the vacuum pumps for a GC/MS should rather be "over-dimensional", i.e. as powerful as possible. This is primarily due to the fact that there is a close correlation between the residual gas pressure in the mass spectrometer and the sensitivity of the instrument. In addition, it is difficult to seal a GC/MS system effectively on account of the many connections and valves, so a higher output than expected is generally necessary.

Typical carrier gas flow rates in capillary gas chromatography of 1 to 2 mL/min require a pump output of approx. 100 to 150 L/s in the high-vacuum range. Either an oil diffusion pump or a simple turbomolecular pump can achieve this performance.

However, if higher carrier gas flow rates are required or chemical ionisation with additional reactant gas is used, a suitable turbomolecular pump with a performance well over 200 L/min is the only appropriate option. In addition, a pre-pump with as high a capacity as possible should also be selected (more than 10 m^3/h). In the case of chemical ionisation it is particularly advantageous for the analyser and the ion source to be in separate components, as up to 1 mbar pressure is created in the ion source, while a high vacuum must be maintained in the mass filter. For this reason high-performance instruments are equipped with two separate high-vacuum pumps, one for the ion source and one for the mass filter area.

At this point the importance of manometers in the mass spectrometer should be emphasised. The most important function of a manometer is to indicate leaks after changing the set-up, replacing the column or maintenance work. In the case of chemical ionisation and the ion trap technique the maintenance of an optimum pressure in the ion source is of key significance for the reproducibility of the analysis, as it directly influences the ionisation yield. Therefore, from the practical point of view, a manometer on the mass spectrometer is an instrument for technical quality assurance.

The "Penning gauging head" (cold cathode) and the "ionisation tube" according to Bayard-Alpert (hot cathode) are typical manometers used for GC/MS. In the Penning gauging head, gas molecules are ionised in a high-voltage field by electrons emitted from a cylindrical cathode and deflected spirally in the magnetic field (increasing the effective path length and thus the probability of ionisation). In the ionisation tube the electrons are emitted from a hot tungsten wire. The electrons migrate several times through a spiral, positively charged grid before landing on its surface. Cations that are formed by impact with residual gas molecules cause an ion flow. The ion flow is proportional to the residual gas concentration in both the Penning gauging heads and in the ionisation tube. The hot cathode measurement is more accurate than the Penning set-up, but the filament is more sensitive to overheating and the tube becomes easily soiled when it is not switched on.

4.3 Ion sources

The ion source is the first typical component group of a GC/MS system. Simply expressed, the components eluted from the analytical separation column or from the interface are ionised in the ion source by electron bombardment. The resulting radical cations and cations are accelerated out of the ion source and focused by electrostatic "lenses" towards the mass filter. In addition to this "classical" electron impact ionisation (EI), chemical ionisation (CI) has also become established as a standard procedure in GC/MS analysis in the fields of occupational and environmental medicine. In this case the ions are formed by charge exchange with a reactant gas that has itself been previously ionised by electron impact. Chemical ionisation can generate cations as well as anions, and therefore a differentiation is made between "positive chemical ionisation" (PCI) and "negative chemical ionisation" (NCI).

4.3.1 Electron impact ionisation (EI)

In principle, the ion source for electron impact ionisation consists of a small hollow space with a volume of less than 1 cm^3 that houses the inlet for the analytical separation column or for the interface, an incandescent filament made of tungsten or rhenium oxide, the target (an anode situated opposite), and an outlet to the mass filter or to the ion focus (generally at an angle of 90° to the sample inlet). The outlet apertures consist of an extractor, a discharger and one or more ion focuses. A potential is applied between the extractor and a repeller situated at the opposite side of the ion source, which causes positively charged molecules to be accelerated towards the system of apertures and out of the ion source (Fig. 4).

The ion source temperature of approx. 200 to 250 °C generally represents a compromise: on the one hand the thermal energy of the molecules should be kept as low as possible to enable efficient ionisation and to minimise the number of collisions between molecules and with the walls within the relatively small volume of the ion

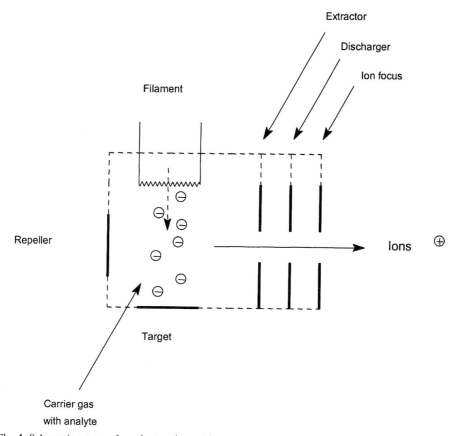

Fig. 4. Schematic set-up of an electron impact ion source

source, on the other hand no "cold" sites should be present to avoid changes to the surface due to substance deposits.

The filament is heated by a current of 200 to 300 µA for ionisation, and high-energy electrons are emitted and accelerated by a voltage towards the opposite electrode. The energy of these electrons is usually given in electron volts (eV), whereby 1 eV is equivalent to the energy taken up by an elemental charge while traversing a potential difference of 1 volt. The emitted electrons collide with the sample molecules on the way to the opposite electrode and can displace an electron from the highest occupied molecular orbital there, thus generating two free electrons at a lower energy and a radical cation from the collision:

$$M + electron\ (70\,eV) \rightarrow M^{*+} + 2\,electrons$$

The ionised molecules are accelerated out of the ion source by electrical repulsion from the repeller anode towards the aperture and lens system, comprising an extractor (cathode opposite the repeller), a discharger (to discharge defocused ions) and one or more ion focuses. The ion beam is focused in the following mass filter by the ion focus.

The first ionisation potential IP of a molecule must be exceeded in order to generate cations. Organic compounds exhibit an IP of 8 to 13 electron volts. Above this threshold the ionisation yield rises continuously and reaches a saturation level at 40 to 60 eV. However, "saturation" in this context means a yield of only about 0.1% of the molecules.

Uncharged molecules are removed from the mass spectrometer by the vacuum pumps. Electron energies of 70 eV are typically used in order to improve the ionisation yield for a wide range of substances. However, such values, which considerably exceed the typical ionisation potentials of organic compounds, cause the excessive impact energy to convert the molecule into excited oscillation states, consequently covalent bonds are broken and fragmentation of the molecule ensues. On principle, the extent of fragmentation can be reduced by lowering the kinetic energy of the impact electrons. However, the ionisation yield falls by several orders of magnitude in this case so that "low-electron volt spectra" are only useful for qualitative purposes, e.g. for identification of the molecule ion.

Interpretation of fragmentation reactions cannot be reviewed in detail here on account of the large number of different mechanisms and special cases. Section 6 gives an overview of more detailed sources of literature.

Mass spectrometry is eminently suitable for quantitative trace analysis thanks to fragmentation:

– Fragmentation reactions are specific for substances and, as a rule, they permit definitive identification of a substance. This also applies with reservations to structural isomers that yield almost identical fragments and must initially be separated by gas chromatography in most cases. While other GC detectors primarily enable identification by means of the retention time, mass spectrometric data provide direct indications of the chemical structure of a compound.

– Fragmentation reactions are readily reproducible. The molecular geometry does not change during the ionisation phase (10^{-15} to 10^{-16} s) (Franck-Condon principle) [75]. Thus the fragmentation pattern of a substance is always identical under constant ionisation conditions. A certain variation is caused only by the thermal energy dispersion of the individual molecules.

4.3.2 Chemical Ionisation (CI)

Chemical ionisation is a procedure for generating ions using less excess kinetic energy than is required for electron impact ionisation. Ions are formed by charge exchange during collision of the sample molecules with a reactant gas previously ionised by electron impact. In contrast to the EI ion source, intermolecular collisions are desirable and are promoted by a closed design and a higher gas pressure (0.1 to 1 mbar). As both positively and negatively charged ions can be generated by the appropriate choice of ionisation conditions, these procedures are known as positive or negative chemical ionisation (PCI or NCI) [76, 77].

4.3.3 Positive Chemical Ionisation (PCI)

Methane, which is frequently utilised as a reactant gas, is used here as an example for the reactions during chemical ionisation. Initially "primary" radical cations are generated (Fig. 5) from methane molecules by electron bombardment in the ion source. Higher potentials of approx. 100 to 150 eV can be applied to the incandescent cathode to enhance the ionisation yield. Reactant gas molecules are largely ionised due to the high methane vapour pressure and the associated "excess of methane".

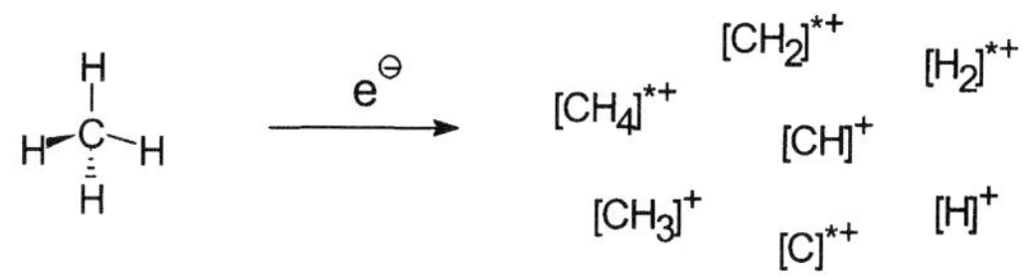

Fig. 5. Generation of radical cations and cations during ionisation of methane

In addition to the primary ions, collisions of the reactant gas molecules with each other result in the generation of "secondary" ions (Fig. 6). Together with the primary ions they form a relatively stable plasma.

$$[CH_4]^{*+} + CH_4 \longrightarrow [CH_5]^+ + CH_3^*$$

$$[CH_3]^+ + CH_4 \longrightarrow [C_2H_5]^+ + H_2$$

$$[CH_2]^{*+} + CH_4 \longrightarrow [C_2H_4]^{*+} + H_2$$

$$[CH_2]^{*+} + CH_4 \longrightarrow [C_2H_3]^+ + H_2 + H^*$$

$$[CH]^+ + CH_4 \longrightarrow [C_2H_2]^{*+} + H_2 + H^*$$

$$[C_2H_5]^+ + CH_4 \longrightarrow [C_3H_7]^+ + H_2$$

$$[C_2H_3]^+ + CH_4 \longrightarrow [C_3H_5]^+ + H_2$$

Fig. 6. Formation of secondary ions during chemical ionisation with methane

The following mechanisms occur due to collisions with sample molecules:

– *Proton transfer*

Sample molecules M can act as Brönstedt bases and accept protons from secondary reactant gas ions RH^+ (e.g. CH_5^+, $C_2H_5^+$, $C_3H_7^+$, NH_4^+, H_3^+). The proton is transferred to the molecule with the highest proton affinity (PA). Therefore the selection of the reactant gas already determines the sample molecules that can be ionised. Table 4 shows the reactant gases most frequently used for chemical ionisation:

Methane is a very efficient reactant gas on account of its relatively low proton affinity (cf. Table 4). In contrast, e.g. isobutane and ammonia lead to considerably "gentler" ionisation. In these cases proton transfer is accompanied by a low excess of energy. Thus fragmentation of the sample molecule is only slight and it occurs with high selectivity.

Table 4. Reactant gases for chemical ionisation (according to [2])

Gas	Proton affinity [kcal/mol]	Reactant ion
Hydrogen	100	H_3^+
Methane	127	CH_5^+
Ethene	160	$C_2H_5^+$
Water vapour	165	H_3O^+
Hydrogen sulphide	171	H_3S^+
Methanol	182	$CH_3OH_2^+$
Isobutane	195	*tert.*-$C_4H_9^+$
Ammonia	207	NH_4^+

Table 5. Proton affinities of selected organic compounds (according to [2, 4])

Compound	Proton affinity [kcal/mol]	Compound	Proton affinity [kcal/mol]
Ethane	121	Methyl sulphide	185
Methyl chloride	165	Nitroethane	185
Trifluoroacetic acid	167	Methyl cyanide	186
Formic acid	175	Toluene	187
Benzene	178	Xylene	187
Cyclopropane	179	Acetic acid	188
Propylene	179	Dimethyl ether	190
Methylcyclopropane	180	Isopropanol	190
Nitromethane	180	2-Butanol	197
Methanol	182	Ethyl acetate	198
Acetaldehyde	185	Acetone	202

Table 5 shows that (with the exception of ethane) all the selected compounds can be chemically ionised using methane, while e.g. methanol is especially suitable for heteroatomic and aromatic compounds.

Ammonia is an extremely selective reactant gas, e.g. for aromatic amines and phthalic acid esters, on account of its high proton affinity. As in the case of hydrogen sulphide and water vapour, it must be considered that ammonia has a corrosive effect on the sensitive surfaces in the ion source and in the mass filter. Therefore these gases cannot be recommended for constant use, and the gas supply installations and pressure reducer must be made of stainless steel.

– *Hydride loss*

When reactant gas ions are formed, species with high hydride (H^-) affinity are created, e.g. CH_3^+, CH_5^+, $C_2H_5^+$. The reactant gas can wrest a hydride ion away from molecules containing hydrogen, leaving a corresponding cation. The exothermal nature of the reaction results in an intensive fragmentation of the $[M-H]^+$-ions. Hydride loss and proton transfer (see above) often occur simultaneously.

– *Cation addition*

As it is thermodynamically favourable, the addition of cations to the sample molecule is a frequently occurring gas phase reaction. This mechanism is particularly prominent in the case of methane. $[C_2H_5^+]$ and $[C_3H_5^+]$ ions are added to form $[M+29]^+$ and $[M+41]^+$ molecule adducts. Therefore together with proton transfer a characteristic addition pattern is observed with an intensity distribution of about $100:10:2$ for $(M+1)$, $(M+29)$ and $(M+41)$, which is observed only in the molecule ion in this form (cf. Fig. 6 b). PCI is therefore an important method for identification of the molecule ions in a compound.

– *Charge transfer*

While the charge donors in the mechanisms described above are protons or hydride ions and a "classical" acid-base reaction takes place, "charge transfer" is
based on an electron transfer, i.e. an "oxidation effect". A noble gas or another
chemically inert compound (e.g. nitrogen, carbon dioxide, carbon monoxide) is
initially ionised with the help of electron impact. The sample molecule receives
its charge by electron transfer from the sample molecule to the reactant gas cation.

This ionisation method is especially "gentle" and results in low fragmentation, as
there is little difference between the recombination energy (e.g. He: 24.6 eV, Ar:
15.8 eV, CO: 14.0 eV, CO_2: 13.8 eV, Xe: 12.1 eV, NO: 9.3 eV, benzene: 9.3 eV) and
the ionisation potential of the (generally organic) compounds. Therefore only a slight
excess energy is transferred to the sample molecule cation. The fragmentation pattern
is similar to low-electron volt spectra, but the ionisation yield is higher.

4.3.4 Negative Chemical Ionisation (NCI)

Negatively charged ions can be detected by reversal of the electrical polarity in the
ion source and at the detector [78, 79]. However, anions are generated only to a limited extent, even during electron impact ionisation, and the ions thus created are
hardly useful for detection because of the large excess of radical cations and cations.
However, chemical ionisation can promote the formation of negatively charged ions
that can be utilised for analysis.

With the exception of the "ion-molecule reactions", the reactant gas primarily serves
to generate slow "thermal" electrons. As in the case of electron impact ionisation the
reactant gas is bombarded with high-energy electrons (>200 eV) that are emitted
from an incandescent filament. Low-energy electrons result from the impact between
the electrons from the incandescent filament and the reactant gas. In the case of
methane the following reaction takes place:

$$CH_4 + electron^-_{(230\,eV)} \rightarrow CH_4^{*+} + 2\,electrons^-_{(thermal)}$$

The energy of the released electrons is only a few electron volts, therefore they can
be "captured" by the sample molecules. Subsequent reactions and fragmentation occur to a considerably lesser extent than in the case of electron impact ionisation.
Negative chemical ionisation can have a considerably higher detection capacity than
electron impact ionisation and PCI:

– NCI exhibits a high selectivity for "electron-trapping" compounds (e.g. halogen-
containing and other heteroatomic compounds) and electron-deficient aromatic
compounds. The sensitivity can be improved by two orders of magnitude compared with EI and PCI, and it lies in the range of that achieved by an electron
capture detector (ECD).

– No reactant gas anions are formed in the case of NCI, and as a consequence background interference is very low, e.g. compared to positive chemical ionisation.

On principle, the gases already mentioned in the section on positive chemical ionisation can also be utilised for NCI, as the high energy of the emitted electrons guarantees a good ionisation yield. However, in contrast to PCI, the identity of the gases is not important for the actual ionisation of the sample molecules, so the reactant gas can be selected on the grounds of safety and the ease of technical installation.
The released electrons can lead to negative ionisation of the sample molecules via various mechanisms:

– *Electron capture*
Ionisation due to electron capture is the quantitatively most important reaction in the NCI mode: thermal electrons are incorporated into unoccupied molecular orbitals and lead to the formation of radical anions:

$$M + \text{electron}^- \rightarrow M^{*-}$$

This reaction is reversible, i.e. in the worst-case scenario the radical anion simply releases the charge it has previously acquired ("spontaneous discharge"). This process proceeds very rapidly, leading to an equilibrium being established between the ionised and the uncharged sample molecules. The higher the electron affinity of the sample molecule, the more the equilibrium lies on the side of the radical anions. For this reason NCI mass spectrometry, like electron capture detection (ECD), is particularly suitable for halogen-containing and heteroatomic compounds. But stabilisation of the anions can also be achieved by fragmentation that is observed as a secondary reaction of NCI:

– *Dissociative electron capture*
Radical anions are stabilised by homolytic bond cleavage in the case of dissociative electron capture:

$$MX + \text{electron}^-_{\text{(thermal)}} \rightarrow M^* + X^-$$

This reaction is also reversible in principle and proceeds rapidly so that a substance-specific equilibrium is established in the ion source.

– *Ion pairing*
Ion pairing results in stabilisation of the radical anion due to heterogenic fragmentation and release of the thermal electron. Therefore it is a combination of "spontaneous discharge" and dissociation:

$$MX + \text{electron}^-_{\text{(thermal)}} \rightarrow M^+ + X^- + \text{electron}^-_{\text{(thermal)}}$$

– *Ion-molecule reactions*

Ion-molecule reactions occur particularly frequently in the presence of oxygen, water or other electron-affinative compounds in the ion source:

$$M + X^- \rightarrow MX^-$$

Initially ionisation of the interfering component (X) occurs and its radical anion or anion associates with the sample molecule. This reaction is also known as "ion adduct formation" and it is an undesirable reaction that competes with electron capture. The different types of ionisation and their effect on the mass spectrum of a compound can be illustrated using N-methylvaline as an example (Figs. 7a to 7c). N-Methylvaline (MEV) is an amino acid that is formed by methylation of the free N-terminus of a haemoglobin chain in the erythrocytes. The parameter can be used for biological monitoring [8]. MEV is cleaved from the globin chain by means of a "modified Edman degradation" and converted to a pentafluorophenylthiohydantoin (PFPTH) derivative [80]. The electron impact mass spectrum of N-methylvaline-PFPTH is shown in Figure 7a, in which the most important fragments are marked. The EI spectrum exhibits

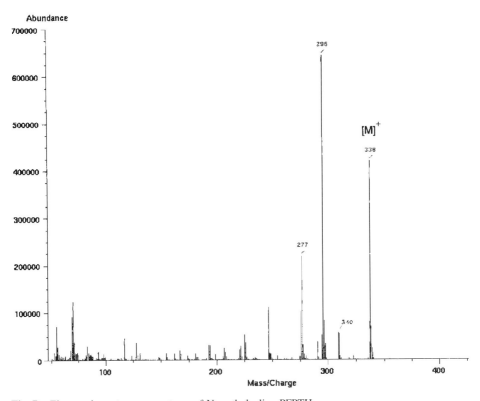

Fig. 7a. Electron impact mass spectrum of N-methylvaline-PFPTH

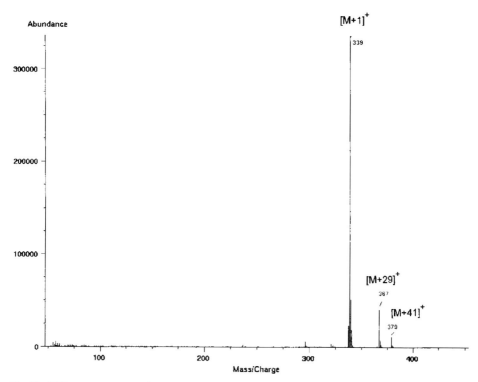

Fig. 7b. PCI mass spectrum of N-methylvaline-PFPTH

marked fragmentation in addition to the molecule ion at 338 m/z. The main fragments can be explained by typical reaction mechanisms (McLafferty rearrangement
(338 → 296), CO cleavage in the α-position to yield a heteroatomic fragment
(338 → 310), loss of a fluoride ion (296 → 277)). The molecule ion 338 m/z exhibits
a peak intensity of about 500,000 relative units.

When the same sample is analysed with positive chemical ionisation using methane
(Fig. 7b), it is evident that the molecule ion hardly fragments. In addition to the pro-
tonated molecule ion [M+1] with a m/z of 339, the only other products of importance to be formed are the adducts [M+29] (molecule ion+C_2H_5) and [M+41]
(molecule ion+C_3H_5). Both adducts confirm that the molecule ion has the atomic
mass or the m/z ratio of 338. However, the intensity of the main signal is only approx. 350,000 relative units, which shows a reduction in the sensitivity compared
with analysis using electron impact ionisation.

The degree of fragmentation in the case of negative chemical ionisation (Fig. 7c) is
similar to that caused by electron impact ionisation. The main signal at 337 m/z is
generated by a ion-molecule reaction (proton transfer). The intensity of this signal is
approx. 1,200,000 relative units.

Compared to electron impact ionisation, the calculated sensitivity is increased by a
factor of 2. In practice, however, the detection limit is improved by a factor of 5 on

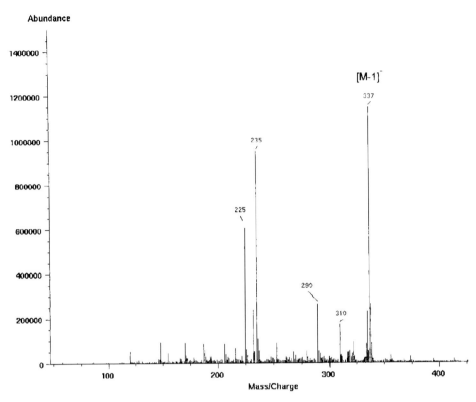

Fig. 7 c. NCI mass spectrum of N-methylvaline-PFPTH

account of the more favourable signal/background ratio for NCI MS in the case of the PFPTH derivatives of alkylated amino acids.

4.4 Mass filter

After the ion source the second most important component group of a mass spectrometer is the mass filter. The charged molecules are deflected from their flight path by means of electromagnetic fields. This deflection depends on the mass or, more exactly expressed, on the mass/charge ratio of the molecule, and leads the ions either to electrically charged surfaces within the mass filter where they discharge and are removed by the high-vacuum pumps, or deflects them towards the detector where they are recorded. The deflection of the flight path can be achieved by electric and magnetic fields with different geometrical designs. In practice several designs have become established: quadrupole, ion trap and sector field instruments. All these designs of mass filters are based on the principle of separation of accelerated ions according to their mass/charge ratio.

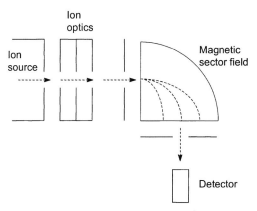

Fig. 8. Deflection of an ion beam by a magnetic sector field

When linearly accelerated, electrically charged molecules reach a magnetic field that is directed at right angles to their flight path and the molecules are deflected in a circular path [72]. This deflection is schematically illustrated in Figure 8. The kinetic energy of these ions is determined by the voltage with which they are accelerated out of the ion source:

$$e\,U = \frac{1}{2}\,mv^2$$

(e = ion charge, U = acceleration voltage, m = molecular mass, v = velocity).
The focused ion beam is bent in the magnetic field H and follows a trajectory in which the strength of the magnetic field is compensated by the centripetal force of the accelerated ions:

$$H\,e\,v = m\,v^2\,r^{-1}$$

The deflection radius is therefore expressed as:

$$r = (m\,v)/(e\,H)$$

It is directly proportional to the mass of the charged ion, its velocity and the strength of the magnetic field. Regardless of the type of ionisation, small molecules usually generate singly charged molecule ions. Thus the ionic charge e is equivalent to the single elemental charge of 1. When combined, these equations give the "basic mass spectrometric equation":

$$m/z = (r^2\,H^2)/(2\,U)\,.$$

There is a constant relationship between the mass/charge ratio and the square of the deflection radius when the acceleration potential and the strength of the magnetic field are kept constant:

$$m/z = r^2 \text{ const.}$$

When $z = 1$ it follows that the deflection radius is proportional to the mass.

This relationship can be utilised for spatial separation of the various ions or charged molecular fragments in the magnetic field of the analyser: only ions with specific m/z ratios are deflected in a circular path in the magnetic sector field, at the end of which the mass spectrometric detector is installed. Any sections of the mass spectrum of a compound or individual substance-characteristic fragments can be investigated by varying the strength of the magnetic field or the acceleration voltage.

The performance of a mass filter is characterised by three factors:

- mass range
- scan speed
- and resolution.

The maximum accessible mass range of the filter primarily depends on the available strength of the magnetic field: while practically all mass spectrometers permit ion detection in the range below 250 to 300 m/z, instruments with weak magnetic fields show an increasing discrimination in the case of higher masses.

For a given sector field geometrical design each magnetic field has a maximum m/z ratio, above which deflection to the detector is no longer possible. Furthermore, the kinetic energy of the molecules in the ion beam exhibits dispersion to a greater or lesser degree: the higher the mass of the ion to be deflected, the more noticeable the differences in kinetic energy. An increasing proportion of the ion beam is not focused with sufficient precision on the detector and the intensity of the signal decreases. This effect causes an increasing distortion in the spectrum in the range of the higher masses, and the relative intensity of the heavier fragments or of the molecule ion seems to diminish.

The stronger the magnetic field, the more precisely the heavier ions with higher dispersion can be deflected. Instruments with an accessible mass range of up to 1000 m/z are especially suitable for use in the fields of occupational and environmental medicine. Organic compounds or their metabolites, even in the form of their volatile derivatives, are seldom heavier than 500 to 600 m/z, so most GC/MS systems meet the stated requirements. Instruments with wider mass ranges up to 2000 m/z are more suitable for applications in the area of protein and DNA analysis. However, in these areas HPLC (high performance liquid chromatography) instruments and special techniques for sample injection are generally required (e.g. matrix-assisted laser desorption ionisation, MALDI), as the compounds to be investigated are no longer volatile enough to be separated by gas chromatography and transported into the mass spectrometer.

The complete mass spectrum of a compound is usually recorded by continually varying the strength of the magnetic field. The result is known as a "scan". This investigation can be carried out only if the intensity of the ion beam remains constant throughout the duration of one or several scans and the concentration of the compound to be investigated is high enough to generate a signal at the detector. While the latter condition is relatively simple to fulfil (a substance quantity of approx. 1 to

10 ng is required in the ion source, which is equivalent to an injection volume of 1 μL at a concentration of 1 to 10 mg/L), the scan speed and scan rate (number of spectral recordings per second) depend on the instrument. They are largely determined by the speed of the magnetic field variation, the response time of the ion detector and the speed of data transfer and data processing.

As a rule, a component separated by gas chromatography is eluted from the separation column after about 5 to 20 seconds. Approx. 15 to 20 individual measurement points or complete scans are required in order to portray a reproducible and quantitatively evaluable peak. As a consequence of these general requirements, the mass filter must perform at a scan rate of at least 1 s^{-1}. Modern GC/MS systems attain scan rates of about 2 to 3 s^{-1} for a mass range of 300 m/z. A scan analysis is graphically depicted by the GC/MS in such a manner that the intensities of the individually recorded masses are added together to give a total signal. Therefore each complete scan yields only one individual value. A seemingly uninterrupted line, similar to a "typical" gas chromatogram and known as a "total ion chromatogram" (TIC), results by continuous recording of several scans per second and by plotting the signals as a function of time.

When closely scrutinised, each "point" of the signal line represents a complete mass spectrum and thus contributes to the identification of the relevant analyte. In addition, individual substance-specific signals can be isolated from the spectral set and can be reconstructed to give a chromatogram. In this way interfering components can be largely suppressed, at least visually.

The narrower the mass range selected, the more frequently a complete scan can be carried out. This is particularly desirable for the definitive identification of a component. However, the sensitivity of the scan technique is low compared with other gas chromatographic detection methods (e.g. flame ionisation). As already mentioned above, the lowest concentration for scan mass spectrometry is about 1 mg/L. The scan technique is inadequate to deal with most investigations in occupational medicine and almost all the questions posed in environmental medicine.

However, in many cases it is not necessary to record complete mass spectra for substance identification. In addition to the gas chromatographic retention time, which represents an important identification characteristic of a compound, the m/z ratio of the molecule ion provides further significant substance-specific information. Moreover, most compounds fragment in a very characteristic and reproducible manner, so that individual ions may be sufficient to enable clear identification of components. This principle is used in the "selected ion monitoring" (SIM) technique.

Only selected ions reach the detector in the SIM mode, i.e. the strength of the magnetic field is varied discontinuously and optimised for the m/z ratio of the ions in question. Two or three individual ions (molecule ion and characteristic or high-intensity fragments) are generally sufficient, together with the retention time, to ensure definitive identification of the component. The advantage of this technique is that the time available to record an ion is distinctly greater than in the scan mode.

Thus at most four seconds per fragment are available for measurement at an elution duration of 12 seconds and with three fragments to be recorded. If this time is divided by the time required for the desired number of measured points, i.e. 20, only 200 milliseconds remain per measurement point and ion. The duration of the switchover from one m/z ratio to the next must be considered (approx. 25 ms), resulting in a total of 175 milliseconds per measurement point and ion.

The measurement time per ion is considerably shorter in the scan mode: recording 20 scans over a range of 300 mass units during an elution duration of 12 seconds leads to a scan rate of 0.6 s^{-1}, i.e. 2 milliseconds of measurement time are available for each ion per scan. This is equivalent to about 1% of the measurement time in the SIM mode. The intensity of an individual ion signal is two orders of magnitude higher in the SIM mode than when a scan is recorded.

Figure 9 (top) shows the gas chromatogram of a urine extract that was investigated to determine a urea herbicide (retention time 17.87 min). The pertinent electron impact mass spectrum (Fig. 9 (middle)) exhibits a main fragment with an m/z ratio of

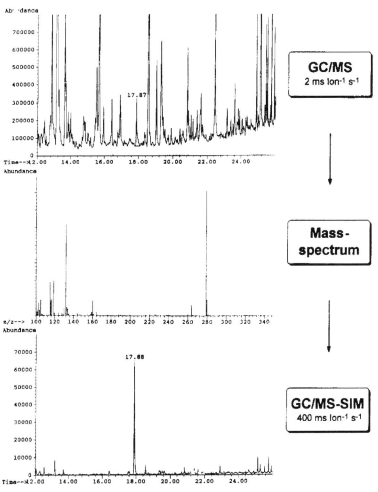

Fig. 9. Switchover from the scan mode to selected ion monitoring

132 in addition to the molecule ion at 279 m/z. A considerably higher selectivity (less interfering components of weaker intensity) and an improved signal/background ratio are achieved by switching the mass spectrometer to a SIM analysis of these two ions (Fig. 9 (bottom)).

The distinct improvement in sensitivity using the SIM technique is accompanied by a relatively slight loss of structural information. The mass range between 160 m/z and 260 m/z yields no relevant information, but requires half of the measurement time. The SIM mode permits the substance to be identified with a sufficiently high degree of certainty on the basis of fewer selected ions.

The following criteria apply when selecting suitable SIM ions:

1. The selectivity of the SIM procedure gradually decreases as the masses become lower (m/z < 100), as many compounds are fragmented in this range. On principle, either the molecule ion or fragments with higher and therefore more substance-specific masses should be selected. Furthermore, many compounds undergo only one or two main fragmentation reactions that are easy to interpret and permit direct conclusions on the nature of the molecule ion (e.g. a-cleavage, McLafferty rearrangement, CO loss, *tert.*-butyl cleavage, etc.).

2. It is advisable to record at least two, preferably three, ions per analyte. On the one hand, analysis of several characteristic fragments enhances the chances of identifying the analyte, on the other the relative ratio of the signal intensities can provide evidence to enable definition of a background interference. As fragmentation reactions are readily reproducible, any deviation of the signal intensities from a specified quotient (e.g. ion A : ion B : ion C) can indicate an interfering component that has not been separated by gas chromatography and may even prevent erroneous identification. Each user must define the limit at which a deviation represents an unacceptable interference to the analysis according to his/her own requirements. The ion trace ratio becomes increasingly inexact, in particular at low concentrations. The signal/background noise ratio (3 : 1), which is frequently defined as the detection limit, is not suitable in this case, particularly as the detection limit must be based on the fragment with the weakest signal.

3. Wherever possible, fragments of strong intensity should be measured for SIM analysis. The detection limit of the method depends directly on the strength of the signal at the detector. Therefore it is advisable to select a "quantifier" (an ion that is used to calculate the analytical result) and "qualifiers" (two or three ions that are measured to confirm the identity of the analyte; the relative intensity ratios serve as a check and they should be less than 10%) from the main fragments.

4. If possible, ions should be chosen that are not subject to any interference from concomitant components which yield the same fragments.

A further advantage of the SIM technique is that the unspecific background is very efficiently suppressed by the high selectivity of the procedure, thus enhancing sensitivity compared with scan recording due to improvement of the signal/background ratio.

In certain investigations it is also important to have a high mass resolution in order to achieve more specific and sensitive detection. The resolution power of a mass spectrometer is defined as the ability to separately detect ions with slight differences in their molecular mass. The resolution R is expressed mathematically as follows

$$R = m/\Delta m$$

Analogous to the definition of gas chromatographic separation, complete mass spectrometric separation is given, when the overlap (the "valley" between two peaks) of two signals of equal height is less than 10% of the peak height. For example, in order to separate masses of 1000 and 1001, a resolution of 1000/1 = 1000 is required. The typical resolution power of low-resolution mass spectrometers is 1000 to 2000, while high-resolution systems achieve R-values of 10,000 or more.

The resolution R of a magnetic sector field instrument is constant over the entire mass range: the distance between two neighbouring signals in the low mass range is therefore relatively large. In contrast, the distance Δm between two signals is constant in quadrupole instruments and ion traps, so the resolution is poorer at lower masses. However, in practice these instruments are adjusted by the manufacturers for constant resolution that is defined by the 10% criterion.

As already mentioned, three technically different types of mass filters capable of different performances are used at present:

4.4.1 Magnetic sector field instruments

The deflection of electrically charged molecules in a magnetic field described above is a "classical" technique of mass spectrometry (Fig. 8). Although the basic principle has been retained, magnetic sector field instruments have become the most powerful of the mass spectrometers as result of many improvements.

However, the molecules accelerated out of the ion source do not move homogeneously in a single defined direction, but are subject to a certain scattering (directional dispersion). Moreover, the kinetic energy of the molecules or molecular fragments has a normal distribution under given external conditions (energy dispersion). While the directional dispersion can be compensated by the "lens effect" of the magnetic sector field, the kinetic energy scatter causes broadening and relative weakening of the ion beam. The mass resolution of a magnetic sector field is also determined by the scatter of the kinetic energy of ions with an identical m/z ratio. Simple instruments use electrostatic apertures for ion focusing. This technique achieves resolutions between 5000 and 10,000 and has the advantage of a relatively simple instrumental set-up. However, it results in considerable loss of sensitivity due to screening of a part of the ion beam.

In "double-focusing" sector field instruments (Fig. 10) a curved electrostatic field is introduced either in front of or after the magnetic field in order to reduce energy dispersion. This field separates the ions with the same m/z ratio but different kinetic energies and also focuses them in the same manner as a lens: the electrostatic field deflects slower ions more strongly than faster ones, thus resulting in spatial separation.

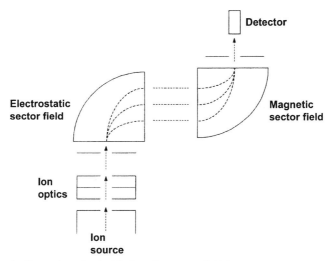

Fig. 10. Schematic illustration of a double-focusing sector field instrument

The directional focusing and the velocity dispersing properties of the sector fields can be harmonised with each other by the sequential arrangement in such a manner that ion beam dispersion is compensated to the greatest possible degree. All ions with the same m/z ratio are focused at the detector regardless of their energy dispersion. Depending on the arrangement of the electrostatic analyser and of the magnetic sector, we differentiate between the Nier-Johnson geometry (first the magnetic field, then the electrostatic analyser; semicircular ionic beam path) and the Mattauch-Herzog geometry (first the electrostatic analyser, then the magnetic field; sigmoid ion beam path). In addition to a higher sensitivity, double-focusing sector field instruments also achieve a considerably higher resolution compared with the simple sector field technique (R = 100,000).

4.4.2 Quadrupole instrument

Quadrupole mass filters utilise the interaction between the ion beam and an oscillating electric field. A quadrupole analyser consists of a total of four parallel rod electrodes that are arranged in a square and point in the longitudinal direction. In each case opposite poles are coupled by a direct voltage U and an alternating voltage with an amplitude of V_0 so that the resulting potential difference V_{total} for each pair of poles is

$$V_{total} = U \pm V_0 \times \cos{(\omega \times t)}$$

(U = direct voltage, V_0 = alternating voltage, ω = angle velocity, t = time) (Fig. 11).

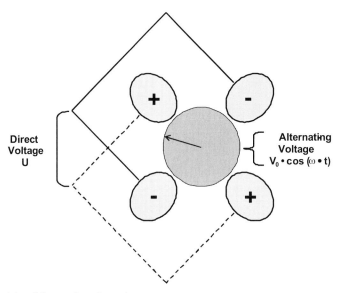

Fig. 11. Principle of the quadrupole analyser

When ions reach the alternating electric field, their flight path is influenced in a complex manner. The resulting flight paths are described by the "Mathieu equations". These are differential equations that describe the oscillation of an ion in a three-dimensional space or, more exactly, in a three-dimensional alternating electric field. Only those ions in resonance with the alternating electric field exhibit a stable flight path and pass through the mass filter. This depends on the mass of the ion and on the selected direct and alternating voltages. To generalise, it can be stated that ions with a small m/z ratio are especially influenced by the alternating voltage, while the heavier ions (high m/z) are subject to deflection by the direct voltage. The ions are not accelerated towards the detector in the quadrupole. Their path is determined by the strength of the direct voltage or alternating voltage, the radius of the alternating electromagnetic field and the frequency of the alternating voltage. Like the cycle frequency of the alternating voltage, the quadrupole dimensions are pre-determined by the design. Therefore in practice the flight path of the ions is determined only by the potentials U and V.

When specific voltages are set, only ions with certain m/z ratios pass through the quadrupole, whereas all other ions either collide with the quadrupole rods and discharge or exhibit an unstable flight path and leave the mass filter between the poles. Quadrupole instruments have several interesting properties:

– They are more compact and cheaper than sector field instruments, and they are very robust.
– As the ions are intended to take the longest possible path through the mass filter, the electrical acceleration voltages in the quadrupole are considerably lower that

in the sector field instrument (5 to 30 volts compared with 5 to 10,000 volts). There is therefore less danger of high voltage surges in the ion source.

– As no magnetic fields are necessary, the scan rate of a quadrupole instrument is higher than that of a sector field instrument. Between 2 and 3 mass spectra of about 300 m/z units can be recorded per second.

However, the operational range of a quadrupole instrument is narrower than that of a sector field instrument. The mass resolution is also lower. The resolution of a quadrupole depends ultimately on the number of oscillations of the ions during their passage through the quadrupole: the higher the number, the better the mass resolution. The lower the acceleration voltage that is set in the ion source, the longer the ions remain in the quadrupole and the more oscillations they exhibit. On the other hand, reduction of the acceleration voltage leads to discrimination of the transmission rate in favour of higher masses, leading to a decrease in sensitivity. Quadrupole instruments are therefore adjusted to a mass resolution of 1000 to 2000 by the manufacturer, which represents a compromise between resolution and sensitivity.

4.4.3 Ion traps

The "ion trap" represents a special form of the quadrupole. The electric field is also generated by applying a direct and an alternating voltage, but the ion trap consists of a central "ring electrode" and two "end cap" electrodes that enclose the ring volume from above and below (Fig. 12).

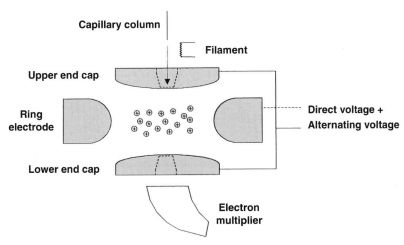

Fig. 12. Principle of the ion trap

Sample molecules are ionised in the resulting hollow space by electrons from an incandescent wire and then accelerated by alternating electric fields in complex trajectories. Thus the ion trap is successively "filled" initially, e.g. with ionised sample molecules that have been previously separated from a mixture by gas chromatography. It is possible to select which ions adopt stable or unstable flight paths within the trap by varying the voltages. In contrast to the quadrupole, in which the ions with stable trajectories impact on the detector and those with unstable trajectories lead to discharge, ions with stable flight paths remain in the trap while the ions of interest are accelerated towards a detector, generally through an aperture in the lower end cap electrode [81].

The most important differences between ion traps and quadrupole or sector field instruments arise from the different procedures for sample ionisation and subsequent detection:

– ions can be "enriched" in the trap, i.e. the time in which the sample molecules flow into the trap with the carrier gas can be varied within certain parameters. As in quadrupole and sector field MS, the areas of components separated by gas chromatography must be defined by several measurement points. No enrichment takes place in instruments that are in continuous operation. The intensity of the subsequently generated ion beam can only be as high as the intensity of the non-ionised sample molecule stream.

– ionisation occurs only when a previously defined optimum "sample concentration" has been reached in the trap. For this purpose the mass spectrometer carries out a brief "pre-scan" before each measurement. In quadrupole and sector field instruments ionisation occurs continuously and independently from the mass concentration in the ion source.

– in the trap all the ions to be detected are sorted very rapidly according to their mass, accelerated towards the detector and recorded. In quadrupole and sector field instruments only one type of ion with a certain m/z ratio is recorded. When a mass spectrum of 1 to 500 m/z is recorded at a scan rate of 2 s^{-1}, 1000 masses per second can be analysed, which is equivalent to a measurement duration of 1 millisecond per ion. The measurement duration per ion in an ion trap mass spectrometer comprises practically the entire ionisation period of the enriched sample (up to 25 milliseconds), as all the enriched ions of a certain m/z ratio are recorded in the subsequent scan.

As a result of the system differences, an ion trap mass spectrometer exhibits considerably higher sensitivity than a quadrupole or a sector field instrument when recording mass spectra. The difference becomes even greater the broader the range scanned because the measurement time per ion diminishes as the scan range increases in continuously operating instruments, while it remains practically unchanged in the case of ion traps. Conversely, an ion trap mass spectrometer permits the recording of complete mass spectra in concentration ranges in which quadrupole and also sector field instruments already have to operate in the selected ion monitoring mode (SIM) (0.01 to 0.1 mg/L). This aspect is particularly advantageous when the component under investigation is present in the sample only in very small concentrations and an unambiguous identification is required. Moreover, ion trap mass spectrometry also permits investigation of unknown samples, i.e. screening of samples on the basis of complete mass spectra. Thus substances that were not originally sought can be detected in the trace range. This is not possible in the case of the SIM technique, which operates on the principle of the selection of previously known substance-characteristic ions.

However, ion trap mass spectrometry also has disadvantages compared with the quadrupole and sector field techniques:

- the typical mass resolution of ion traps is between 500 and 1000. Thus more exact analysis, e.g. of isotope distribution, usually proves difficult.
- like the correct selection of the sample concentration, optimisation of the duration of sample enrichment in the ion trap and ionisation depend to a large extent on the experience of the analyst. The theory and practice of ion trap mass spectrometry are more complicated than in the case of the quadrupole technique, especially in combination with chemical ionisation or tandem mass spectrometry (see below).
- ion traps show poorer repeatability than quadrupoles. Whereas the repeatability in quadrupole or sector field analysis is less than 2% (multiple injection of one sample), it can be up to 10% in ion trap MS, depending on the consistency of the sample. This imprecision is caused by the complex processes in the ion trap and the complicated pressure and flow control technique. One possible way of compensating for these effects is to use isotope-labelled compounds as internal standards: provided they have the same chemical structure, these analytes behave exactly like the unlabelled target compounds in the mass spectrometer. Fluctuations in the signal intensity can be compensated mathematically. However, a prerequisite of this method is the availability of isotope-labelled compounds.

4.4.4 Tandem mass spectrometry (MS/MS)

As described above, ion trap mass spectrometry permits very efficient suppression of matrix constituents or interfering components. However, the sensitivity of this procedure is not considerably better than that of quadrupole MS in the SIM-mode, while the reproducibility is rather poorer. A considerable improvement in the signal/background ratio and thus in sensitivity can be achieved by means of tandem mass spectrometry (tandem MS, MS/MS). Moreover, this technique is being increasingly used as a detector in conjunction with high performance liquid chromatography (HPLC).

On principle, quadrupole mass spectrometers, sector field instruments and ion traps are all suitable for tandem MS. In the first case described above the individual steps of the mass spectrometric analysis take place sequentially at different sites in the mass spectrometer (separation of the parent ions in the first quadrupole, absorption of impact energy in the collision cell, separation of the daughter ions in the second quadrupole or in the magnetic sector field). This technique is therefore known as "tandem in space". In ion trap instruments the processes take place in chronological order at the same site, i.e. in the trap. Accordingly, this technique is called "tandem in time" (Fig. 13).

Tandem in space

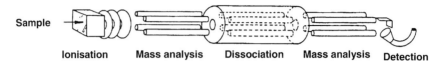

Sample → Ionisation   Mass analysis   Dissociation   Mass analysis   Detection

Tandem in time

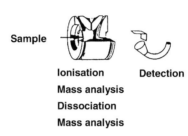

Sample →

Ionisation   Detection
Mass analysis
Dissociation
Mass analysis

Fig. 13. Principle of tandem mass spectrometry with sequentially arranged quadrupoles ("tandem in space") and ion traps ("tandem in time")

In tandem mass spectrometry with quadrupoles the sample molecules transported by the carrier gas into the ion source are initially ionised and accelerated into the first mass filter. There one ion with a specific m/z ratio or a certain m/z range is separated as already described for the SIM or scan technique. However, the selected ions are not detected after leaving the quadrupole, but enter a second quadrupole, in which they are induced to further fragment due to high energy impact with an inert gas (reduction of the mean free path due to increase in pressure).

This process is normally known as "collision-induced dissociation" (CID) or "collision-activated decomposition" (CAD) and generates "daughter ions". Daughter ions are formed from molecule ion fragments that are called parent ions or precursor ions in this context. The spectrum of the daughter ions is recorded in the second mass filter (scan mode). Alternatively, the analysis can also be restricted to individual, characteristic daughter ions (selected ion monitoring mode). Tandem mass spectrometry therefore enables the separation of undesirable components as part of a mass selective clean-up in the first step and allows selection of substance-specific fragments or mass ranges in the second step.

The sequential arrangement of two quadrupoles for separation according to the m/z ratio permits free variation of the scan and SIM techniques:

– A combination of the first quadrupole running in the SIM mode and the second quadrupole in the scan mode enables clarification or confirmation of the structure of a compound present in the trace concentration range in a complex matrix. A parent ion (e.g. a molecule ion or a characteristic fragment) is initially isolated, then further fragmentation is induced by collision, leading to the formation of daughter ions. The scan of these reaction products (daughter ion scan) generally yields substance-specific structural information.

– The SIM-SIM combination permits detection of substance-specific daughter ions with the highest possible intensity, i.e. it is especially suitable for highly specific and highly sensitive detection. In trace analysis it is normally used to detect an individual daughter ion (selected reaction monitoring, SRM) or a few fragments (multiple reaction monitoring, MRM).

– Only certain daughter ions that have been cleaved from a parent ion with a similar structure (parent ion scan) are recorded using the scan-SIM combination. Thus substance groups with certain characteristic structures or fragmentation reactions can be identified.

– The scan-scan combination is used for a "neutral loss scan", i.e. it can be used to identify all the parent ions from which neutral fragments of a certain mass are lost. For this purpose both quadrupoles are operated in the scan mode, but the mass range is shifted to correspond with the neutral fragment, e.g. a McLafferty product. The detector then displays only daughter ions whose parent ion shows a loss of the relevant neutral fragment.

In general, a "gentle" initial ionisation, e.g. by means of PCI or NCI, is preferable in tandem mass spectrometry. The intensive fragmentation that ions undergo as a result of electron impact ionisation leads to a low intensity of the individual ions. As interfering components are mainly separated by the first quadrupole, a lower fragmentation rate with higher ion intensities is more favourable. The collision cell in continuously operating tandem mass spectrometers is frequently composed of a quadrupole or octopole rod system in which only one alternating electrical voltage is applied to focus the ion beam. For this reason these instruments are also known as "triple quadrupole" systems. As a rule, a noble gas (helium, argon, xenon) or nitrogen with a pressure of about 10^{-4} mbar is used as the reactant gas for the collision. The energies transferred during collision are in the range of 50 to 100 eV.

In ion traps the ion selection processes and the collisions within the trap are staggered in time. After the primary ionisation, the ions with masses higher and lower than those desired are then removed from the source by appropriate adjustment of the voltage. An alternating voltage that is in resonance with the circulation frequency of the ions is subsequently applied between the end caps. The ions are kinetically excited in this manner and fragment when they collide with the carrier gas atoms (usually helium). The resulting daughter ions are retained in the trap and can be subsequently detected in the scan or SIM mode.

For quadrupole and sector field instruments as well as ion traps the yield of the CID process is of key importance. While the impact energy and the pressure have to be exactly optimised in the continuously operating instruments to avoid e.g. dispersion effects at excessive pressures, the duration and the strength of the resonance voltage must be optimised in the ion trap.

Compared with the triple quadrupole instrument, the ion trap technique exhibits a higher fragmentation yield and a better daughter ion transmission to the detector. Furthermore, complete mass spectra can be recorded, as in the conventional ion trap technique, even in the trace range, while triple quadrupole instruments must already be operated in the SIM mode for comparable sensitivity. The accessibility of a broader mass range and the possibility of performing the parent ion and neutral loss scans are the most important advantages of the triple quadrupole instruments. In addition, as already mentioned when the conventional technique was described, the reproducibility of the tandem technique is considerably better in the case of quadrupole and sector field instruments. However, these shortcomings of the ion trap technique can be partly compensated by isotope-labelled standards, and their use is particularly recommended in this case.

Tandem MS instruments have rarely been used for detection in routine analysis of xenobiotics in biological material. Therefore no example is yet to be found in the Deutsche Forschungsgemeinschaft's collection of methods in this series. The first application of the MS/MS technique was a procedure to assay polycyclic musk compounds (PMC) in blood completed in 2003 [82].

This example demonstrates the superior signal/background ratio of tandem MS compared with single quadrupole mass spectrometry when used for the analysis of samples from complex matrices (Fig. 14). The signal/background ratio increases from 217 to 1048, thus the sensitivity is enhanced by a factor of five.

This method also shows another important advantage of tandem MS over the conventional quadrupole technique, the possibility of recording complete fragment mass spectra, even close to the detection limit (Fig. 15).

To date ion trap tandem MS has not been used for certain applications, e.g. chlorinated compounds, because only slight further fragmentation of the parent ions was possible. However, the ion trap technique can be distinctly superior to the quadrupole technique in the case of readily fragmentable analytes such as the polycyclic musk compounds. If the polycyclic musk compounds are enriched by a factor of 50 during processing, detection limits of 0.1 µg/L can be achieved using the ion trap technique in the MS/MS mode. This is only a fifth of the detection limit achieved by MS/MS detection with a triple quadrupole mass spectrometer and with the same sample preparation (Figs. 16, 17).

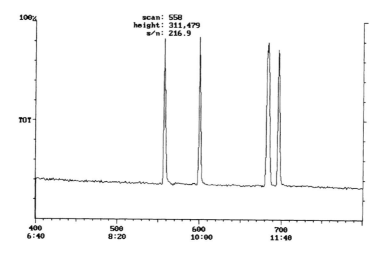

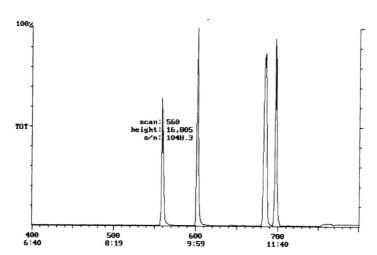

Fig. 14. Total ion chromatogram in the MS mode (above) and total ion chromatogram of the daughter ions of selected parent ions in the MS/MS mode (below) of a mixture of the polycyclic musk compounds ADBI (4-acetyl-1,1-dimethyl-6-*tert*.-butyldihydroindene), AHDI (6-acetyl-1,1,2,3,3,5-hexamethyldihydroindene), HHCB (1,3,4,6,7,8-hexahydro-4,6,6,7,8,8-hexamethylcyclopenta-[g]-2-benzopyrane) and AHTN (7-acetyl-1,1,3,4,4,6-hexamethyltetrahydronaphthalene) (in the order of elution, concentration: 1 mg/L (equivalent to 20 µg/L in blood)), detected by an ion trap mass spectrometer (Varian Saturn 3)

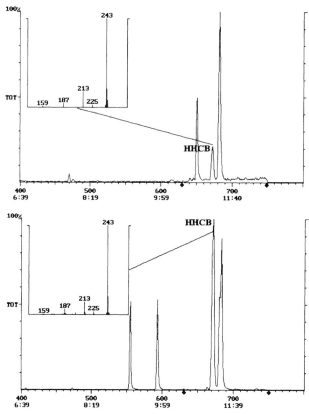

Fig. 15. Daughter ion chromatograms and mass spectra after MS/MS (above: HHCB spectrum of a processed blood sample (0.12 µg/L), below: HHCB spectrum of a calibration standard (4 µg/L)

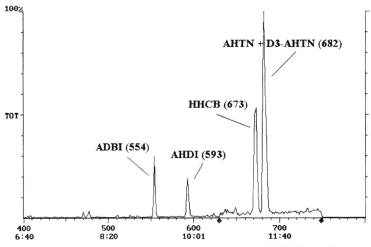

Fig. 16. Daughter ion chromatogram (full scan) following ion trap MS/MS of a blood sample spiked with 1 µg/L each of AHDI, HHCB and AHTN (D$_3$-AHTN: internal standard: 4 µg/L) (fragmentation on the y-axis)

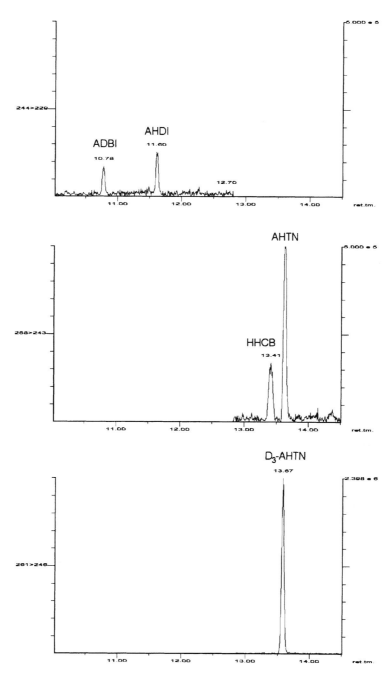

Fig. 17. Daughter ion chromatograms (full scan) following triple quadrupole MS/MS of blood sample spiked with 5 µg/L each of ADBI, AHDI, HHCB and AHTN (D$_3$-AHTN: internal standard: 5 µg/L) (fragmentation on the y-axis)

4.5 Detectors

Ions generated in the ion source and selected in the mass filter are normally detected by a secondary electron multiplier (SEM) or a photon multiplier.

All electron multipliers function according to the same principle (Fig. 18). A beam of positive ions that hits an impact plate coated with copper/beryllium oxide (conversion dynode) causes the release of "primary electrons". These electrons are accelerated by an electrical potential (1 to 3 kV) towards a further dynode and release further electrons ("secondary electrons"). Dynodes (about 10 to 15 dynodes) arranged in sequence generate an electron cascade that is converted as an electrical potential by an amplifier to a measurable signal (Fig. 18a). The amplification achieved by an electron multiplier is in the magnitude of 10^4 to 10^7 secondary electrons per primary electron. In the case of a photon multiplier the electrons pass a plate containing phosphorus that emits photons in the direction of a photocathode. The release of electrons from the alkaline metal coating there also results in a measurable current as a signal.

So-called "channeltrons" are shaped like a horn with the larger aperture pointing towards the mass filter (Fig. 18b). An insulating coating of lead oxide enables a large potential difference to be applied between both ends of the multiplier. In this case the electron cascade travels continuously from the entrance of the channeltron to the amplifier. This design is more compact and cheaper than the electron multiplier with individual dynodes and is therefore very frequently used.

Changes in the multiplier voltage influence the measurement signal to a large extent: a reduction of 200 volts can cause the intensity to diminish by 90%. This is particu-

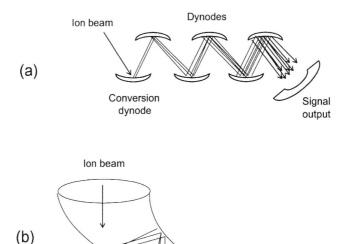

Fig. 18. Principle of the electron multiplier (a=dynodes, b=channeltron)

larly interesting if the samples to be investigated have relatively high unavoidable concentrations that would lead to overloading of the multiplier or to a signal outside the linear range (about 5 to 6 orders of magnitude). On the other hand an increase in the multiplier voltage in order to enhance the measurement signal or to improve the detection limit is advisable only if the background noise is low or can be further reduced by the selection of the general mass spectrometric conditions (suppression of the background signal, ion selection, measurement time per ion, etc.); otherwise amplifying the signals also increases the background noise.

Depending on the frequency of use, the operational life of an SEM is about 1 to 3 years. As electron multipliers have a finite number of secondary electrons that can be released, the number decreases with each cascade. The state of the multiplier can be relatively easily assessed by the voltage that has to be set by the GC/MS system during autotuning to reach the standard specifications. Normally the electron multiplier voltage in a new instrument is about 1000 to 1200 volts. This voltage rises to a maximum of 3000 volts with increasing use of the electron multiplier. In this case the multiplier must be replaced. However, a rise in the voltage can also be caused by a contaminated ion source that gives a poor ionisation yield. In this case the electron multiplier adjusts the voltage to a high value that amplifies even small ion beams to a signal of a previously set intensity. If the voltage is readjusted to the range below 1500 volts after the ion source has been cleaned, the multiplier is still functional. If the voltage remains high despite optimal maintenance of the ion source, the electron multiplier must be replaced. Wherever possible, only "clean" samples with low concentrations should be injected in order to prolong the useful life of a multiplier. In addition, data recording can be limited to the retention time range of interest in the gas chromatogram. Most GC/MS systems permit the multiplier to be switched off before and after this time interval.

5 Summary

Gas chromatography in combination with mass spectrometry is one of the key detection techniques used in occupational and environmental medicine. Lower exposure to xenobiotic substances at the workplace and increasing interest in exposure from the environment have fuelled the need for both sensitive and also substance-specific methods for biological monitoring in concentrations as low as the ppt range.

On the one hand, GC/MS meets these requirements with its broad dynamic working range of several orders of magnitude and its high reproducibility (detector stability, ionisation yield, fragmentation pattern). On the other, the mass spectrometer provides considerably more information on the chemical structure of the relevant analytes than other detectors normally used with gas chromatography. Complete mass spectra, even at low concentrations in the ppm range, can be recorded by suitable selection of the instrumental parameters, thus enabling clear identification of substances. As a rule, the detection limit of a method can be improved by up to two orders of magnitude by limiting the analysis to a few specific molecule or fragment ions in the selected ion monitoring mode, which usually poses no problems. At present four mass spectrometric techniques are mainly used in occupational and environmental medicine: single quadrupole MS, sector field MS, triple quadrupole MS ("tandem in space") and ion trap MS ("tandem in time)". From a technical point of view the difference between these mass spectrometers lies above all in their resolution power and their sensitivity. Both properties are especially important in the analysis of isomer mixtures (e.g. polychlorinated biphenyls, dioxins and furans) and in very low concentration ranges, in which analytical detection is limited by interfering components of the matrix or other analytes present in excess.

Sector field mass spectrometry is remarkable for its very high resolution, while the main advantage of ion traps lies in their ability to record complete mass spectra, even in the trace range. Triple quadrupole instruments permit a very efficient suppression of undesirable signals or matrix components, and their robustness is comparable with that of quadrupole instruments.

Quadrupole GC/MS, which is regarded as a standard technique on account of its current widespread application, has reached its limits in the above-mentioned cases. However, quadrupole instruments are suitable for most investigations in occupational and environmental medicine on account of their considerably simpler instrumental technology, their high ionisation and mass filter stability, and the possibility of "classical" spectra interpretation. This is especially true as the detection power and selectivity can be further enhanced by the use of chemical ionisation. The CI mode permits a "gentle" ionisation of sample molecules, so fragmentation reactions can be suppressed to a large extent. In the case of analysis of compounds containing halogens or aromatic substances, sensitivity can be improved by detecting negatively charged ions to include the range accessible to an electron capture detector.

The selection of the optimum mass spectrometer for an analysis in occupational or environmental medicine principally depends on the particular application: the methods already published in this series can and should provide important pointers to the development of further biological monitoring parameters on the basis of GC/MS.

6 References

[1] *J. Angerer, W.A. König, G. Machata, H. Muffler, K.H. Schaller* and *E. Schulte:* Gas chromatographic methods for the determination of organic substances in biological material. In: J. Angerer and K.H. Schaller (eds.) Analyses of Hazardous Substances in Biological Materials. Volume 3, pp. 1–44. VCH Verlagsgesellschaft (1991).

[2] *H.J. Hübschmann:* Handbuch der GC/MS. Grundlagen und Anwendung, 1st edition. Wiley-VCH (1996).

[3] *M. Schallies:* Flüssigkeits- und Dünnschichtchromatographie. In: H. Naumer and W. Heller (eds.) Untersuchungsmethoden in der Chemie, 2nd edition. Georg Thieme Verlag (1986).

[4] *M. Oehme:* Praktische Einführung in die GC/MS-Analytik mit Quadrupolen. Grundlagen und Anwendungen. In: W. Dünges (ed.) Handbibliothek Chemie. Hüthig GmbH, Heidelberg (1996).

[5] *G. Müller:* Thiodiessigsäure (Bis(carboxymethyl)sulfid). Bestimmung in Harn. In: J. Angerer and K.H. Schaller (eds.) Analysen in Biologischem Material. Loose-leaf collection, 6th issue. VCH Verlagsgesellschaft (1982).

[6] *H. Pauschmann:* Gaschromatographie (GC). In: H. Naumer and W. Heller (eds.) Untersuchungsmethoden in der Chemie, 2nd edition. Georg Thieme Verlag (1986).

[7] *G. Müller* and *E. Jeske:* S-Phenylmercapturic acid in urine. In: J. Angerer and K. H. Schaller (eds.) Analyses of Hazardous Substances in Biological Materials. Volume 5, pp. 143–162. VCH Verlagsgesellschaft (1996).

[8] *N. van Sittert, J. Angerer, M. Bader, M. Blaszkewicz, D. Ellrich, A. Krämer* and *J. Lewalter:* N-2-Cyanoethylvaline, N-2-hydroxyethylvaline, N-methylvaline in blood. In: J. Angerer and K. H. Schaller (eds.) Analyses of Hazardous Substances in Biological Materials. Volume 5, pp. 181–210. VCH Verlagsgesellschaft (1996).

[9] *W. Krämer, W. Merz* and *W. Ziemer:* Chlorophenoxycarboxylic acids in urine (4-chloro-2-methylphenoxyacetic acid, 2,4-dichlorophenoxyacetic acid, 4-chloro-2-methylphenoxypropionic acid, 2,4-dichlorophenoxypropionic acid). In: J. Angerer and K. H. Schaller (eds.) Analyses of Hazardous Substances in Biological Materials. Volume 5, pp. 77–96. VCH Verlagsgesellschaft (1996).

[10] *J. Angerer, W. Butte, H. W. Hoppe, G. Leng, J. Lewalter, R. Heinrich-Ramm* and *A. Ritter:* Pyrethroid metabolites in urine (*cis*-3-(2,2-dichlorovinyl)-2,2-dimethylcyclopropane-1-carboxylic acid, *trans*-3-(2,2-dichlorovinyl)-2,2-dimethylcyclopropane-1-carboxylic acid, *cis*-3-(2,2-dibromovinyl)-2,2-dimethylcyclopropane-1-carboxylic acid, 3-phenoxybenzoic acid, 4-fluoro-3-phenoxybenzoic acid). In: J. Angerer and K. H. Schaller (eds.) Analyses of Hazardous Substances in Biological Materials. Volume 6, pp. 231–254. Wiley-VCH (1999).

[11] *H. W. Hoppe:* Pentachlorophenol in urine and serum/plasma. In: J. Angerer and K. H. Schaller (eds.) Analyses of Hazardous Substances in Biological Materials. Volume 6, pp. 189–210. Wiley-VCH (1999).

[12] *J. Angerer:* Chlorophenols (2,4-dichlorophenol, 2,5-dichlorophenol, 2,6-dichlorophenol, 2,3,4-trichlorophenol, 2,4,5-trichlorophenol, 2,4,6-trichlorophenol, 2,3,4,6-tetrachlorophenol). In: J. Angerer and K. H. Schaller (eds.) Analyses of Hazardous Substances in Biological Materials. Volume 7, pp. 143–170. Wiley-VCH (2001).

[13] *J. Lewalter* and *W. Gries:* Haemoglobin adducts of aromatic amines: aniline, o-, m- and p-toluidine, o-anisidine, p-chloroaniline, *α*- and *β*-naphthylamine, 4-aminodiphenyl, benzidine, 4,4′-diaminodiphenylmethane, 3,3′-dichlorobenzidine. In: J. Angerer and K. H. Schaller (eds.) Analyses of Hazardous Substances in Biological Materials. Volume 7, pp. 191–220. Wiley-VCH (2001).

[14] *J. Lewalter, G. Leng* and *D. Ellrich:* N-benzylvaline after exposure to benzylchloride in blood. In: J. Angerer and K. H. Schaller (eds.) Analyses of Hazardous Substances in Biological Materials. Volume 8, pp. 35–52. Wiley-VCH (2003).

[15] *M. Müller:* Cotinine in urine. In: J. Angerer and K. H. Schaller (eds.) Analyses of Hazardous Substances in Biological Materials. Volume 8, pp. 53–67. Wiley-VCH (2003).

[16] *J. Lewalter, G. Skarping, D. Ellrich* and *U. Schoen:* Hexamethylene diisocyanate (HDI) and hexamethylenediamine (HDA) in urine. In: J. Angerer and K. H. Schaller (eds.) Analyses of Hazardous Substances in Biological Materials. Volume 8, pp. 119–132. Wiley-VCH (2003).

[17] *M. Bader, T. Göen* and *J. Angerer:* 1-(4-(1-Hydroxy-1-methylethyl)-phenyl)-3-methylurea (HMEPMU) as a metabolite of isoproturon in urine. In: J. Angerer and K. H. Schaller (eds.) Analyses of Hazardous Substances in Biological Materials. Volume 8, pp. 151–166. Wiley-VCH (2003).

[18] *H. W. Hoppe* and *T. Weiss:* Organochlorine compounds in whole blood and plasma. In: J. Angerer and K. H. Schaller (eds.) Analyses of Hazardous Substances in Biological Materials. Volume 8, pp. 187–220. Wiley-VCH (2003).

[19] *M. Ball:* Dioxins, furans and WHO PCB in whole blood. In: J. Angerer and K. H. Schaller (eds.) Analyses of Hazardous Substances in Biological Materials. Volume 8, pp. 85–118. Wiley-VCH (2003).

[20] *K. Hauff* and *R. Schierl:* Oxazaphosphorines: Cyclophosphamide and ifosfamide in urine. In: J. Angerer and K. H. Schaller (eds.) Analyses of Hazardous Substances in Biological Materials. Volume 8, pp. 221–238. Wiley-VCH (2003).

[21] *K. Kaaria, A. Hirvonen, H. Norppa, P. Piirila, H. Vainio* and *C. Rosenberg:* Exposure to 2,4- and 2,6-toluene diisocyanate (TDI) during production of flexible foam: determination of airborne TDI and urinary 2,4- and 2,6-toluenediamine. Analyst 126 (7), 1025–1031 (2001).

[22] *G. Sabbioni, R. Hartley* and *S. Schneider:* Synthesis of adducts with amino acids as potential dosimeters for the biomonitoring of humans exposed to toluenediisocyanate. Chem Res Toxicol 14, 1573–1583 (2001).

[23] *T. Weiss* and *J. Angerer:* Simultaneous determination of various aromatic amines and metabolites of aromatic nitro compounds in urine for low level exposure using gas-chromatography-mass spectrometry. J. Chromatogr. B 778, 179–192 (2002).

[24] *C. J. Sennbro, C. H. Lindh, H. Tinnerberg, C. Gustavsson, M. Littorin, H. Welinder* and *B. A. Jönsson:* Development, validation and characterisation of an analytical method for the quantification of hydrolysable urinary metabolites and plasma protein adducts of 2,4- and 2,6-toluene diisocyanate, 1,5-naphthalene diisocyanate and 4,4′-methylenediphenyl diisocyanate. Biomarkers 8, 204–217 (2003).

[25] *J. Dmitrovic, S. C. Chan* and *S. H. Chan:* Analysis of pesticides and PCB congeners in serum by GC/MS with SPE sample cleanup. Toxicol Letters 134, 253–258 (2002).

[26] *C. Thomsen, L. S. Haug, H. Leknes, E. Lundanes, G. Becher* and *G. Lindström:* Comparing electron ionization high-resolution and electron capture low-resolution mass spectrometric determination of polybrominated diphenyl ethers in plasma, serum and milk. Chemosphere 46, 641–648 (2002).

[27] *R. Turci, F. Bruno* and *C. Minoia:* Determination of coplanar and non-coplanar polychlorinated biphenyls in human serum by gas chromatography with mass spectrometric detection: electron impact or electron-capture negative ionization? Rapid Commun Mass Spectrom 17, 1881–1888 (2003).

[28] *J. F. Focant, J. W. Cochran, J. M. Dimandja, E. DePauw, A. Sjodin, W. E. Turner* and *D. G. Patterson:* High-throughput analysis of human serum for selected polychlorinated biphenyls (PCBs) by gas chromatography-isotope dilution time-of-flight mass spectrometry (GC-ID-TOFMS). Analyst 129, 331–336 (2004).

[29] *J. Hardt* and *J. Angerer:* Determination of dialkyl phosphates in human urine using gas chromatography-mass spectrometry. J. Anal. Toxicol. 24, 678–684 (2000).

[30] *M. J. Perry, D. C. Christiani, J. Mathew, D. Degenhardt, J. Tortorelli, J. Strauss* and *W. C. Sonzogni:* Urinalysis of atrazine exposure in farm pesticide applicators. Toxicol. Ind. Health 16, 285–290 (2001).

[31] *K. Wittke, H. Hajimiragha, L. Dunemann* and *J. Begerow:* Determination of dichloroanilines in human urine by GC-MS, GC-MS-MS and GC-ECD as markers of low-level pesticide exposure. J. Chromatogr. B. 755, 215–228 (2001).

[32] *U. Heudorf* and *J. Angerer:* Metabolites of organophosphorous insecticides in urine specimens from inhabitants of a residential area. Environ. Res. 86, 80–87 (2001).

[33] *G. Leng, U. Ranft, D. Sugiri, W. Hadnagy, E. Berger-Preiss* and *H. Idel:* Pyrethroids used indoors – biological monitoring of exposure to pyrethroids following an indoor pest control operation. Int. J. Hyg. Environ. Health 206, 85–92 (2003).

[34] *J. B. Laurens, X. Y. Mbianda, J. H. Spies, J. B. Ubbink* and *W. J. Vermaark:* Validated method for quantitation of biomarkers for benzene and its alkylated analogues in urine. J. Chromatogr. B. 774, 173–185 (2002).

[35] *A. Barbieri, A. Accorsi, G. B. Raffi, L. Nicoli, F. S. Violante:* Lack of sensitivity of urinary trans,trans-muconic acid in determining low-level (ppb) benzene exposure in children. Arch. Environ. Health 57, 224–228 (2004).

[36] *G. A. Jacobson* and *S. McLean:* Biological monitoring of low level occupational xylene exposure and the role of recent exposure. Ann. Occup. Hyg. 47, 331–336 (2003).

[37] *A. Accorsi, A. Barbieri, G. B. Raffi* and *F. S. Violante:* Biomonitoring of exposure to nitrous oxide, sevoflurane, isoflurane and halothane by automated GC/MS headspace analysis. Int. Arch. Occup. Environ. Health 74, 541–548 (2001).

[38] *A. Gentili, A. Accorsi, A. Pigna, V. Bachiocco, I. Domenichini, S. Baroncini* and *F. S. Violante:* Exposure of personnel to sevoflurane during paediatric anaesthesia: influence of professional role and anaesthetic procedure. Eur. J. Anaesthesiol. 21, 638–645 (2004).

[39] *T. Nakao, O. Aozasa, S. Ohta* and *H. Miyata:* Assessment of human exposure to PCDDs, PCDFs and Co-PCBs using hair as a human pollution indicator sample I: development of an

analytical method for human hair and evaluation for exposure assessment. Chemosphere 48, 885–896 (2002).

[40] *I. Zwirner-Baier, K. Deckart, R. Jäckh* and *H. G. Neumann:* Biomonitoring of aromatic amines VI: determination of hemoglobin adducts after feeding aniline hydrochloride in the diet of rats for 4 weeks. Arch. Toxicol. 77, 672–677 (2003).

[41] *C. Rosenberg, K. Nikkila, M. L. Henriks-Eckerman, K. Peltonen* and *K. Engstrom:* Biological monitoring of aromatic diisocyanates in workers exposed to thermal degradation products of polyurethanes. J. Environ. Monit. 4, 711–716 (2002).

[42] *A. Fidder, D. Noort, A. G. Hulst, L. P. DeJong* and *H. P. Benschop:* Biomonitoring of exposure to lewisite based on adducts to haemoglobin. Arch. Toxicol. 74, 207–214 (2000).

[43] *C. Prado, J. Garrido* and *J. F. Periago:* Urinary benzene determination by SPME/GC-MS. A study of variables by fractional factorial design and response surface methodology. J. Chromatogr. B 804, 255–261 (2004).

[44] *K. Semchuk, H. McDuffie, A. Senthilselvan, A. Cessna* and *D. Irvine:* Body mass index and bromoxynil exposure in a sample of rural residents during spring herbicide application. J. Toxicol. Environ. Health A 67, 1321–1352 (2004).

[45] *S. Szucs, L. Toth, J. Legoza, A. Sarvary* and *R. Adany:* Simultaneous determination of styrene, toluene and xylene metabolites in urine by gas chromatography/mass spectrometry. Arch. Toxicol 76, 560–569 (2002).

[46] *N. J. van Sittert, H. J. Megens, W. P. Watson* and *P. J. Boogaard:* Biomarkers of exposure to 1,3-butadiene as a basis for cancer risk assessment. Toxicol. Sci. 56, 189–202 (2000).

[47] *P. Begemann, R. J. Sram* and *H. G. Neumann:* Hemoglobin adducts of epoxybutene in workers occupationally exposed to 1,3-butadiene. Arch. Toxicol. 74, 680–687 (2001).

[48] *T. Göen, G. Korinth* and *H. Drexler:* Butoxyethoxyacetic acid, a biomarker of exposure to water-based cleaning agents. Toxicol. Letters 134, 295–300 (2002).

[49] *D. L. Hughes, D. J. Ritter* and *R. D. Wilson:* Determination of 2,4-dichlorophenoxyacetic acid (2,4-D) in human urine with mass selective detection. J. Environ. Sci. Health B 36, 755–764 (2001).

[50] *L. M. Gonzalez-Reche, T. Schettgen* and *J. Angerer:* New approaches to the metabolism of xylenes: verification of the formation of phenylmercapturic acid metabolites of xylenes. Arch. Toxicol. 77, 80–85 (2003).

[51] *F. Sandner, J. Fornara, W. Dott* and *J. Hollender:* Sensitive biomonitoring of monoterpene exposure by gas chromatography-mass spectrometric measurement of hydroxy terpenes in urine. J. Chromatogr. B 780, 225–230 (2002).

[52] *A. Accorsi, S. Valenti, A. Barbierei, G. B. Raffi* and *F. S. Violante:* Enflurane as an internal standard in monitoring halogenated volatile anaesthetics by headspace gas chromatography-mass spectrometry. J. Chromatogr. A 985, 259–264 (2003).

[53] *S. Fustinoni, L. Campo, C. Colosio, S. Birindelli* and *V. Foa:* Application of gas chromatography-mass spectrometry for the determination of urinary ethylenethiourea in humans. J. Chromatogr. B 814, 251–258 (2005).

[54] *B. Åkesson* and *B. A. G. Jönsson:* Biological monitoring of N-methyl-2-pyrrolidone using 5-hydroxy-N-methyl-2-pyrrolidone in plasma and urine as the biomarker. Scand. J. Work Environ. Health 26, 213–218 (2000).

[55] *P. Akrill, J. Cocker* and *S. Dixon:* Dermal exposure to aqueous solutions of N-methyl-2-pyrrolidone. Toxicol. Letters 134, 265–269 (2002a).

[56] *B. A. G. Jönsson* and *B. Åkesson:* Human experimental exposure to N-methyl-2-pyrrolidone (NMP): toxicokinetics of NMP, 5-hydroxy-N-methyl-2-pyrrolidone, N-methylsuccinimide and 2-hydroxy-N-methylsuccinimide (2-HMSI), and biological monitoring using 2-HMSI as a biomarker. Int. Arch. Occup. Environ. Health 76, 267–274 (2003).

[57] *S. Letzel, T. Göen, M. Bader, T. Kraus* and *J. Angerer:* Exposure to nitroaromatic explosives and health effects during disposal of military waste. Occup. Environ. Med. 60, 483–488 (2003).

[58] *P. Akrill, R. Guiver* and *J. Cocker:* Biological monitoring of nitroglycerin exposure by urine analysis. Toxicol. Letters 134, 271–276 (2002b).

[59] *N. F. van Nimmen, K. L. Poels* and *H. A. Veulemans:* Highly sensitive gas chromatographic-mass spectrometric screening method for the determination of picogram levels of fentanyl, sufetanil and alfetanil and their major metabolites in urine of opioid exposed workers. J. Chromatogr. B 804, 375–387 (2004).

[60] *M. Moreno-Frias, M. Jimenez-Torres, A. Garrido-Frenich, J. L. Martinez-Vidal, F. Olea-Serrano* and *N. Olea:* Determination of organochlorine compounds in human biological samples by GC-MS/MS. Biomed. Chromatogr. 18, 102–111 (2004).

[61] *A. Stambouli, M. A. Bellimam, N. El Karni, T. Bouayoun* and *A. El Bouri:* Optimization of an analytical method for detection paraphenylenediamine (PPD) by GC/MS-ion trap in biological liquids. Forensic Sci. Int. 146 Supplement 1, S87–S92 (2004).

[62] *J. P. Aston, R. L. Ball, J. E. Pople, K. Jones* and *J. Cocker:* Development and validation of a competitive immunoassay for urinary S-phenylmercapturic acid and its application in benzene biological monitoring. Biomarkers 7, 103–112 (2002).

[63] *S. Liu* and *J. D. Pleil:* Human blood and environmental media screening method for pesticides and polychlorinated biphenyl compounds using liquid extraction and gas chromatography-mass spectrometry analysis. J. Chromatogr. B 769, 155–167 (2002).

[64] *J. L. Martinez-Vidal, M. Moreno-Frias, A. Garido-Frenich, F. Olea-Serrano* and *N. Olea:* Determination of endocrine disrupting pesticides and polychlorinated biphenyls in human serum by GC-ECD and GC-MS/MS and evaluation of contributions to the uncertainty of the results. Anal. Bioanal. Chem. 372, 766–775 (2002).

[65] *S. Waidyanatha, Y. Zheng* and *S. M. Rappaport:* Determination of polycyclic aromatic hydrocarbons in urine of coke oven workers by headspace solid phase microextraction and gas chromatography-mass spectrometry. Chem. Biol. Interact. 145, 165–174 (2003).

[66] *C. J. Smith, W. Huang, C. J. Walcott, W. Turner, J. Grainger* and *D. G. Patterson:* Quantification of monohydroxy-PAH metabolites in urine by solid-phase extraction with isotope dilution-GC-MS. Anal. Bioanal. Chem. 372, 216–220 (2002).

[67] *L. Elflein, E. Berger-Preiss, A. Preiss, M. Elend, K. Levsen* and *G. Wünsch:* Human biomonitoring of pyrethrum and pyrethroid insecticides used indoors: determination of the metabolites E-cis/trans-chrysanthemumdicarboxylic acid in human urine by gas chromatography-mass spectrometry with negative chemical ionization. J. Chromatogr. B 795, 195–207 (2003).

[68] *J. Hardt* and *J. Angerer:* Biological monitoring of workers after the application of insecticidal pyrethroids. Int. Arch. Occup. Environ. Health 76, 492–498 (2003).

[69] *T. Schettgen, T. Weiss* and *J. Angerer:* Biological monitoring of phenmedipham: determination of m-toluidine in urine. Arch. Toxicol. 75, 145–149 (2001).

[70] *J. R. Barr, A. R. Woolfitt, V. L. Maggio* and *D. G. Patterson:* Measurement of toxaphene congeners in pooled human serum collected in three U.S. cities using high-resolution mass spectrometry. Arch. Environ. Contam. Toxicol. 46, 551–556 (2004).

[71] *M. Imbriani, Q. Niu, S. Negri* and *S. Ghittori:* Trichloroethylene in urine as biological exposure index. Ind. Health 39, 225–230 (2001).

[72] *M. Hesse, H. Meier* and *B. Zeeh:* Spektroskopische Methoden in der organischen Chemie, 2nd ed. Georg Thieme Verlag, Stuttgart (1984).

[73] *C. H. Suelter* and *J. T. Watson (eds.):* Biomedical applications of mass spectrometry, 1st ed. Wiley Interscience (1990).

[74] *F. W. McLafferty* and *F. Turecek:* Interpretation von Massenspektren, 1st ed. Spektrum Akademischer Verlag (1995).

[75] *G. Wedler:* Lehrbuch der Physikalischen Chemie, 2nd ed. Wiley-VCH, Weinheim (1985).

[76] *W. J. Richter* and *H. Schwarz:* Chemische Ionisation – ein stark an Bedeutung gewinnendes massenspektrometrisches Analysenverfahren. Angewandte Chemie 90, 449–469 (1978).

[77] *A. G. Harrison:* Chemical Ionization Mass Spectrometry, 2nd ed. CRC Press, Boca Raton, USA (1992). ISBN: 0-8493-4254-6

[78] *M. McKeown:* Instrumentation for negative ion detection. Environ. Health Perspect. 36, 97–102 (1980).

[79] *G. C. Stafford:* Instrumental aspects of positive and negative ion chemical ionization mass spectrometry. Environ. Health Perspect. 36, 85–88 (1980).

[80] *M. Bader, J. Lewalter* and *J. Angerer:* Analysis of N-alkylated amino acids in human hemoglobin: evidence for elevated N-methylvaline levels in smokers. Int. Arch. Occup. Environ. Health 67, 237–242 (1995).

[81] *R. G. Cooks, G. L. Glish, S. A. McLuckey* and *R. E. Kaiser:* Ion Trap Mass Spectrometry. Chemical and Engineering News 3, 26–41 (1991).

[82] *W. Butte* and *A. Schmidt:* Bestimmung mehrkerniger (polycyclischer) Moschusverbindungen im Blut mit GC/MS/MS. Manuscript approved and in the process of editorial revision in the "Analyses in Biological Materials" group.

Authors: *M. Bader, W. Butte, H. W. Hoppe, G. Leng*

Substances

Essential Biomonitoring Methods. DFG, Deutsche Forschungsgemeinschaft
Copyright © 2006 WILEY-VCH Verlag GmbH & Co. KGaA, Weinheim
ISBN: 3-527-31478-4

Alcohols and Ketones

Application	Determination in blood and urine
Analytical principle	Capillary gas chromatography Headspace technique
Completed in	February 1996

Summary

The concentrations of alcohols and ketones in blood and in urine due to occupational exposure can be reliably determined using this analytical method.

The highly volatile alcohols and ketones contained in both blood and urine are determined by means of capillary gas chromatography using the headspace technique. For this purpose the blood or urine samples are warmed to 40 °C or 50 °C in airtight crimp top vials. After distribution of the alcohols and ketones between the liquid and vapour phases has reached equilibrium, an aliquot of the headspace is withdrawn and analysed by gas chromatography. A flame ionization detector (FID) serves as a detector. Calibration curves are obtained by analysing blood and urine samples to which known quantities of alcohols and ketones have been added. The resulting peak areas are plotted as a function of the concentrations used.

Methanol

Blood

Within-series imprecision:	Standard deviation (rel.)	$s_w = 2.1\,\%$
	Prognostic range	$u = 4.7\,\%$
	At a concentration of 23.8 mg methanol per litre blood and where $n = 10$ determinations	
Between-day imprecision:	Standard deviation (rel.)	$s = 4.6\,\%$
	Prognostic range	$u = 11.3\,\%$
	At a concentration of 23.8 mg methanol per litre blood and where $n = 6$ days	
Inaccuracy:	Recovery rate	$r = 96 - 98.4\,\%$
Detection limit:	0.6 mg methanol per litre blood	

Essential Biomonitoring Methods. DFG, Deutsche Forschungsgemeinschaft
Copyright © 2006 WILEY-VCH Verlag GmbH & Co. KGaA, Weinheim
ISBN: 3-527-31478-4

Urine

Within-series imprecision:	Standard deviation (rel.) $s_w = 4\%$
	Prognostic range $u = 8.9\%$
	At a concentration of 6.1 mg methanol per litre urine and where $n = 10$ determinations
Between-day imprecision:	Standard deviation (rel.) $s = 7.6\%$
	Prognostic range $u = 18.6\%$
	At a concentration of 6.3 mg methanol per litre urine and where $n = 6$ days
Inaccuracy:	Recovery rate $r = 96.4 - 104.1\%$
Detection limit:	0.6 mg methanol per litre urine

Ethanol

Blood

Within-series imprecision:	Standard deviation (rel.) $s_w = 2.1\%$
	Prognostic range $u = 4.5\%$
	At a concentration of 7.9 mg ethanol per litre blood and where $n = 10$ determinations
Between-day imprecision:	Standard deviation (rel.) $s = 6.9\%$
	Prognostic range $u = 16.9\%$
	At a concentration of 7.9 mg ethanol per litre blood and where $n = 6$ days
Inaccuracy:	Recovery rate $r = 96.4 - 110.9\%$
Detection limit:	1.3 mg ethanol per litre blood

Urine

Within-series imprecision:	Standard deviation (rel.) $s_w = 2.9\%$
	Prognostic range $u = 6.5\%$
	At a concentration of 6.3 mg ethanol per litre urine and where $n = 10$ determinations
Between-day imprecision:	Standard deviation (rel.) $s = 4.4\%$
	Prognostic range $u = 10.8\%$
	At a concentration of 6.3 mg ethanol per litre urine and where $n = 6$ days
Inaccuracy:	Recovery rate $r = 97.1 - 99.7\%$
Detection limit:	0.8 mg ethanol per litre urine

Acetone

Blood

Within-series imprecision: Standard deviation (rel.) $s_w = 1.9\,\%$
Prognostic range $u = 4.2\,\%$
At a concentration of 0.8 mg acetone per litre blood and where $n = 10$ determinations

Between-day imprecision: Standard deviation (rel.) $s = 3.1\,\%$
Prognostic range $u = 7.6\,\%$
At a concentration of 0.8 mg acetone per litre blood and where $n = 6$ days

Inaccuracy: Recovery rate $r = 102.1 - 105.7\,\%$

Detection limit: 0.2 mg acetone per litre blood

Urine

Within-series imprecision: Standard deviation (rel.) $s_w = 2.6\,\%$
Prognostic range $u = 5.8\,\%$
At a concentration of 1.3 mg acetone per litre urine and where $n = 10$ determinations

Between-day imprecision: Standard deviation (rel.) $s = 1.2\,\%$
Prognostic range $u = 2.9\,\%$
At a concentration of 1.3 mg acetone per litre urine and where $n = 6$ days

Inaccuracy: Recovery rate $r = 99.9 - 100.9\,\%$

Detection limit: 0.1 mg acetone per litre urine

2-Propanol (Isopropanol)

Blood

Within-series imprecision: Standard deviation (rel.) $s_w = 2.8\,\%$
Prognostic range $u = 6.2\,\%$
At a concentration of 3.9 mg 2-propanol per litre blood and where $n = 10$ determinations

Between-day imprecision: Standard deviation (rel.) $s = 2.7\,\%$
Prognostic range $u = 6.6\,\%$
At a concentration of 3.9 mg 2-propanol per litre blood and where $n = 6$ days

Inaccuracy: Recovery rate $r = 97.4 - 105.7\,\%$

Detection limit: 0.6 mg 2-propanol per litre blood

Urine

Within-series imprecision:	Standard deviation (rel.) $\quad s_w = 1.4\,\%$ Prognostic range $\qquad\qquad u = 3.1\,\%$ At a concentration of 6.3 mg 2-propanol per litre urine and where $n = 10$ determinations
Between-day imprecision:	Standard deviation (rel.) $\quad s = 1.4\,\%$ Prognostic range $\qquad\qquad u = 3.4\,\%$ At a concentration of 6.3 mg 2-propanol per litre urine and where $n = 6$ days
Inaccuracy:	Recovery rate $\qquad\qquad\qquad r = 94.5-99.5\,\%$
Detection limit:	0.4 mg 2-propanol per litre urine

1-Propanol

Blood

Within-series imprecision:	Standard deviation (rel.) $\quad s_w = 2.1\,\%$ Prognostic range $\qquad\qquad u = 4.7\,\%$ At a concentration of 4 mg 1-propanol per litre blood and where $n = 10$ determinations
Between-day imprecision:	Standard deviation (rel.) $\quad s = 5.9\,\%$ Prognostic range $\qquad\qquad u = 6.1\,\%$ At a concentration of 4 mg 1-propanol per litre blood and where $n = 6$ days
Inaccuracy:	Recovery rate $\qquad\qquad\qquad r = 99.8-106.6\,\%$
Detection limit:	0.8 mg 1-propanol per litre blood

Urine

Within-series imprecision:	Standard deviation (rel.) $\quad s_w = 1.1\,\%$ Prognostic range $\qquad\qquad u = 2.5\,\%$ At a concentration of 6.4 mg 1-propanol per litre urine and where $n = 10$ determinations
Between-day imprecision:	Standard deviation (rel.) $\quad s = 1.8\,\%$ Prognostic range $\qquad\qquad u = 4.4\,\%$ At a concentration of 6.4 mg 1-propanol per litre urine and where $n = 6$ days
Inaccuracy:	Recovery rate $\qquad\qquad\qquad r = 96.3-98.3\,\%$
Detection limit:	0.4 mg 1-propanol per litre urine

2-Butanone (Methyl ethyl ketone)

Blood

Within-series imprecision:	Standard deviation (rel.)	$s_w = 1.6\,\%$
	Prognostic range	$u = 3.6\,\%$
	At a concentration of 0.8 mg 2-butanone per litre blood and where $n = 10$ determinations	

Between-day imprecision:	Standard deviation (rel.)	$s = 5.9\,\%$
	Prognostic range	$u = 14.4\,\%$
	At a concentration of 0.8 mg 2-butanone per litre blood and where $n = 6$ days	

Inaccuracy:	Recovery rate	$r = 99.6 - 101.3\,\%$

Detection limit:	0.1 mg 2-butanone per litre blood

Urine

Within-series imprecision:	Standard deviation (rel.)	$s_w = 1.3\,\%$
	Prognostic range	$u = 2.9\,\%$
	At a concentration of 1.3 mg 2-butanone per litre urine and where $n = 10$ determinations	

Between-day imprecision:	Standard deviation (rel.)	$s = 2.6\,\%$
	Prognostic range	$u = 6.4\,\%$
	At a concentration of 1.3 mg 2-butanone per litre urine and where $n = 6$ days	

Inaccuracy:	Recovery rate	$r = 95.4 - 96.5\,\%$

Detection limit:	0.08 mg 2-butanone per litre urine

2-Butanol

Blood

Within-series imprecision:	Standard deviation (rel.)	$s_w = 1.5\,\%$
	Prognostic range	$u = 3.3\,\%$
	At a concentration of 8 mg 2-butanol per litre blood and where $n = 10$ determinations	

Between-day imprecision:	Standard deviation (rel.)	$s = 0.8\,\%$
	Prognostic range	$u = 5.9\,\%$
	At a concentration of 8 mg 2-butanol per litre blood and where $n = 6$ days	

Inaccuracy:	Recovery rate	$r = 96.8 - 100.2\,\%$

Detection limit:	0.4 mg 2-butanol per litre blood

Urine

Within-series imprecision: Standard deviation (rel.) $s_w = 0.8\%$
                              Prognostic range $u = 1.8\%$
                              At a concentration of 6.2 mg 2-butanol per litre
                              urine and where $n = 10$ determinations

Between-day imprecision: Standard deviation (rel.) $s = 0.8\%$
                            Prognostic range $u = 1.9\%$
                            At a concentration of 6.2 mg 2-butanol per litre
                            urine and where $n = 6$ days

Inaccuracy: Recovery rate $r = 90.9 - 98.6\%$

Detection limit: 0.2 mg 2-butanol per litre urine

2-Methyl-1-propanol (Isobutanol)

Blood

Within-series imprecision: Standard deviation (rel.) $s_w = 3.7\%$
                              Prognostic range $u = 10\%$
                              At a concentration of 8 mg 2-methyl-1-propanol
                              per litre blood and where n = 10 determinations

Between-day imprecision: Standard deviation (rel.) $s = 4.7\%$
                            Prognostic range $u = 11.5\%$
                            At a concentration of 8 mg 2-methly-1-propanol
                            per litre blood and where $n = 6$ days

Inaccuracy: Recovery rate $r = 75.7 - 100.6\%$

Detection limit: 0.4 mg 2-methyl-1-propanol per litre blood

Urine

Within-series imprecision: Standard deviation (rel.) $s_w = 1.1\%$
                              Prognostic range $u = 2.5\%$
                              At a concentration of 6.3 mg 2-methyl-1-propanol
                              per litre urine and where $n = 10$ determinations

Between-day imprecision: Standard deviation (rel.) $s = 0.7\%$
                            Prognostic range $u = 1.7\%$
                            At a concentration of 6.3 mg 2-methyl-1-propanol
                            per litre urine and where $n = 6$ days

Inaccuracy: Recovery rate $r = 89 - 97.6\%$

Detection limit: 0.2 mg 2-methyl-1-propanol per litre urine

1-Butanol

Blood

Within-series imprecision: Standard deviation (rel.) $s_w = 3.7\,\%$
                               Prognostic range $u = 8.2\,\%$
                               At a concentration of 8 mg 1-butanol per litre
                               blood and where $n = 10$ determinations

Between-day imprecision: Standard deviation (rel.) $s = 3.8\,\%$
                               Prognostic range $u = 9.3\,\%$
                               At a concentration of 8 mg 1-butanol per litre
                               blood and where $n = 6$ days

Inaccuracy: Recovery rate $r = 106\,\%$

Detection limit: 0.8 mg 1-butanol per litre blood

Urine

Within-series imprecision: Standard deviation (rel.) $s_w = 1.1\,\%$
                               Prognostic range $u = 2.5\,\%$
                               At a concentration of 6.3 mg 1-butanol
                               per litre urine and where $n = 10$ determinations

Between-day imprecision: Standard deviation (rel.) $s = 1.8\,\%$
                               Prognostic range $u = 4.4\,\%$
                               At a concentration of 6.3 mg 1-butanol per litre
                               urine and where $n = 6$ days

Inaccuracy: Recovery rate $r = 98.9\,\%$

Detection limit: 0.3 mg 1-butanol per litre urine

4-Methyl-2-pentanone (Methyl isobutyl ketone)

Blood

Within-series imprecision: Standard deviation (rel.) $s_w = 0.9\,\%$
                               Prognostic range $u = 2\,\%$

                               At a concentration of 0.8 mg 4-methyl-2-pentanone
                               per litre blood and where $n = 10$ determinations

Between-day imprecision: Standard deviation (rel.) $s = 5.6\,\%$
                               Prognostic range $u = 13.7\,\%$
                               At a concentration of 0.8 mg 4-methyl-2-pentanone
                               per litre blood and where $n = 6$ days

Inaccuracy: Recovery rate $r = 95.8 - 100.2\ \%$

Detection limit: 0.05 mg 4-methyl-2-pentanone per litre blood

Urine

Within-series imprecision: Standard deviation (rel.) $s_w = 1.1\ \%$
                                Prognostic range $u = \ \ 2.5\ \%$
                                At a concentration of 1.3 mg 4-methyl-2-pentanone
                                per litre urine and where $n = 10$ determinations

Between-day imprecision: Standard deviation (rel.) $s = \ \ 3.6\ \%$
                                Prognostic range $u = 8.8\ \%$
                                At a concentration of 1.3 mg 4-methyl-2-pentanone
                                per litre urine and where $n = 6$ days

Inaccuracy: Recovery rate $r = 94.6 - 96\ \%$

Detection limit: 0.03 mg 4-methyl-2-pentanone per litre urine

2-Hexanone

Blood

Within-series imprecision: Standard deviation (rel.) $s_w = 1.8\ \%$
                                Prognostic range $u = \ \ 4\ \%$
                                At a concentration of 0.8 mg 2-hexanone
                                per litre blood and where $n = 10$ determinations

Between-day imprecision: Standard deviation (rel.) $s = \ \ 4.7\ \%$
                                Prognostic range $u = 12.1\ \%$
                                At a concentration of 0.8 mg 2-hexanone per litre
                                blood and where $n = 5$ days

Inaccuracy: Recovery rate $r = 96 - 100.8\ \%$

Detection limit: 0.07 mg 2-hexanone per litre blood

Urine

Within-series imprecision: Standard deviation (rel.) $s_w = 0.8\ \%$
                                Prognostic range $u = \ \ 1.8\ \%$
                                At a concentration of 1.2 mg 2-hexanone
                                per litre urine and where $n = 10$ determinations

Between-day imprecision: Standard deviation (rel.) $s = \ \ 2.9\ \%$
                                Prognostic range $u = 7.1\ \%$
                                At a concentration of 1.2 mg 2-hexanone per litre
                                urine and where $n = 6$ days

Inaccuracy:	Recovery rate	$r = 93.6-98.7\%$
Detection limit:	0.03 mg 2-hexanone per litre urine	

Alcohols

Alcohols are used as organic solvents, cleaning fluids and as starting materials for chemical syntheses (e. g. ester synthesis). In addition, ethanol is a widely consumed stimulant.

Straight-chain alcohols form a homologous series. According to the degree of substitution of the carbon atom bearing the OH group they are classified as primary, secondary or tertiary alcohols. Compounds with more than one OH group are called diols, triols, etc.

The primary alcohols of lower molecular weight are highly volatile liquids. They exhibit narcotic effects which become stronger as the number of carbon atoms increases, i. e. as their lipophilic properties increase. Other effects also intensify in a comparable manner, e. g. their antimicrobial effect, haemolytic effect or their lethality in the case of acute poisoning [1]. The MAK and BAT values of several alcohols are summarized in Table 1 [2].

Table 1. MAK and BAT values of alcohols

	MAK value			BAT value
	$[mL/m^3]$ [ppm]	$[mg/m^3]$		$[mg/L]$ (material investigated)
Methanol	200	260	*H*	30 (urine)
Ethanol	1000	1900		
1-Propanol	–	–		
2-Propanol	400	980		50 (blood, urine)
1-Butanol	100	300		
2-Methyl-1-propanol	100	300		
2-Butanol	100	300		

H: indicates the risk of absorption through the skin

Methanol

Methanol (methyl alcohol) is produced from synthesis gas (CO and H_2) in a high or low pressure process. Methanol has a variety of uses in chemical synthesis, as a solvent, a cleaning fluid, an antifreeze and as a fuel additive (antiknock agent).

Besides intake of methanol via the respiratory tract, occupational exposure leads to significant absorption of methanol through the skin. Methanol has been marked with an "H" in the list of MAK and BAT values to indicate that it is readily absorbable through the skin [2]. After absorption, methanol tends to be distributed in the aqueous body fluids [3]. In contrast to ethanol, a considerable proportion (30–60 %) of the absorbed methanol is exhaled from the lungs due to its relatively slow oxidation by alcohol dehydrogenase. A lesser quantity of methanol is also excreted unchanged through the kidneys. The remaining methanol is oxidized to formic acid and especially to carbon dioxide in the organism (cf. Figure 1). The first rate-determining step in the breakdown of methanol is the oxidation of methanol to formaldehyde which is catalyzed by alcohol dehydrogenase. The formaldehyde is not accumulated in the organism; instead it is rapidly further oxidized to formic acid. The generated formic acid (or formate) is subsequently metabolized to carbon dioxide by means of a mechanism which is dependent on tetrahydrofolic acid. Only a small amount of the formic acid (<5 % of the absorbed methanol) is excreted via the kidneys. In the case of acute poisoning, therefore, considerable accumulation of formic acid occurs in the human organism. This can lead to a reduction of the bicarbonate concentration in blood and subsequently to acidosis (reduction of the blood pH to values below 7). In the course of acute methanol poisoning pronounced symptoms of narcosis are observed one day after intake of methanol. Two to three days later massive metabolic acidosis occurs, accompanied by initial, usually reversible visual disorders. Depending on the seriousness of the case, irreversible damage to the eyes can occur 5 to 6 days after intake of methanol. There are significant correlations between the methanol content of the blood and the narcotic phase as well as between the formic acid content in the blood and the occurrence of visual disorders. Death as a result of intake of a single dose of 100–250 mL, in rare cases of only 30 mL of methanol, is caused by respiratory failure, oedema of the CNS or of the lungs, circulatory collapse or uraemia. These conditions can be preceded by a state of intoxication, vomiting, shortage of breath and delirium. These symptoms of poisoning can also appear after methanol vapour has been inhaled.

Chronic poisoning due to repeated intake of small amounts of methanol below the acutely toxic level can cause damage to sight and hearing.

Biological monitoring of chronic exposure to methanol can be carried out by determination of methanol itself in blood and urine, but also by monitoring the excretion of its metabolite formic acid. As a rule it is preferable to determine the methanol level in urine, as it is a more sensitive diagnostic parameter than the methanol concentration in blood. This is also reflected in the BAT value. The formic acid level is more closely associated with

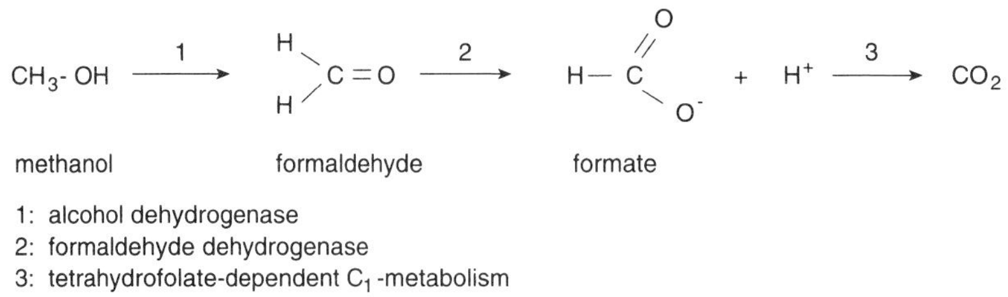

1: alcohol dehydrogenase
2: formaldehyde dehydrogenase
3: tetrahydrofolate-dependent C_1-metabolism

Figure 1. Oxidation of methanol in the human body

the toxic effect of acidosis than the excretion of methanol, but because of its high and fluctuating content it is not sufficiently specific for monitoring low levels and individual exposure to methanol. Under conditions found in practice, it should be noted that an accumulation of methanol in urine is observable. Even though no occupational exposure has taken place traces of methanol can be present in the urine, e. g. as a result of drinking fruit juices.

Concentrations of up to 3.8 mg/L methanol were found in the urine of 49 persons who were not exposed to the substance at the workplace (cf. Figure 2) [4].

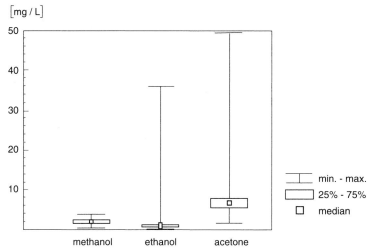

Figure 2. Methanol, ethanol and acetone concentrations in the urine of a normal group (n = 49)

Ethanol

Ethanol can be manufactured from ethylene by direct catalytic hydration at higher temperatures and pressures up to 250 bar.

It is used as a starting material in chemical syntheses, as a solvent, as a disinfectant and for medical purposes [5].

Ethanol is mainly absorbed (up to approx. 80 %) in the upper part of the small intestine.

When passage through the stomach is rapid, (when the stomach is empty) absorption is especially fast (about 50 % in 15 min).

A small amount of ethanol is excreted unchanged via the kidneys (1–2 %) and via the lungs (2–3 %). The major proportion of the absorbed ethanol is metabolized, whereby the elimination rate (0.1 g/kg body weight per hour [men] or 0.085 g/kg body weight per hour [women]) is independent of the ethanol concentration. 10 % of the ethanol consumed is excreted in the urine as glucuronide and sulphate derivatives. In addition, alcohol dehydrogenase and aldehyde dehydrogenase metabolize ethanol to acetic acid. Figure 3 shows an overview of the enzymatic oxidation of ethanol to acetic acid.

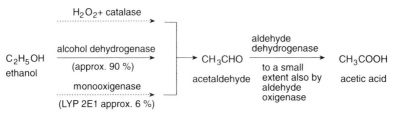

Figure 3. Overview of the enzymatic oxidation of ethanol to acetic acid

Ethanol exhibits both depressive and stimulative effects on the central nervous system. The primary effect of ethanol is to influence the cell membrane accompanied by changes to the membrane fluidity. The underlying mechanisms have not yet been finally clarified. In small doses, ethanol causes a slight increase in blood pressure. All doses lead to increased respiration and urine excretion, and a reduction in muscle performance. At an alcohol concentration of 1.4 per thousand acute poisoning sets in. The symptoms include psychomotor excitation, nausea and vomiting as well as reduction of the body temperature. In addition, the blood sugar level falls. Above the level of 2 per thousand the narcotic effect predominates causing symptoms of paralysis and serious impairment of the sense of balance. A deep coma, which may result in death, has been described at alcohol concentrations of 3.5 to 4 per thousand.

Frequent consumption of ethanol can lead to addiction (chronic alcoholism).

A clear correlation has been established between the increased consumption of alcohol during pregnancy and the incidence of deformity in the newly born [6]. The children show retardation of growth, characteristic changes in facial features and intelligence deficiency.

Moreover, epidemiological studies have shown a correlation between alcohol consumption and the incidence of tumours in the mouth, throat, oesophagus and the liver. A possible relationship between alcohol consumption and the formation of carcinomas in other organs, such as the pancreas and the small intestine, is under discussion [7].

Ethanol concentrations of up to 36 mg/L were found in the urine of 49 people who were not occupationally exposed to the substance (cf. Figure 2) [4].

Propanols

1-Propanol is employed as a disinfectant and as a solvent in cosmetics and for the manufacture of paints, plastics and cellulose. 2-Propanol is used for cosmetic substances and as a solvent, extraction agent and preserving agent.

The most important consequence of propanol poisoning is its effect on the central nervous system. Irritation to the mucous membranes is more pronounced than with ethanol. 2-Propanol is oxidized by alcohol dehydrogenase to acetone in the organism. Besides its narcotic effect, acute poisoning causes acidosis and ketonuria.

Butanols

1-Butanol, 2-methyl-1-propanol and 2-butanol are colourless, flammable liquids which are moderately miscible with water, but completely miscible in organic solvents. They are mainly used for the manufacture of paints, solvents and flotation agents, but they are also used in the perfume industry.

In humans butanols mainly cause irritation to the eyes, the respiratory tract and the skin. At higher concentrations their narcotic effect is of central importance.

Ketones

Ketones are organic compounds with the general formula $R^1R^2C = O$, whereby R^1 and R^2 are alkyl and/or aryl groups or R^1 and R^2 can form a ring.

Ketones are relatively volatile, with boiling points only a little above those of the corresponding alkanes. Up to C_5 the aliphatic ketones are soluble in water and they are widely used in the chemical industry as solvents. In contrast, the aliphatic-aromatic ketones and the aromatic ketones are high-boiling liquids or solids. Their main use is for chemical synthesis [5].

Due to their volatility ketones are predominantly taken into the body at the workplace by inhalation. Furthermore, considerable amounts of ketones can be absorbed into the body through the intact skin. The intake of ketones causes irritation of the mucous membranes of the nose, the throat and the eyes [8, 9].

Table 2 shows the MAK and BAT values of some ketones [2].

Table 2. MAK and BAT values of some ketones

	MAK value		BAT value
	[mL/m³] [ppm]	[mg/m³]	[mg/L] (material investigated)
Acetone	500	1200	40 (urine)
2-Butanone	200	590	5 (urine)
4-Methyl-2-pentanone	100	400	3.5 (urine)
2-Hexanone	5	21	5 (urine)

Exhalation of the absorbed ketones via the lungs is the most important elimination route. In contrast, only small amounts of the unchanged solvents are excreted in the urine [9]. Despite this fact, determination of the excretion of ketones in the urine seems preferable to determination of the blood level as a parameter for biological monitoring. This is also reflected by the established BAT values [2]. One possible explanation is the – relatively – good solubility of the short-chain ketones in water and their resulting distribution through the aqueous body fluids. This also leads to a high inhalative retention rate and a slow establishment of equilibrium between the ambient air and the blood concentration at the normal duration of exposure at the workplace. This has been shown in various field studies, e. g. for butanone [10, 11]. Therefore, greatly differing blood/urine concentration ratios are found, which has so far precluded the establishment of a correlating blood concentration, e. g. for butanone.

The metabolism of ketones generally proceeds via an oxidative hydroxylation and subsequent reduction to the secondary alcohol. In some cases, e. g. in the case of 2-hexanone and 4-methyl-2-pentanone, oxidative formation of γ-diketones is possible. Recent findings have shown that these compounds are responsible for the so-called gamma diketone polyneuropathy [9].

In contrast, the ketones most frequently used in industry, such as acetone and 2-butanone seem to be only mildly toxic [8].

Author: *J. Angerer, J. Gündel*
Examiners: *R. Heinrich-Ramm, M. Blaszkewicz*

Alcohols and Ketones

Application   Determination in blood and urine

Analytical principle Capillary gas chromatography
                          Headspace technique

Completed in   February 1996

Contents

Essential Biomonitoring Methods. DFG, Deutsche Forschungsgemeinschaft
Copyright © 2006 WILEY-VCH Verlag GmbH & Co. KGaA, Weinheim
ISBN: 3-527-31478-4

1 General principles

The highly volatile alcohols and ketones contained in both blood and urine are determined by means of capillary gas chromatography using the headspace technique. For this purpose the blood or urine samples are warmed to 40 °C or 50 °C in airtight crimp top vials. After distribution of the alcohols and ketones between the liquid and vapour phases has reached equilibrium, an aliquot of the headspace is withdrawn and analysed by gas chromatography. A flame ionization detector (FID) serves as a detector.

Calibration curves are obtained by analysing blood and urine samples to which known quantities of alcohols and ketones have been added. The peak areas are plotted as a function of the concentrations used.

2 Equipment, chemicals and solutions

2.1 Equipment

Gas chromatograph with split-splitless injector, flame ionization detector, integrator and plotter

Device for automatically injecting headspace samples (headspace autosampler)

1000 μL syringe for gas chromatography (e. g. from Hamilton)

Automatic pipettes (e. g. Eppendorf Multipette)

Microlitre pipettes, adjustable for dosages between 10 and 100, but also between 100 and 1000 μL (e. g. from Eppendorf)

10, 20, 25, 50, 100 and 500 mL volumetric flasks

20 mL crimp top vials with teflon-coated butyl rubber stoppers and aluminium crimp caps as well as crimping tongs for sealing and opening them.

The crimp top vials and especially the stoppers must be heated in the drying cupboard at 100 °C for more than 24 hours. It is advisable to heat the stoppers for as long as a week.

Disposable syringes containing potassium EDTA as an anticoagulant (e.g potassium EDTA Monovetten® from Sarstedt)

2.2 Chemicals

Chemicals of the highest commercially available purity are always used.

Methanol for trace analysis (e. g. from Merck)

Ethanol p.a. (e. g. from Merck)

1-Propanol p.a. (e. g. from Merck)

2-Propanol p.a. (e. g. from Merck)

1-Butanol p.a. (e. g. from Merck)

2-Butanol p.a. (e. g. from Merck)

2-Methyl-1-propanol for trace analysis (e. g. from Merck)

Acetone for trace analysis (e. g. from Merck)

2-Butanone p.a. (e. g. from Merck)

4-Methyl-2-pentanone p.a. (e. g. from Merck)

2-Hexanone p.a. (e. g. from Merck)

Ultrapure water (ASTM type 1) or double distilled water

Sodium chloride p.a. (e. g. from Merck)

Defibrinated sheep's blood (e. g. from Froscheck, Mülheim/Ruhr)

K_2-EDTA

Purified nitrogen (99.999 %)

Hydrogen/compressed air

2.3 Solutions

0.9 % Sodium chloride solution:
4.5 g sodium chloride are dissolved in ultrapure water in a 500 mL volumetric flask. The flask is filled up to the mark with ultrapure water.

2.4 Calibration standards

2.4.1 Calibration standards for the determination in blood

Starting solution (in 0.9 % sodium chloride solution):
About 1 mL ethanol, 500 µL each of methanol, 2-propanol, 1-propanol, 2-methyl-1-propanol, 1-butanol, 2-butanol and 100 µL each of acetone, 2-butanone, 4-methyl-2-pentanone and 2-hexanone are pipetted into a 10 mL volumetric flask one after another. The mass of added alcohols and ketones is calculated using their specific weights and checked gravimetrically. The volumetric flask is filled up to its nominal volume with 0.9 % sodium chloride solution to give a solution containing the individual alcohols and ketones in concentrations between 7.91 and 79.36 g/L. Table 3 shows the volumes of the individual alcohols and ketones which were added, as well as their corresponding masses and their concentrations in the starting solution.

Table 3. Preparation and Concentrations of the alcohols and ketones in 0.9 % NaCl solution and in blood

Substance	Volume of the individual alcohols and ketones	Equivalent mass in 10 mL 0.9 % NaCl solution	Concentration of the starting solution in 0.9 % NaCl solution	Concentration of stock solution I in 0.9 % NaCl solution	Concentration of stock solution II in blood
	[µL]	[g]	[g/L]	[g/L]	[mg/L]
Methanol	500	0.3935	39.35	3.93	393
Ethanol	1000	0.7936	79.36	7.94	794
2-Propanol	500	0.3928	39.28	3.93	393
1-Propanol	500	0.4020	40.20	4.02	402
2-Methyl-1-propanol	500	0.4030	40.30	4.03	403
1-Butanol	500	0.4065	40.65	4.07	407
2-Butanol	500	0.4055	40.55	4.06	406
Acetone	100	0.0791	7.91	0.79	79
2-Butanone	100	0.0805	8.05	0.81	81
4-Methyl-2-pentanone	100	0.0800	8.00	0.80	80
2-Hexanone	100	0.0830	8.30	0.83	83

Stock solution I (in 0.9 % sodium chloride solution):
1 mL of the starting solution is pipetted into a 10 mL volumetric flask and the flask is filled up to the mark with 0.9 % sodium chloride solution. The concentrations of the individual alcohols and ketones are given in Table 3, column 5.

Stock solution II (in blood):
2 mL of stock solution I are pipetted into a 20 mL volumetric flask. The flask is subsequently filled up to the mark with blood. The concentrations of the individual alcohols and ketones are shown in Table 3, column 6.

Calibration standards of the alcohols and ketones in blood:
Between 0.5 and 5 mL of the alcohols and ketones dissolved in blood (stock solution II) are pipetted into a 50 mL volumetric flask which is then filled to the mark with blood (cf. Table 4). After thorough mixing of the calibration solutions, aliquots (each 2 mL) are filled into the crimp top vials. The vials are immediately covered with teflon-coated rubber septa and sealed with aluminium crimp caps. These calibration standards can be stored in this form in the deep-freezer. No changes in the concentrations could be determined after a storage period of one year.

2.4.2 Calibration standards for the determination in urine

Starting solution in ultrapure water:
After pipetting 250 µL methanol into a 25 mL volumetric flask, about 500 µL of each of the other alcohols and approximately 100 µL of each ketone are added with pipettes. The volumetric flask is filled up to mark with ultrapure water. The resulting solution contains the individual alcohols and ketones in concentrations between 3.16 and 16.26 g/L. Table 5 gives the volumes of the individual alcohols and ketones which are added, their corresponding masses and their concentrations in the starting solution.

Stock solution:
2 mL of the starting solution are pipetted into a 50 mL volumetric flask. Then the flask is filled up to the mark with urine. The concentrations of the individual alcohols and ketones are given in Table 5, column 5.

Calibration standards of the alcohols and ketones in urine:
Between 1 and 10 mL of the alcohols and ketones dissolved in urine are pipetted into a 100 mL volumetric flask which is then filled to the mark with urine (cf. Table 6). After thorough mixing of the calibration standards, aliquots (each 2 mL) are filled into the crimp top vials. The vials are immediately covered with teflon-coated rubber septa and sealed with aluminium crimp caps. These calibration standards can be stored in this form in the deep-freezer. No changes in the concentrations could be ascertained after a storage period of one year.

Table 4. Pipetting scheme for the preparation of the calibration standards in blood

Calibration standard No.	Volume of stock solution II in blood [mL]	Final volume of the calibration standards [mL]	Concentration of the calibration standards [mg/L]										
			Methanol	Ethanol	2-Propanol	1-Propanol	2-Methyl-1-propanol	1-Butanol	2-Butanol	Acetone	2-Butanone	4-Methyl-2-pentanone	2-Hexanone
1	0.50	50	3.94	7.94	3.93	4.02	4.03	4.07	4.06	0.79	0.81	0.80	0.83
2	1.00	50	7.88	15.88	7.86	8.04	8.06	8.14	8.12	1.58	1.62	1.60	1.66
3	2.50	50	19.70	39.70	19.65	20.10	20.15	20.35	20.30	3.95	4.05	4.00	4.15
4	5.00	50	39.40	79.40	39.30	40.20	40.30	40.70	40.60	7.90	8.10	8.00	8.30

Table 5. Preparation and concentrations of the alcohols and ketones in water and urine

Substance	Volume of the individual alcohols and ketones	Equivalent mass in 25 mL water	Concentration of the starting solution	Concentration of the stock solution
	[μL]	[g]	[g/L]	[mg/L]
Methanol	250	0.1950	7.80	312.0
Ethanol	500	0.3968	15.87	634.8
2-Propanol	500	0.3927	15.71	628.4
1-Propanol	500	0.4020	16.08	643.2
2-Methyl-1-propanol	500	0.4030	16.12	644.8
1-Butanol	500	0.4065	16.26	650.4
2-Butanol	500	0.4055	16.22	648.8
Acetone	100	0.079	3.16	126.4
2-Butanone	100	0.0805	3.22	128.8
4-Methyl-2-pentanone	100	0.0800	3.20	128.0
2-Hexanone	100	0.0830	3.32	132.8

Table 6. Pipetting scheme for the preparation of the calibration standards in urine

Calibration standard	Volume of stock solution in urine	Final volume of the calibration standards	Concentration of the calibration standards										
No.	[mL]	[mL]	[mg/L]										
			Methanol	Ethanol	2-Propanol	1-Propanol	2-Methyl-1-propanol	1-Butanol	2-Butanol	Acetone	2-Butanone	4-Methyl-2-pentanone	2-Hexanone
1	1.0	100	3.12	6.35	6.28	6.43	6.44	6.50	6.48	1.26	1.29	1.28	1.33
2	2.0	100	6.24	12.70	12.56	12.86	12.88	13.00	12.96	2.52	2.58	2.56	2.66
3	5.0	100	15.60	31.75	31.40	32.15	32.20	32.50	32.40	6.30	6.45	6.40	6.65
4	10.0	100	31.20	63.50	62.80	64.30	64.40	65.00	64.80	12.60	12.90	12.80	13.30

3 Specimen collection and sample preparation

For the purpose of occupational health surveillance the assay sample (blood or urine) for determination of the alcohols and ketones is generally collected at the end of the exposure period. If the substance in question has been assigned a BAT value, the sampling time can be found in the current list of MAK and BAT values [2]. In order to prevent contamination of the blood sample by solvents, the arm of the test person is cleansed with 3 % aqueous hydrogen peroxide solution or with soap and water rather than the usual disinfectant before withdrawal of the blood specimen. The blood sample is collected using a disposable syringe which already contains an anticoagulant (e. g. EDTA Monovette from Sarstedt).

Approximately 5 mg of the anticoagulant potassium EDTA in solid form are placed in the crimp top vials and they are carefully covered with the teflon-coated rubber septa and sealed with the aluminium crimp caps.

2 mL of the blood sample are injected into a crimp top vial prepared as described above immediately after collection. The liquid is then vigorously swirled around to dissolve the anticoagulant in the vessel. Coagulation of the blood sample is thus prevented. Filling the urine samples into crimp top vials has also proved advantageous. The samples can be transported in this state, no further precautions are necessary. Samples can be stored for several days in the refrigerator until the analysis is carried out. If longer storage is necessary deep freezing (-18 °C) has proved the best method of storing the samples. Before analysis the samples are brought to room temperature.

The crimp top vials are placed in the headspace autosampler. The samples are incubated at a temperature of 40 °C (blood) or 50 °C (urine) for at least 1 hour. This ensures that equilibrium of the assay alcohols and ketones between the aqueous phase and the vapour phase has been reached.

4 Operational parameters for gas chromatography

Column:	Material:	Fused silica
	Length:	25 m
	Inner diameter:	0.32 mm
	Stationary phase:	Porapak Q
	Film thickness:	10 µm
Detector:	Flame ionization detector	
Temperatures:	Column:	4 min at 100 °C, increase: 10 °C per min until 140 °C; then 2 min isothermal,

	increase 10 °C per min until 170 °C then 6 min isothermal, increase 3 °C per min until 190 °C then 8 min isothermal, increase 10 °C per min until 220 °C and 10 min isothermal
Injection block:	230 °C
Detector:	270 °C
Carrier gas:	Purified nitrogen with a pre-column pressure of 1 bar (1000 hPa), flow rate 1 mL/min
Split:	5 mL/min
Make up gas:	Purified nitrogen, 30 mL/min
Sample volume:	1 mL

The following retention times (Table 7) of the individual alcohols and ketones in blood or urine serve only as a guide.

Table 7. Recorded retention times

Analyte	Retention time [min]
Methanol	5.80
Ethanol	7.60
Acetone	8.80
2-Propanol	9.20
1-Propanol	13.80
2-Butanone	16.20
2-Butanol	17.30
2-Methyl-1-propanol	20.10
1-Butanol	23.00
4-Methyl-2-pentanone	28.00
2-Hexanone	33.20

Figure 4 shows the gas chromatogram of a calibration standard in urine.

5 Analytical determination

After incubating the crimp top vials at 40 °C (blood) or 50 °C (urine) for 1 hour, 1000 µL of the vapour in the headspace are injected into the gas chromatograph using a headspace auto-sampler.

6 Calibration and calculation of the analytical result

A sample of each of the calibration standard solutions prepared as described in Section 2.3 is analysed as instructed (cf. Section 4). A blood or urine sample, respectively, which has not been spiked with alcohols and ketones serves to determine the baseline level of the specimens. These samples are kept deep-frozen in sealed airtight crimp top vials until analysis. The resulting peak areas for the individual alcohols and ketones are plotted as a function of the corresponding concentrations used. If baseline levels are ascertained they must be previously subtracted. The peaks areas in the chromatogram, in which the peaks are very sharp (cf. Figure 4), should be evaluated using an integrator.

A complete new calibration curve need not be plotted for each analytical series. It is sufficient to determine a calibration standard with every analytical series. A new calibration curve should be plotted if systematic deviation of the quality control results is detected.

The characteristic retention times of the individual alcohols and ketones are recorded with the peak areas or peak heights. Any baseline levels which may occur are subtracted. With these values the equivalent concentration in mg per litre blood or urine is read off the appropriate calibration curve.

In the concentration range given for the calibration standards there is a linear relationship between the gas chromatographic peak signal or its peak area and the concentration of the corresponding alcohol or ketone.

7 Standardization and quality control

Quality control of the analytical results is carried out according to TRgA 410 [12] and the Special Preliminary Remarks in this series. At present no control material containing a specified quantity of alcohols and ketones is commercially available. Therefore internal control of the results is carried out using control material prepared in the laboratory. For this purpose defibrinated sheep's blood or pooled urine is spiked with defined amounts of alcohols and ketones. A supply of control material for six months is prepared, aliquots are filled into ampoules and kept deep-frozen. The concentration of the alcohols and ketones in this control material should lie in the middle of the most frequently occurring concentration range. The expected value and the tolerance range of this quality control material is determined in a pre-analytical period (one analysis of the control material is carried out on each of 20 different days) [13].

8 Reliability of the method

8.1 Precision

In order to determine the precision in the series, animal blood or urine from people who were not occupationally exposed to the alcohols and ketones was spiked with defined quantities of the alcohols and ketones. Solutions were obtained containing the alcohols and ketones in concentrations between 0.8 and 23.8 mg per litre blood, or between 1.2 and 6.5 mg per litre urine respectively. These solutions were processed and analyzed ten times as described in Sections 3 and 4. The relative standard deviations were between 0.9 and 4.5 % or between 0.8 and 4 % respectively. The corresponding prognostic ranges were between 2.0 and 10 % or between 1.8 and 8.9 % respectively. The individual results are shown in Table 8.

Table 8. Precision in the series for the determination of alcohols and ketones in blood or urine respectively.

Analyte	Blood				Urine			
	Conc. [mg/L]	n	s_w [%]	u [%]	Conc. [mg/L]	n	s_w [%]	u [%]
Methanol	23.8	10	2.1	4.7	6.1	10	4.0	8.9
Ethanol	7.9	10	2.0	4.5	6.3	10	2.9	6.5
Acetone	0.8	10	1.9	4.2	1.3	10	2.6	5.8
2-Propanol	3.9	10	2.8	6.2	6.3	10	1.4	3.1
1-Propanol	4.0	10	2.1	4.7	6.4	10	1.1	2.5
2-Butanone	0.8	10	1.6	3.6	1.3	10	1.3	2.9
2-Butanol	8.0	10	1.5	3.3	6.2	10	0.8	1.8
2-Methyl-1-propanol	8.0	10	4.5	10.0	6.3	10	1.1	2.5
1-Butanol	8.0	10	3.7	8.2	6.5	10	1.1	2.5
4-Methyl-2-pentanone	0.8	10	0.9	2.0	1.3	10	1.1	2.5
2-Hexanone	0.8	10	1.8	4.0	1.2	10	0.8	1.8

The between-day precision was determined by processing and analyzing the blood and urine samples prepared according to Section 3 and 4 on five to six different days. Relative standard deviations between 2.4 and 6.9 % or between 0.7 and 7.6 % respectively were calculated. The equivalent prognostic ranges were between 5.9 and 16.9 % or between 1.7 and 18.6 % respectively. The individual results are given in Table 9.

Table 9. Between-day precision for the determination of alcohols and ketones in blood or urine respectively

Analyte	Blood				Urine			
	Conc. [mg/L]	n	s_w [%]	u [%]	Conc. [mg/L]	n	s_w [%]	u [%]
Methanol	23.8	6	4.6	11.3	6.3	6	7.6	18.6
Ethanol	7.9	6	6.9	16.9	6.3	6	4.4	10.8
Acetone	0.8	6	3.1	7.6	1.3	6	1.2	2.9
2-Propanol	3.9	6	2.7	6.6	6.3	6	1.4	3.4
1-Propanol	4.0	6	2.5	6.1	6.4	6	1.8	4.4
2-Butanone	0.8	6	5.9	14.4	1.3	6	2.6	6.4
2-Butanol	8.0	6	2.4	5.9	6.2	6	0.8	1.9
2-Methyl-1-propanol	8.0	6	4.7	11.5	6.3	6	0.7	1.7
1-Butanol	8.0	6	3.8	9.3	6.5	6	1.8	4.4
4-Methyl-2-pentanone	0.8	6	5.6	13.7	1.3	6	3.6	8.8
2-Hexanone	0.8	6	4.7	12.1	1.2	6	2.9	7.1

8.2 Accuracy

Recovery experiments were carried out by the examiners of the method to check its accuracy. For this purpose samples of sheep's blood or urine were spiked with specific quantities of the alcohols and ketones. The results of the analysis of these samples were compared with the values read off the appropriate calibration curves. The recovery rate was calculated as the ratio of the measured result with the values for the calibration standards. Table 10 shows the recovery rates.

Table 10. Recovery rates of the alcohols and ketones in blood or urine

Analyte	Blood Recovery rate [%]	Urine Recovery rate [%]
Methanol	96.0 – 98.4	96.4 – 104.1
Ethanol	96.4 – 100.9	97.1 – 99.7
Acetone	102.1 – 105.7	99.9 – 100.9
2-Propanol	97.4 – 105.7	94.5 – 99.5
1-Propanol	99.8 – 106.6	96.3 – 98.3
2-Butanone	99.6 – 101.3	95.4 – 96.5
2-Butanol	96.8 – 100.2	90.9 – 98.6
2-Methyl-1-propanol	75.7 – 100.6	89.0 – 97.6
1-Butanol	106.0	98.9
4-Methyl-2-pentanone	95.8 – 100.2	94.6 – 96.0
2-Hexanone	96.0 – 100.8	93.6 – 98.7

8.3 Detection limit

The detection limits were calculated as three times the signal/background ratio (see Table 11).

Table 11. Detection limits

Analyte	Blood Detection limit [mg/L]	Urine Detection limit [mg/L]
Methanol	0.6	0.6
Ethanol	1.3	0.8
Acetone	0.2	0.1
2-Propanol	0.6	0.4
1-Propanol	0.8	0.4
2-Butanone	0.1	0.08
2-Butanol	0.4	0.2
2-Methyl-1-propanol	0.4	0.2
1-Butanol	0.8	0.3
4-Methyl-2-pentanone	0.05	0.03
2-Hexanone	0.07	0.03

8.4 Sources of error

a) Pre-analytical phase

It is essential to adhere to the sample collection time stipulated in the list of MAK and BAT values [2]. This is usually at the end of the exposure period.

Coagulation of the blood sample impedes reproducible distribution of the alcohols and ketones between the biological matrix and the vapour phase at equilibrium. Thus, the blood sample should be vigorously swirled round after filling it into the crimp top vial to ensure it is thoroughly mixed with the anticoagulant. Disposable syringes containing solid EDTA (e. g. EDTA Monovettes from Sarstedt) are recommended for sample collection.

In order to prevent loss of the highly volatile alcohols or ketones during transport and storage, it is important to fill the samples into the crimp top vials and seal them immediately after sample collection. It should be impossible to turn the seals by hand. To ensure a tight seal the crimping tongs must be appropriately adjusted,

The use of Vacutainers for the analysis of highly volatile alcohols and ketones is not recommended.

Exogenous contamination can influence the results of this method to determine alcohols and ketones. In order to prevent the pollution of the samples with solvents, it is necessary to avoid using the usual disinfectant for cleansing the arm of the test person before withdrawal of the blood specimen. It is advisable to use a 3 % solution of hydrogen peroxide. Thorough cleaning with soap and water has also proved sufficient. Thus, interference in the chromatogram from overlapping of the solvents contained in swabs can be avoided.

Furthermore, the crimp top vials and the teflon-coated butyl rubber stoppers should be thoroughly heated before use. This should be carried out in the drying cupboard at 100 °C for three days. Any traces of solvents contained in the butyl rubber stoppers or adhering to the glass walls are thus expelled.

However, carefully filling the urine into crimp top vials after collection does not seem to be absolutely necessary to preserve the water-soluble analytes in urine. Thus, no difference in concentration was detected for a water-soluble 2-butanone urine sample which was transferred from a screw-topped plastic bottle to a crimp top vial only when it reached in the laboratory compared with the same sample which was immediately filled into a crimp top vial after collection [11].

b) Analytical phase

It is of decisive importance for the reliability and accuracy of the analytical results that the distribution equilibrium between the liquid and the vapour phase has definitely been reached. Experience has shown that for alcohols and ketones the equilibrium has been established after 1 hour. Nevertheless, the user of this method should check how long attainment of equilibrium takes under the operational conditions of the apparatus used.

Meanwhile the use of an autosampler has become standard practice in headspace analysis.

10 Discussion of the method

Gas chromatographic headspace analysis using a capillary column is the current standard method of determining highly volatile organic solvents in blood and urine. Carrying out the analysis is simple because no sample preparation is necessary. As no analytical background interference occurs, very low detection limits can be achieved.

Eleven different alcohols and ketones can be determined in blood and urine using this method. Those substances which have already been assigned BAT values in urine (4-methyl-2-pentanone 3.5 mg/L, 2-butanone 5 mg/L, 2-hexanone 5 mg/L, methanol 30 mg/L, acetone 40 mg/L) are detectable in concentrations between $^1/_{50}$ and $^1/_{400}$ of the current (1995) BAT value in urine [2].

The precision of the method was determined using a headspace autosampler. At the concentrations used here, relative standard deviations of under 10 % can thus be achieved.

In order to attain a reproducible distribution equilibrium between the biological matrix and the vapour phase, it is essential to prevent coagulation of the blood sample. Thus, in addition to the anticoagulant in the disposable syringes, EDTA is also added to the headspace vials as a precaution, To ensure that the distribution equilibrium is reached, the headspace vials should be incubated at 40 °C for at least 1 hour. When the alcohols and ketones are determined in urine, incubation for at least one hour at 50 °C is necessary. Within a relatively wide range the analytical results are normally independent of the amount filled into the headspace vials. Erroneous results occur only when the vials have been clearly overfilled.

Based on the calibration curves for methanol, ethanol, acetone, 2-butanone, 4-methyl-2-pentanone, 2-propanol and 1-propanol, one of the examiners of the method was able to demonstrate that the addition of solid salts (e. g. 0.5 g Na_2SO_4 per litre urine) to the urine sample can increase the concentration of the volatile analyte in the vapour phase up to 4 to 11 times (salting-out effect [14] cf. Figure 5.). However, a distinctly lower reproducibility was observed for the analysis of the propanol isomers, which was apparent in the poorer linearity of the calibration curves.

Instruments used:

Gas chromatograph 3300 with data system DS 600 from Varian and an autosampler Dani HSS 3950

11 References

[1] *W. Forth, D. Henschler* and *W. Rommel* (eds.): Allgemeine und spezielle Pharmakologie und Toxikologie. 6th edition, Wissenschaftsverlag, 1992.
[2] *Deutsche Forschungsgemeinschaft*: List of MAK and BAT values 1995. Maximum Concentrations at the Workplace and Biological Tolerance Values for Working Materials. Commission for the

Investigation of Health Harzards of Chemical Compounds in the Work Area. Report No 31. VCH Verlagsgesellschaft, Weinheim 1995.

[3] *S. Moeschlin* (ed.): Klinik und Therapie der Vergiftungen. Georg Thieme Verlag, Stuttgart 1986.

[4] *R. Heinrich-Ramm*. Personal communication.

[5] *Römpp Chemie Lexikon*. 9th extended and revised edition, Georg Thieme Verlag, Stuttgart 1989.

[6] *O. E. Pratt*: Alcohol and the developing fetus. Br. Med. Bull. *38(1)*, 48–53 (1982).

[7] *R. R. Watson* (ed.): Alcohol and cancer. CRC Press, Boca Raton 1992.

[8] *H. Greim* (ed.): Gesundheitsschädliche Arbeitsstoffe. Toxikologisch-arbeitsmedizinische Begründungen von MAK-Werten. Deutsche Forschungsgemeinschaft. VCH Weinheim, 20th issue (1994).

[9] *G. D. George* and *F. E. Clayton* (eds.): Patty's Industrial Hygiene and Toxicology. Fourth Edition, Vol. 2, Part C, John Wiley & Sons, Inc. New York 1994.

[10] *J. Angerer*: 2-Butanon. In: H. Greim, and G. Lehnert (eds.): Biologische Arbeitsstoff-Toleranz-Werte (BAT-Werte) und Expositionsäqivalente für krebserzeugende Arbeitsstoffe (EKA). Arbeitsmedizinisch-toxikologische Begründungen. Deutsche Forschungsgemeinschaft. VCH Weinheim, 5th issue (1990).

[11] *R. Heinrich-Ramm, J. Gebhard, V. Kassebart* and *D. Szadkowski*: Biological Monitoring bei Butanonexposition während Lackierarbeiten an Flugzeugen. In: G. Triebig and O. Stelzer (eds.): Verhandlungen der Deutschen Gesellschaft für Arbeitsmedizin und Umweltmedizin. Gentner Verlag, Stuttgart 1993.

[12] *TRgA 410*: Statistische Qualitätssicherung. In: Technische Regeln und Richtlinien des BMA zur Verordnung über gefährliche Stoffe. Bek. des BMA vom 03.04.1979, Bundesarbeitsblatt 6/1979, p. 88 ff.

[13] *J. Angerer* and *K. H. Schaller*: Erfahrungen mit der statistischen Qualitätskontrolle im arbeitsmedizinisch-toxikologischen Laboratorium. Arbeitsmed. Sozialmed. Präventivmed. *1*, 33–35 (1977).

[14] *G. Machata* and *J. Angerer*: Head-Space-Technik (Dampfraumanalyse) Sammelmethode. In: H. Greim (ed.): Analytische Methoden zur Prüfung gesundheitsschädlicher Arbeitsstoffe. Vol. 2: Analysen in biologischem Material, 7th issue, VCH Verlagsgesellschaft, Weinheim 1983.

Author: *J. Angerer, J. Gündel*
Examiners: *R. Heinrich-Ramm, M. Blaszkewicz*

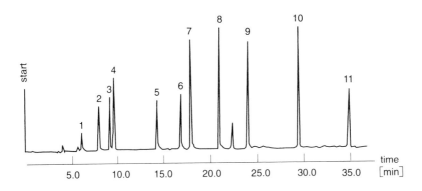

1. Methanol (12.5 mg/L)
2. Ethanol (12.5 mg/L)
3. Acetone (2.6 mg/L)
4. 2-Propanol (12.5 mg/L)
5. 1-Propanol (12.9 mg/L)
6. 2-Butanone (2.6 mg/L)

7. 2-Butanol (12.7 mg/L)
8. 2-Methyl-1-propanol (12.7 mg/L)
9. 1-Butanol (12.8 mg/L)
10. 4-Methyl-2-pentanone (2.4 mg/L)
11. 2-Hexanone (2.4 mg/L)

Figure 4. Gas chromatogram of a spiked urine sample

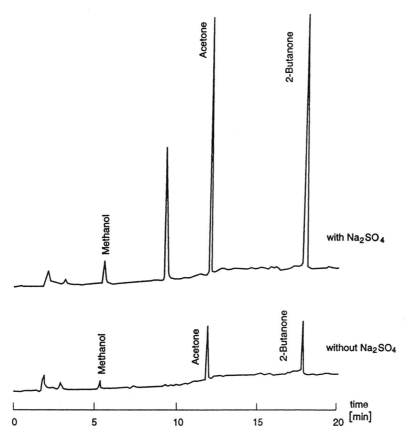

Figure 5. Gas chromatogram of a urine sample with and without the addition of 0.5 g Na_2SO_4 per litre urine

Antimony, Lead, Cadmium, Platinum, Mercury, Tellurium, Thallium, Bismuth, Tungsten, Tin

Application   Determination in urine

Analytical principle   Inductively coupled plasma quadrupole mass
                          spectrometry (Quadrupole ICP-MS)

Completed in   August 1998

Summary

Using the quadrupole ICP mass spectrometry (Q-ICP-MS) method described here, antimony, lead, cadmium, platinum, mercury, tellurium, thallium, bismuth, tungsten and tin present in urine due to occupational exposure can be simply, sensitively and specifically determined. With the exception of platinum and bismuth, the ecological concentration range can also be detected. After UV digestion of the urine samples, an internal standard is added and the samples are introduced into the ICP-MS by means of a pneumatic nebulizer. Evaluation is carried out using the standard addition procedure.

Antimony

Within-series imprecision: Standard deviation (rel.) $s_w = 14.8$ or 1.1%
                            Prognostic range $u = 26.0$ or 2.0%
                            at concentrations of 0.1 or 2.37 µg antimony per litre
                            urine and where $n = 10$ determinations

Between-day imprecision: Standard deviation (rel.) $s = 3.1\%$
                            Prognostic range $u = 5.6\%$
                            at a concentration of 1.1 µg antimony per litre urine
                            and where $n = 10$ days

Accuracy: Recovery rate $r = 112.7\%$

Detection limit: 30 ng antimony per litre urine

Essential Biomonitoring Methods. DFG, Deutsche Forschungsgemeinschaft
Copyright © 2006 WILEY-VCH Verlag GmbH & Co. KGaA, Weinheim
ISBN: 3-527-31478-4

Lead

Within-series imprecision: Standard deviation (rel.) $s_w = 2.0$ or 0.4%
Prognostic range $u = 3.8$ or 0.7%
at concentrations of 0.46 or 2.48 µg lead per litre urine
and where $n = 10$ determinations

Between-day imprecision: Standard deviation (rel.) $s = 3.7\%$
Prognostic range $u = 6.0\%$
at a concentration of 1.5 µg lead per litre urine
and where $n = 10$ days

Accuracy: Recovery rate $r = 100.9\%$

Detection limit: 30 ng lead per litre urine

Cadmium

Within-series imprecision: Standard deviation (rel.) $s_w = 4.8$ or 0.9%
Prognostic range $u = 9.0$ or 1.5%
at concentrations of 0.21 or 2 µg cadmium per litre urine
and where $n = 10$ determinations

Between-day imprecision: Standard deviation (rel.) $s = 2.4\%$
Prognostic range $u = 4.2\%$
at a concentration of 1 µg cadmium per litre urine
and where $n = 10$ days

Accuracy: Recovery rate $r = 95\%$

Detection limit: 20 ng cadmium per litre urine

Platinum

Within-series imprecision: Standard deviation (rel.) $s_w = 0.8\%$
Prognostic range $u = 1.5\%$
at a concentration of 2 µg platinum per litre urine
and where $n = 10$ determinations

Between-day imprecision: Standard deviation (rel.) $s = 2.4\%$
Prognostic range $u = 4.5\%$
at a concentration of 1 µg platinum per litre urine
and where $n = 10$ days

Accuracy: Recovery rate $r = 93.8\%$

Detection limit: 10 ng platinum per litre urine

Mercury

Within-series imprecision: Standard deviation (rel.) $s_w = 16.2$ or 1.5%
Prognostic range $u = 29.3$ or 2.8%
at concentrations of 0.13 or 2.26 µg mercury per litre
urine and where $n = 10$ determinations

Between-day imprecision: Standard deviation (rel.) $s = 8\%$
Prognostic range $u = 14\%$
at a concentration of 1.1 µg mercury per litre urine
and where $n = 10$ days

Accuracy: Recovery rate $r = 106\%$

Detection limit: 30 ng mercury per litre urine

Tellurium

Within-series imprecision: Standard deviation (rel.) $s_w = 9.1$ or 1.6%
Prognostic range $u = 17.2$ or 3.0%
at concentrations of 0.04 or 3.37 µg tellurium per litre
urine and where $n = 10$ determinations

Between-day imprecision: Standard deviation (rel.) $s = 3.6\%$
Prognostic range $u = 6.5\%$
at a concentration of 1.5 µg tellurium per litre urine
and where $n = 10$ days

Accuracy: Recovery rate $r = 96.4\%$

Detection limit: 10 ng tellurium per litre urine

Thallium

Within-series imprecision: Standard deviation (rel.) $s_w = 4.4$ or 0.8%
Prognostic range $u = 8.1$ or 1.5%
at concentrations of 0.12 or 2.1 µg thallium per litre urine
and where $n = 10$ determinations

Between-day imprecision: Standard deviation (rel.) $s = 3.7\%$
Prognostic range $u = 7.0\%$
at a concentration of 1.1 µg thallium per litre urine
(addition of 1 µg/L thallium) and where $n = 10$ days

Accuracy: Recovery rate $r = 94\%$

Detection limit: 5 ng thallium per litre urine

Bismuth

Within-series imprecision: Standard deviation (rel.) $s_w = 0.5\%$
Prognostic range $u = 0.9\%$
at a concentration of 2 µg bismuth per litre urine
and where $n = 10$ determinations

Between-day imprecision: Standard deviation (rel.) $s = 1.8\%$
Prognostic range $u = 3.2\%$
at a concentration of 1.0 µg bismuth per litre urine
and where $n = 10$ days

Accuracy: Recovery rate $r = 79.7\%$

Detection limit: 5 ng bismuth per litre urine

Tungsten

Within-series imprecision: Standard deviation (rel.) $s_w = 4.5$ or 0.3%
Prognostic range $u = 8.1$ or 0.5%
at concentrations of 0.18 or 2.54 µg tungsten per litre
urine and where $n = 10$ determinations

Between-day imprecision: Standard deviation (rel.) $s = 3.6\%$
Prognostic range $u = 6.8\%$
at a concentration of 1.2 µg tungsten per litre urine
and where $n = 10$ days

Accuracy: Recovery rate $r = 116.5\%$

Detection limit: 20 ng tungsten per litre urine

Tin

Within-series imprecision: Standard deviation (rel.) $s_w = 9.8$ or 0.6%
Prognostic range $u = 18.1$ or 1.0%
at concentrations of 0.39 or 2.28 µg tin per litre urine
and where $n = 10$ determinations

Between-day imprecision: Standard deviation (rel.) $s = 4.2\%$
Prognostic range $u = 7.8\%$
at a concentration of 1.3 µg tin per litre urine
and where $n = 10$ days

Accuracy: Recovery rate $r = 95.5\%$

Detection limit: 50 ng tin per litre urine

Antimony

Antimony is a brittle, shiny metal with an atomic mass of 121.75 and an atomic number of 51. It belongs to group Vb in the periodic table, and its chemical, physical and toxicological properties closely resemble those of its neighbour arsenic. It occurs in oxidation states III and V. Antimony is a relatively rare metal with a concentration in the Earth's crust of approximately 0.3 mg/kg. The worldwide production is about 60000 t/a [1].

As antimony alloys are extremely hard and corrosion-resistant, the metal is used industrially together with lead, copper and tin to form alloys for various products such as rechargeable batteries, type-metal, bearings, ammunition, pipes and cables. It is also employed in the ceramics and glassware industry (antimony trioxide as a plaining agent), as a pigment for the manufacture of dyes and paints and in the textile industry (flame-retardant equipment), etc.

Organic compounds of antimony, especially sodium stibogluconate, stibenyl, stibocaptate and stibophen are used to treat parasitic diseases and infections, especially in tropical medicine.

Antimony is released into the environment, especially as a result of the combustion of fossil fuels (approximately 5000–10000 t/a worldwide), and is therefore detectable in low concentrations in practically all environmentally relevant matrices.

The toxicity of antimony compounds depends on both the oxidation state and the solubility of the relevant compound. Thus its trivalent compounds are approximately 10 times more toxic than its pentavalent compounds. The lethal dose for organic antimony compounds is about 100 mg/kg body weight.

The toxicology of the various antimony compounds is summarized in Henschler [3], Elinder and Friberg [4] and Merian and Stemmer [5]. Inhalation of dust and vapour containing antimony represents the most important route of intake at the workplace. Its absorption into the blood via the lungs is relatively swift. Once there, it is bound to the erythrocytes and blocks the activity of certain enzymes. For the most part antimony is excreted in the urine (80%) and the faeces (20%) after 1–3 days [6].

A MAK value of 0.5 mg/m^3 (1998) – measured as the inspirable aerosol portion – has been stipulated for metallic antimony [2]. Antimony trioxide has been assigned to group 2 of the carcinogenic working materials [2]. The MAK value for antimony hydride is 0.1 mL/m^3 or 0.52 mg/m^3 (1998) [2]. In some isolated cases, measurements have shown concentrations up to 3 mg/m^3 at workplaces [7]. Knowledge of the concentrations in the blood and urine of exposed persons is very incomplete due to insufficient analytical detection, as the detection limits of approximately 0.5 µg per litre urine achieved by hydride atomic absorption spectrometry [8] or graphite furnace atomic absorption spectrometry [8] are not low enough to include the concentration range of interest to environmental medicine.

ICP-MS opens up new possibilities of investigating these questions by attaining detection limits of 0.03 µg per litre urine (quadrupole ICP-MS: Q-ICP-MS) and approximately 0.005 µg per litre urine (sector field ICP-MS: SF-ICP-MS) [9]. Initial investigations of samples from the general public indicate that a reference value for the antimony concentration in urine could be in the order of 50 ng per litre [9].

Lead

Lead (atomic weight 207.2) is a soft, greyish-blue metal which is extracted from lead ores by smelting. Its melting point is 327 °C, its boiling point 1750 °C and its density is 11.34 kg/L.

In the vaporous state (vaporization from 550 °C) lead is oxidized to lead oxide (PbO) in the air. Lead can form divalent and tetravalent compounds, it is readily soluble in nitric acid and is passivated by orthophosphoric acid, hydrochloric acid and sulphuric acid. The content of lead in the Earth's crust is very low at 16 ppm. The extractable reserves of lead are estimated at approximately 23 million tonnes worldwide [10].

For further information regarding lead the reader is referred to the general toxicological section of the "Lead in blood" method published in volume 2 of "Analyses of Hazardous Substances in Biological Materials" [11].

Cadmium

Cadmium (atomic number 48, relative atomic mass 112.41) is a malleable, relatively soft metal with a melting point of 321 °C and a boiling point of 767 °C. With a content of approximately $5\times10^{-5}\%$ in the Earth's crust, cadmium belongs to the rare metals. Sphalerite (ZnS) contains about 0.1–0.5% and smithsonite up to 5% of the metal. Greenockite is the most important mineral.

The most important technical use of cadmium is in the manufacture of batteries. Between 25% and 30% of the total annual cadmium production is employed for the protection of iron and similar metals against corrosion. A similar amount is used in cadmium-containing pigments and cadmium soaps, which function as stabilisers for PVC. In addition, small quantities of cadmium are present in alloys used for making bearings and solder [12]. Among other routes, cadmium is introduced into the environment by means of the combustion of fossil fuels, via waste and sewage sludge, but also through the use of phosphate fertilizers.

Several reviews dealing with the absorption, distribution, retention and excretion of cadmium in the human organism have been published [13–17]. The current state of knowledge of the absorption and kinetics of cadmium can be summarized as follows: Depending on their solubility and particle size, 20 to 30% of the respirable cadmium (particle size <5 μm) are absorbed. The gastro-intestinal absorption rate is less than 10%, but exhibits large inter-individual variation [17]. The individual fluctuation in the level of the iron depot seems to be a significant influencing factor in this case. Iron deficiency can increase the normal absorption rate of cadmium by a factor of more than three [18, 19]. Therefore, the oral absorption rate is higher as a rule for women than men. Cadmium retained in the lungs is released from this site with a half-life of about 5 days while its concentration increases simultaneously in other organs [20]. With regard to the mechanisms which lead to the distribution of absorbed cadmium in the human organism, it is known that the metal becomes bound to the low-molecular protein metallothionein, probably during its first passage through the liver. A certain detoxification is associated with the formation of this cadmium-metallothionein complex, whereby an increased formation of metallothionein can be induced by chronic exposure to cadmium [13].

The target organ for the chronic intake of cadmium is the kidney. The metal accumulates there and initially causes damage to the renal tubuli. In the case of higher internal levels, the function of the glomeruli is also impaired. Enzymuria and proteinuria occur as a consequence of this kidney damage. In the past, early symptoms of a kidney dysfunction appeared as a result of exposure to the then prevailing levels in the environment. Furthermore, the contribution of tobacco smoke to cadmium exposure is not inconsiderable [21]. Cadmium and its compounds cadmium chloride, cadmium oxide, cadmium sulphate, cadmium sulphide and other bioavailable compounds (in the form of inspirable dust/ aerosols) have proved carcinogenic in animal studies. Therefore the Deutsche Forschungsgemeinschaft's Commission for the Investigation of Health Hazards of Chemical Compounds in the Work Area (Commission for Working Materials) has assigned cadmium and the compounds mentioned above to group 2 of the carcinogenic working materials [2].

The technical exposure limit (TRK value) is 0.03 mg/m^3 in areas where batteries are manufactured, for the thermal extraction of zinc, lead and copper, and for the welding of alloys containing cadmium. The TRK value (1998) for other workplaces is 0.015 mg/m^3 – measured as the inspirable aerosol portion in each case [2].

In principle, determination of the cadmium concentration in both blood and urine is suitable for estimating the inner exposure to cadmium. While the cadmium level in blood reflects a recent exposure to cadmium, the determination of the cadmium concentration in urine opens up the possibility of quantifying this metal in the entire organism. The "Human Biomonitoring" Commission of the Umweltbundesamt (Ministry for the Environment) and the Arbeitsstoffkommission (Commission for Working Materials) have set a reference value (human biomonitoring value: HBM) and a threshold value

Table 1. Reference values and threshold values of cadmium for the general population and for occupationally exposed persons.

Designation	Group of persons	Value
TRK value	Workers	0.03 or 0.015 mg/m^3
Reference value	Children (6–12 years old)	0.5 µg/L whole blood 0.5 µg/g creatinine or 0.5 µg/L urine
	Adults (non-smokers 25–69 years old)	1.0 µg/L whole blood 1.0 µg/g creatinine or 1.0 µg/L urine
HBM I value	Children, adolescents and young adults (<25 years old)	1 µg/g creatinine
	Adults (>25 years old)	2 µg/g creatinine
HBM II value	Children, adolescents and young adults (<25 years old)	3 µg/g creatinine
	Adults (>25 years old)	5 µg/g creatinine
BAT value	Workers	15 µg/L urine 15 µg/L whole blood

(biological tolerance value: BAT) for the evaluation of an inner cadmium stress for the general population and for people exposed to cadmium at the workplace (cf. Tab. 1) [2, 22]:

Platinum

At a concentration of 5 µg/kg in the upper 16 km of the Earth's crust (approximately as much as palladium and gold), platinum is the 76th most abundant element [12]. Platinum can occur in oxidation states from 0 to +VI, whereby bivalent and tetravalent compounds are the most common. In solutions it occurs exclusively in the form of coordination compounds (platinates). The relative atomic masses of the naturally occurring isotopes of platinum are 190 (relative occurrence: 0.01%), 192 (0.79%), 194 (32.9%), 195 (33.8%), 196 (25.3%) and 198 (7.2%) [12].

As finely dispersed platinum exhibits excellent catalytic properties, it is used in large-scale technical processes, such as the manufacture of nitric acid from ammonia, the oxidation of ammonia to produce fertilizers, in petrochemical processes, for example hydrocracking, isomerization, aromatization, hydrogenation, etc., and in catalytic converters to reduce the pollution from the exhaust gases of automobiles [12].

Moreover, certain platinum compounds (cis-platinum, carboplatinum) have been successfully applied in the therapy of cancer. In addition, alloys containing precious metals which are used in dental health can contain a maximum of about 20% platinum according to the manufacturers' declarations [23].

In 1988 125 t of platinum were produced worldwide, of which 31% was used in the western world to manufacture automobile catalytic converters, 29% for the manufacture of jewellery and 14% in the chemical and petrochemical industry [12].

Automobile catalytic converters, such as the three-way catalytic converters commonly mounted on cars in Germany, consist of a ceramic carrier material coated with aluminium oxide, bearing a total amount of about 2–3 g of platinum or palladium and rhodium as well as small proportions of other catalytically active precious metals [24]. Platinum is now increasingly being replaced by palladium for cost reasons. The ratio of platinum to rhodium or palladium to rhodium is approximately 5:1. Under normal running conditions the above-mentioned precious metals are released into the air with the exhaust fumes as suspended particles and vapours as a result of mechanical abrasion and thermal stress. Thus, a distinct rise in the platinum content of dust deposits and in plants compared with its natural occurrence has been observed, particularly in the vicinity of roads carrying high volumes of traffic [25, 26]. According to an investigation carried out by the National Academy of Sciences [27], platinum is emitted from automobile catalytic converters predominantly in its metallic form and as platinum dioxide (PtO_2) bound to the carrier material Al_2O_3. In particles with a mean aerodynamic diameter of more than 5 µm, König et al. [28] found platinum concentrations which were equivalent to a concentration of between 3.3 and 39.0 ng/m^3 in the exhaust fumes. In the case of particles with a mean diameter between 0.1 and 20 µm, platinum concentrations between 43 and 88 ng/m^3 were found in exhaust fumes by Innacker and Malessa [29]. The mean platinum emission

from an engine with a three-way catalytic converter is given as 15 ng/m^3 at a speed
of 100 km/h [24].

According to Alt et al. [30], the total platinum content in suspended dust samples
was 0.6 to 130 μg/kg, of which a proportion of 30–43% is soluble in 0.07 mol/L
HCl. The platinum content of the air was between 0.02 and 5.1 pg/m^3, i.e. about 4
orders of magnitude below the level in automobile exhaust fumes.

The main route of intake of platinum and its compounds is by inhalation, oral intake
is low in comparison. It is excreted in the urine, whereby 20–45% is eliminated within 24 h after its intake [31]. Platinum oxides and soluble platinum compounds can
have a sensitizing and allergic effect on humans following dermal exposure and
when these compounds are inhaled [31–33]. These effects were observed exclusively
in persons who were exposed at the workplace. Workers in platinum refining plants
and in the production of catalytic converters are primarily affected. However, elevated internal exposure to platinum was also found in other occupational groups,
such as dental technicians and hospital personnel who had contact with cytostatic
agents. Table 2 contains the results of platinum determinations in the urine of various
groups of people who are exposed to platinum at their workplace. Recent investigations [34–36] have shown that people who are regularly exposed to heavy traffic,
such as road construction workers on motorways, workers employed in motorway
maintenance as well as bus drivers and taxi drivers do not excrete elevated platinum
concentrations in their urine.

At present little is known for sure about the possible effects on humans, animals and
plants as a result of the increasing emission of platinum into the environment. Similarly, there is little information on the possible consequences of the release of platinum from dental alloys containing precious metals.

Messerschmidt et al. [40], Begerow et al. [34, 41–43], Schramel et al. [44], Schierl
et al. [35, 38], Nygren et al. [45], and Philippeit and Angerer [46] have carried out
investigations to determine platinum concentrations in the urine of persons who had
not been exposed to platinum at the workplace. The results are summarized in Table
3.

As the table demonstrates, there is excellent agreement between the results for the
physiological platinum concentrations in urine published by Messerschmidt et al.
[40], Schierl et al. [35, 38], Schramel et al. [44], Begerow et al. [41–43], and by Philippeit and Angerer [46]. In the light of the other results, the platinum concentrations
published by Nygren et al. [45] must be regarded as distinctly too high.

In contrast to exposure to automobile emissions, the release of platinum from dental
alloys can make a considerable contribution to the total exposure of the affected person to the metal. It was shown that the platinum excretion in urine was increased by
several hundred percent when a commercially available, frequently used dental alloy
with a high proportion of gold (Pt content: 9.0%) was inserted [48]. Before insertion
of the dental alloy, the platinum excretion of the three investigated test persons was
between 1.0 and 7.4 ng/L, and thus in the range of the background environmental exposure (cf. Tab. 3). After insertion of the artificial denture with a high gold content,
a distinct increase in the urinary excretion of platinum was observed in all three test
persons, and the elevated levels were maintained throughout the three-month investi-

Table 2. Results of platinum determination in the urine of people who are occupationally exposed to the metal [ng/L].

Occupation	n	Mean value	Range	Reference
Production of catalytic converters	19	950	23–9200	[37]
Recycling of catalytic converters	5	320	20–630	[37]
Handling of platinum nozzles	16	214	10–2900	[37]
Production of catalytic converters	34		16–6270[1]	[38]
Hospital personnel	21		<1.8–34.4	[39]
Dental technicians	27	25.7	0.8–167.8	[34]
Road construction workers	17	0.9	0.2–4.4	[34]
Bus drivers	29	2.8	1.0–40	[35]
MOT testers	13	2.2	0.5–21.0	[35]
Motorway maintenance	18		<1.0–6.6	[36]
Taxi drivers	10	1.3	1.0–28	[35]

[1] ng/g creatinine

Table 3. Results of investigations to determine the platinum content in the urine of the general public [ng/L]

Group of persons	n	Mean value	Range	Reference
Adults	14	1.1	0.5–14.3	[40]
Adults	16	1.7	0.5–7.7	[42]
Adults	21	1.8	0.5–7.7[1]	[43]
Children	262	1.8	0.2–19	[43]
Trainees	17	1.1	0.3–2.2	[34]
Adults	10	5.4	1.2–35	[44]
Adults	12	6.3	2.1–17.4	[34]
Adults	12		1–12[1]	[38]
Adults	21	126		[45]
Adults (without gold fillings in their teeth)	20	1.2	0.9–6.6	[46]
Adults (with gold fillings in their teeth)	26	23.1	0.9–151.2	[46]

[1] ng/g creatinine

gation period. In the first few days after insertion the platinum excretion increased on average by a factor of 12 to values between 10.5 and 59.6 ng/L, three months after inserting the artificial denture the mean platinum concentrations in urine were still 7 times higher than the original level. In vitro experiments clearly confirmed the release of platinum from this type of alloy [47]. In this context, Philippeit and Angerer were also able to detect elevated platinum concentrations in the urine of persons with artificial dentures containing a high concentration of gold [46]. In this case the mean platinum excretion in the urine of 26 people was 23.1 ng/L (cf. Tab. 3) [46].

Mercury (Hg)

Mercury is a silvery-white metal which is insoluble in water. It has a high surface tension and its density is 13.6 g/mL. It is the only metal which is liquid at room temperature. Hg and its chemical compounds are of great importance as working materials and also as pollutants of the environment. Among its many uses, mercury is filled into barometers and thermometers, it is employed to extract gold and silver from sand containing these precious metals. It is used in neon tubes and mercury vapour lamps, as a cathode material in chlorine-alkali electrolysis and in the manufacture of dry batteries [12]. The two decisive action mechanisms of mercury poisoning are the denaturization of proteins at the application site, and inhibition of enzymes as a result of divalent mercury ions reacting with free enzymatic thiol groups. The site of action and the toxic effects depend on the differing kinetic behaviour of the individual compounds. It is primarily the central nervous system which exhibits symptoms of poisoning by elemental mercury and its organic compounds, whereas damage to the kidney region (glomeruli, tubuli) is caused by bivalent mercury ions.

It is advisable to determine the mercury levels in various biological matrices, especially in blood and urine, in order to estimate the degree of exposure to mercury at the workplace and in the environment. The concentration of mercury in biological materials depends on its route of intake, on the chemical bonding state of the element and on the duration of the exposure and the degree to which the substance is incorporated. The mercury concentration in the blood and urine is a measure of the dose absorbed during the previous weeks [48–50].

The "Human Biomonitoring" Commission of the Umweltbundesamt (Ministry for the Environment) and the Arbeitsstoffkommission (Commission for Working Materials) have set reference values and threshold values for the evaluation of inner mercury stress for the general population and for people exposed at the workplace [2, 51]. The reference value for children and adults without amalgam fillings is 1.4 µg Hg per litre urine or 1.0 µg Hg per gram creatinine. The value can be up to four times higher when the person has amalgam fillings. The relevant reference value for the determination of mercury in blood is 1.5 µg Hg per litre for children or 2.0 µg Hg per litre blood for adults. The BAT value (Biological Tolerance Value for Working Materials) can be used as a reference for occupational medical and toxicological assessment of the mercury level in the blood and urine of people who are exposed to mercury. The current BAT value (1998) is 25 µg per litre blood or 100 µg per litre urine in the case of exposure to inorganic compounds and metallic mercury [2]. The mercury content of whole blood is measured for the evaluation of occupational exposure to organic mercury compounds, in particular the alkyl mercury compounds. In this case the current BAT value is 100 µg per litre blood [2]. On a collective basis, these BAT values for inorganic compounds or metallic mercury correlate with the current (1998) MAK values of 0.012 mL/m^3 and 0.1 mg/m^3. The Commission for the Investigation of Health Hazards in the Work Area has assigned organic mercury compounds to group 3 of the carcinogenic working materials. This means that, despite justified concern, organic mercury compounds may have a carcinogenic effect on humans, it is impossible make a definitive assessment at present due to lack of information. However, in vitro experiments and animal studies have provided indications of a carcinogenic effect, but there are insufficient grounds for assigning these compounds to another category [2].

Tellurium

Like sulphur and selenium, tellurium (Te) is a chalcogen. Its atomic number is 52 and it has an atomic mass of 127.61.

Elemental tellurium exists in two polymorphic forms: the silvery-white, shiny hexagonal rhombohedral crystals, and amorphous tellurium which is a fine, black powder. Both elemental tellurium and tellurium dioxide are barely soluble.

Tellurium and selenium are chemically similar, but tellurium has more metallic properties. It occurs in oxidation states II, IV and VI, the tetravalent compounds being the most stable. Tellurium is a very rare element. Its concentration in the Earth's crust is only a few μg/kg.

Tellurium is used as an additive in the steel, chemical, electrical and electronic industries. The worldwide production of tellurium is in the magnitude of 1000 t/a [1].

The degree to which tellurium compounds are absorbed varies to a large extent. While acute poisoning has been observed at the workplace, nothing is known about the chronic effects of small doses.

Tellurium hydride (H_2Te) and tellurium hexafluoride (TeF_6), both of them colourless and extremely toxic gases, are of toxicological interest, as are several other compounds, such as tellurium dioxide (TeO_2) and salts derived from orthotelluric(VI) acid (H_6TeO_6) and telluric(IV) acid (H_2TeO_3).

The MAK value (1998) for tellurium and its compounds has been set at 0.1 mg/m^3. However, this value should not be applied to tellurium hexafluoride and tellurium hydride, as the concentration limits at which these compounds can be regarded as harmless are not known. The concentration of these substances should not exceed 0.01 mg/m^3 under any circumstances [2].

ICP-MS opens up new possibilities of investigating these questions by achieving detection limits of 0.01 μg/L urine (Q-ICP-MS) and approximately 0.001 μg/L urine (SF-ICP-MS).

Initial investigations of urine samples from the general public indicate that a reference value for the tellurium concentration could be in the order of 50 ng per litre urine [9].

Thallium

Thallium (Tl) has an atomic mass of 204.37 and its atomic number is 81. It is classified as a heavy metal on account of its density (11.83 g/cm^3).

In many ways its chemical behaviour resembles that of lead, its neighbour in the periodic table. The similarity between the ionic radii of Tl(I) ions and potassium ions is of great significance for the physiological characteristics of this element.

Thallium occurs in the oxidation states +I and +VI in its compounds, the former being its more stable form.

Although thallium is employed industrially in small quantities only, it has numerous uses [1]. Thus, its sulphide, arsenide, selenide and telluride are used in semiconductor technology. In the glass industry it is used as an additive in the manufacture of low-melting, extremely durable special glass with a high refractive index. Due to their high transparency in the infrared range, mixed crystals of thallium halides are employed in the manufacture of lenses and prisms.

Products containing thallium, particularly thallium(I) sulphate, are still employed today as pesticides against rats and insects in many countries.

Several review articles presenting the clinic symptoms and pharmacokinetics of thallium have been published [52, 53].

Thallium and its inorganic compounds are absorbed into the human body through the lungs and the gastro-intestinal tract. Intake is rapid and almost complete. It enters the bloodstream through which it is swiftly transported to the interstitial and intracellular spaces in various organs and tissues.

In humans thallium acts as a general cytotoxin and is thus highly poisonous when taken in acute oral doses. At present there is still little knowledge about chronic thallium poisoning in humans.

The MAK value (1998) for soluble thallium compounds is currently 0.1 mg/m^3, based on the inspirable dust fraction [2].

ICP-MS, which achieves detection limits of approximately 0.005 µg per litre urine (Q-ICP-MS) and 0.0005 µg per litre urine (SF-ICP-MS) for thallium, is vastly superior to the previous determination methods (GFAAS, ICP-OES, voltammetry).

Initial investigations of urine samples from the general population indicate that a reference value for the thallium concentration could be in the order of 100 ng per litre urine [9]. For further details regarding thallium the reader is referred to the general toxicological section of the "Thallium in Urine" method in Volume 5 of "Analyses of Hazardous Substances in Biological Materials" [54].

Bismuth

Bismuth (Bi) is a greyish-white shiny metal with a melting point of 271 °C, a boiling point of 1560 °C, a density of 9.8 g/cm^3 and an atomic mass of 208.98.

The chemical behaviour of bismuth is similar to that of Pb, As and Sb. As it occurs considerably less frequently (its mean concentration in the Earth's crust is 0.19 mg/kg), no environmental damage due to bismuth has yet been discovered. Dissolved bismuth compounds are rapidly converted to insoluble compounds.

Bismuth is used for the manufacture of numerous readily fusible alloys, in quenching baths for steel production, for the silvering of mirrors and in dental health. Bi compounds have a multitude of uses in industrial processes and products, such as in the manufacture and reprocessing of nuclear fuel rods, in battery cathodes, in semiconductors and for catalysts in the chemical industry. The *Merck Index* [55] lists 39 bismuth compounds, 17 of which are employed in the pharmaceutical industry and in many cosmetic products.

The consequences of intoxication range from stomatitis, local pigmentation, erythema to kidney damage. In addition, cases of bismuth encephalopathy have been described throughout the world. Such cases are probably the result of intoxication from insoluble Bi salts which are deposited in the brain. In contrast, soluble bismuth salts, which are used in the therapy of various diseases (stomach preparations, syphilis therapy), are rapidly excreted.

Nothing is known about a general risk from bismuth and its compounds. Until now no bismuth compounds have been included in the list of MAK values [2].

Tungsten

Tungsten (atomic mass 183.8) belongs to the transition metals. At 3410 °C it has the highest melting point of all the metals. The most frequently produced radioisotopes are [181]W, [185]W and [187]W. Tungsten can occur in oxidation states from 0 to VI and it is closely related chemically to molybdenum [12].

For further details regarding tungsten the reader is referred to the general toxicological section of the "ICP Collective Method" in Volume 5 of "Analyses of Hazardous Substances in Biological Materials" [56].

Tin

The atomic mass of tin is 118.9, its atomic number is 50 and it exists in 3 allotropic forms (α, β and γ forms). Metallic tin is mainly produced by reduction of its dioxide (SnO_2).

Tin occurs in oxidation states +II and +IV in its compounds. It forms numerous anionic complexes with halides and ligands containing oxygen, especially in its tetravalent state. As its position in group IV of the periodic table would indicate, tin forms organometallic compounds with covalent C-Sn bonds. Pure tin is utilized in the form of tin foil and as a thin coating to prevent the corrosion of iron (tinplate for cans). The most frequent use for elemental tin is for alloys such as bronze, soft solder, lettertype metal, white metal (bearing metal), Wood's alloy etc. Elemental tin is resistant to water and air at room temperature.

$SnCl_2$ is frequently used as a reducing agent. SnO is employed in the manufacture of enamel, SnO_2 is a constituent of opalescent glass, while various inorganic compounds of tin are used for dyeing [1].

The organotin compounds are of particular significance. They are added to paints as fungicides, disinfectants and anti-fouling agents. They function as stabilizers in plastics, catalysts for olefin polymerization and auxiliary reagents in the production of foamed polyurethanes.

The relatively low toxicity of orally ingested metallic and inorganic tin is probably due to the fact that it is not readily absorbable. In contrast, many different organotin compounds are very toxic, compounds containing short-chain alkyl groups being especially dangerous. Toxicity increases with the number of alkyl groups in the compound [57]. Oedema of the white cerebral matter and damage to nerve cells in certain regions of the brain can be caused by exposure to these substances.

Authors: *P. Schramel, I. Wendler*
Examiners: *L. Dunemann, M. Fleischer, H. Emons*

Antimony, Lead, Cadmium, Platinum, Mercury, Tellurium, Thallium, Bismuth, Tungsten, Tin

Application   Determination in urine

Analytical principle Inductively coupled plasma quadrupole mass
                         spectrometry (Quadrupole ICP-MS)

Completed in   August 1998

Contents

Essential Biomonitoring Methods. DFG, Deutsche Forschungsgemeinschaft
Copyright © 2006 WILEY-VCH Verlag GmbH & Co. KGaA, Weinheim
ISBN: 3-527-31478-4

1 General principles

After UV digestion of the urine samples, an internal standard is added and the samples are introduced into the ICP-MS by means of a pneumatic nebulizer. The evaluation is carried out using the standard addition procedure.

2 Equipment, chemicals and solutions

2.1 Equipment

ICP mass spectrometer (quadrupole or SF-ICP-MS) with autosampler, PC and printer

UV digestion device with 20 mL quartz vessels (e.g. UV 1000 from Kürner)

Microlitre pipette, adjustable between 10 and 100 µL (e.g. from Eppendorf)

Microlitre pipette, adjustable between 100 and 1000 µL (e.g. from Eppendorf)

Millilitre pipette, adjustable between 1 and 5 mL (e.g. from Eppendorf) and/or 1 and 10 mL (e.g. from Rainin)

10, 20, 100 and 1000 mL Volumetric flasks

100 mL Measuring pipettes

Quartz glass or plastic sample vials: approx. 20 mL (depending on the autosampler)

2.2 Chemicals

Antimony standard solution (1 g/L) in the form of Sb_2O_3 in 5% HCl (e.g. from Spex)

Lead standard solution (1 g/L) in the form of $Pb(NO_3)_2$ in 5% HNO_3 (e.g. from Spex)

Cadmium standard solution (1 g/L) in the form of Cd in 5% HNO_3 (e.g. from Spex)

Platinum standard solution (1 g/L) in the form of Pt in 10% HCl (e.g. from Spex)

Mercury standard solution (1 g/L) in the form of Hg in 10% HNO_3 (e.g. from Spex)

Tellurium standard solution (1 g/L) in the form of Te in 5% HNO_3 (e.g. from Spex)

Thallium standard solution (1 g/L) in the form of $TlNO_3$ in 5% HNO_3 (e.g. from Spex)

Bismuth standard solution (1 g/L) in the form of Bi in 10% HNO_3 (e.g. from Spex)

Tungsten standard solution (1 g/L) in the form of $(NH_4)_2WO_4$ in 2% HNO_3+5% HF (e.g. from Spex)

Tin standard solution (1 g/L) in the form of Sn in 20% HCl (e.g. from Spex)

Rhodium standard solution (1 g/L) in the form of Rh in 10% HCl (e.g. from Spex)

Iridium standard solution (1 g/L) in the form of Ir in 2–5% HCl (e.g. from Spex)

Ultrapure water (equivalent to ASTM type 1) or double-distilled water

Argon (welding argon) for ICP

65% HNO_3 (subboiling distilled or "Suprapur" from Merck)

2.3 Solutions

1 M HNO_3 (to clean the glassware):
About 500 mL ultrapure water are placed in a 1000 mL volumetric flask, then 70 mL of the 65% HNO_3 are added with a pipette. The flask is subsequently filled to its nominal volume with ultrapure water while the contents are swirled gently.

1.4 M HNO_3:
About 500 mL ultrapure water are placed in a 1000 mL volumetric flask, then 100 mL of the 65% HNO_3 are added with a pipette. The flask is subsequently filled to its nominal value with ultrapure water while the contents are swirled gently.

These solutions can be stored for several months at $4\,°C$.

2.4 Solution of the internal standard (rhodium and iridium)

1 mL each of the rhodium and iridium standard solutions is pipetted into a 100 mL volumetric flask. The volumetric flask is subsequently filled to its nominal volume with ultrapure water (10 mg/L).

2.5 Standard addition solution

Starting solution:
0.1 mL of the standard solutions of antimony, lead, cadmium, platinum, mercury, tellurium, thallium, bismuth, tungsten and tin are transferred to a 100 mL volumetric flask with a pipette. The flask is then filled to its nominal volume with 1.4 M HNO_3 (1 mg/L).

Standard addition solution:

1 mL of the starting solution is pipetted into a 10 mL volumetric flask. The volumetric flask is subsequently filled to its nominal volume with 1.4 M HNO_3 (0.1 mg/L).

These solutions must be freshly prepared daily.

3 Specimen collection and sample preparation

3.1 Specimen collection

As is the case for all trace element analyses, it is essential to ensure that the reagents are of the highest possible purity and that the vessels are thoroughly clean. This also applies to sample collection.

To prevent a possible exogenous contamination, the plastic vessels for sample collection must be cleaned before use by leaving them filled with 1 M HNO_3 for at least 2 hours, rinsing them thoroughly with ultrapure water and drying them. For determination in the range of the detection limit the cleansing effect can be improved by warming the nitric acid.

The urine should be collected and stored in polyethylene vessels, whereby it is always advisable to collect urine over a 24-hour period. If the determination cannot be carried out immediately, the urine can be stored in the refrigerator for about 1 week at approximately $+4\,°C$, but the urine must be acidified (approx. 10 mL HNO_3 per litre urine). If longer storage is necessary, it is advisable to keep the samples in the deep-freezer at $-18\,°C$.

Prior to further processing, the urine samples are thawed and brought to room temperature.

3.2 UV digestion

Before an aliquot is withdrawn for UV digestion, the samples are thoroughly shaken to ensure they are homogeneous.

Then 4 mL urine, 1 mL concentrated HNO_3, 11 mL H_2O and 4 mL H_2O_2 are pipetted into the 20 mL digestion vessels (dilution 1:5). These are placed into the UV digestion device and exposed to UV light for approximately 1 hour. The colourless to slightly yellow solutions are brought to room temperature.

A reagent blank is included in each analytical series. Ultrapure water is used instead of urine in this case.

Between 4.80 and 4.95 mL of the digested sample are pipetted into a 20 mL auto-sampler vial, then the solution of the internal standard and the standard addition solution are added in accordance with the pipetting scheme shown in Table 4.

Table 4. Pipetting scheme for the standard addition procedure.

Sample		Internal standard	Standard addition solution	Desig-nation	Added metal concentration, based on the urine volume used
Urine	Water				
[mL]	[mL]	[mL]	[mL]		[µg/L]
4.95	–	0.05	–	ADD1	–
4.90	–	0.05	0.05	ADD2	5
4.85	–	0.05	0.1	ADD3	10
4.80	–	0.05	0.15	ADD4	15
–	4.95	0.05	–	Blank value	–

Standard addition solution: 0.1 µg/L

4 Operational parameters for ICP-MS

4.1 Plasma settings

Power supply: 1.2 kW

Sample introduction: peristaltic pump, output <1 mL/min

Nebulizer: cross flow or Meinhard

Nebulizer chamber: Scott type (quartz glass or Rayton®)
Zyklon chamber (advisable – due to less memory effects)

Plasma conditions: combustion gas 15 L/min
nebulizer gas approx. 0.7–0.8 L/min (must be optimized)
plasma gas 0.8 L/min

The given plasma conditions serve only as a guide. The operational parameters must be optimally adjusted for each individual instrument used.

4.2 MS parameters (Q-ICP-MS)

Table 5. ICP-MS parameters.

Sweeps/Reading	20
Reading/Replicate	6
Number of replicates	1
Points across peak	normal
Resolution	peak hop
Scanning mode	0
Baseline time (ms)	replicat
Transfer frequency	+
Polarity	

Table 6. MS program parameters.

Element	Mass	Times [ms]	
		Replicate	Dwell
Rh*	103	2000	100
Cd	111	2000	100
Sn	118	2000	100
Sb	121	2000	100
Te	126		
W	182	2000	100
Ir*	193	2000	100
Pt	195	2000	100
Hg	202	2000	100
Tl	205	2000	100
Pb	208	2000	100
Bi	209	2000	100

*: internal standard

Element equations:
Rh 103 = Rh 103
Cd 111 = Cd 111
Sn 118 = Sn 118
Sb 121 = Sb 121
Te 126 = Te 126–0.003404×Xe 129
W 182 = W 182
Ir 193 = Ir 193
Pt 195 = Pt 195
Hg 202 = Hg 202
Tl 205 = Tl 205
Pb 208 = Pb 208
Bi 209 = Bi 209

Manual settings:
Plasma flow: 15 L/min   RF power: 1200 Watts
Nebulizer flow: 0.75 L/min   CEM voltage: 3.7 kV
Auxiliary flow: 0.8 L/min   Sample uptake: 0.9 mL/min

5 Analytical determination

The ADD1 to ADD4 solutions are introduced into the plasma and analyzed (see Figs. 1 and 2).

The analytical determination is carried out by quadrupole MS. Four measurement points are thus obtained for the evaluation of the analytical result.

6 Calibration and calculation of the analytical result

The concentrations of the metals in the urine sample are obtained from a graph with the aid of the standard addition procedure. The reagent blank value is subtracted from the peak heights of the unspiked and the three spiked samples, which are then divided by the corresponding value for the appropriate internal standard, and the resulting quotients are plotted as a function of the metal concentrations. Linear graphs are obtained and their point of intersection with the concentration axis gives the concentration of the metal in µg per litre urine in each case.

The more modern generation of instruments is equipped with computer-supported evaluation programs which perform the evaluation automatically.

The linear operational range extends to 150 µg of the metals per litre urine. The urine samples must be further diluted when the metal concentrations are not within the linear range of the graph.

7 Standardization and quality control

Quality control of the analytical results is carried out in accordance with the guidelines of the Bundesärztekammer (German Medical Association) [58, 59] and the special preliminary remarks in Volume 1 of the "Analyses of Hazardous Substances in Biological Materials". Material containing antimony, lead, cadmium, mercury and thallium is commercially available for internal quality control, e.g. "Control urine" from Recipe, Munich.

As no control material is commercially available for the remaining metals which can be determined by this method, it must be prepared in-house in the laboratory. For this purpose, urine is spiked with a defined quantity of the metals. Aliquots of this solution can be stored in the deep-freezer for up to a year and used for quality control. The concentration of this control material should lie in the middle of the most frequently occurring concentration range. The theoretical value and the tolerance range for this quality control material is determined in the course of a pre-analytical period (one analysis of the control material on 20 different days) [58, 60].

External quality control can be realized by participation in round-robin experiments. The round-robin experiments carried out to test analysis in occupational and environmental medicine in Germany include antimony, lead, cadmium, mercury and thallium in the concentration range of interest to occupational medicine and cadmium, platinum and mercury in the concentration of interest to environmental medicine in the quality control programme [61, 62].

8 Reliability of the method

8.1 Precision

The precision in the series was checked in 10 analyses of pooled urine from people who had not been exposed to the metals at the workplace. The following relative standard deviations and prognostic ranges were found (cf. Tab. 7).

In addition, the ten metals were added to the same pooled urine so that the concentration of each was 2 µg per litre, and the spiked sample was processed and analyzed 10 times. The following standard deviations and prognostic ranges were obtained (cf. Tab. 8).

The precision from day to day was tested using the same pooled urine after adding 1 µg of each of the metals to 1 L urine and the sample was processed and analyzed on 10 different days. The following standard deviations and prognostic ranges were obtained (cf. Tab. 9).

Table 7. Precision in the series.

Element	Concentration	Standard deviation (rel.)	Prognostic range
	[µg/L]	[%]	[%]
Cd	0.21	4.8	9.0
Sn	0.39	9.8	18.1
Sb	0.10	14.8	26.0
Te	0.04	9.1	17.2
W	0.18	4.5	8.1
Pt	–	–	–
Hg	0.13	16.2	29.3
Tl	0.12	4.4	8.1
Pb	0.46	2.0	3.8
Bi	–	–	–

Table 8. Precision in the series after addition of 2 µg/L each ($n = 10$).

Element	Concentration	Standard deviation (rel.)	Prognostic range
	[µg/L]	[%]	[%]
Cd	2.10	0.9	1.5
Sn	2.28	0.6	1.0
Sb	2.37	1.1	2.0
Te	3.37	1.6	3.0
W	2.54	0.3	0.5
Pt	2.00	0.8	1.5
Hg	2.26	1.5	2.8
Tl	2.10	0.8	1.5
Pb	2.48	0.4	0.7
Bi	2.00	0.5	0.9

Table 9. Precision from day to day ($n = 10$ days).

Element	Concentration	Standard deviation (rel.)	Prognostic range
	[µg/L]	[%]	[%]
Cd	1.21	2.4	4.2
Sn	1.39	4.2	7.8
Sb	1.10	3.1	5.6
Te	1.04	3.6	6.5
W	1.18	3.6	6.8
Pt	1.00	2.4	4.5
Hg	1.13	8.0	14.0
Tl	1.12	3.7	7.0
Pb	1.46	3.7	6.0
Bi	1.00	1.8	3.2

8.2 Accuracy

Recovery experiments were performed to check the accuracy of the method. For this purpose pooled urine was spiked with 2 µg of each of the metals and analyzed (cf. Tab. 10).

Table 10. Recovery rates.

Element	Concentration (theoretical)	Concentration (found)	Recovery rate
	[µg/L]	[µg/L)	[%]
Cd	2.10	1.99	95.0
Sn	2.28	2.17	95.5
Sb	2.37	2.67	112.7
Te	3.37	3.25	96.4
W	2.54	2.96	116.5
Pt	2.00	1.90	94.9
Hg	2.26	2.39	106.0
Tl	2.10	1.97	93.8
Pb	2.48	2.50	100.9
Bi	2.00	1.59	79.7

Furthermore, the method described here was checked with the aid of commercially available control material. The following results were obtained (cf. Tab. 11).

Table 11. Results obtained using commercially available control material. (Note: the control material had different concentration levels, i.e. level I and II).

Element/level	Unit	Reference value	Range	Found ($n=5$)
Sb/I	μg/L	12.2	9.5–14.9	12.3±0.6
Sb/II		45.5	36.1–51.9	45.5±1.0
Pb/I	μg/L	37.5	30.2–44.6	39.2±1.2
Pb/II		64.5	53.7–75.3	66.1±2.0
Cd/I	μg/L	11.1	9.0–13.2	11.1±0.4
Cd/II		21.7	17.8–25.7	20.9±1.0
Hg/I	μg/L	12.3	9.8–14.9	14.5±1.5
Hg/II		120	92–147	122±5
Tl/I	μg/L	3.1	1.8–4.6	3.6±0.3
Tl/II		17.9	13.5–22.3	21.1±1.5

8.3 Detection limits

Under the analytical conditions given here the detection limits in urine samples were calculated as three times the standard deviation of the background signal at the mass of the given isotope (cf. Tab. 12).

Table 12. Calculated detection limits.

Element	Isotope m/e [amu]	Detection limit Urine [μg/L]
Sb	121	0.03
Pb	208	0.03
Cd	111	0.02
Pt	195	0.01
Hg	202	0.03
Te	126	0.01
Tl	205	0.005
Bi	209	0.005
W	182	0.02
Sn	118	0.05

8.4 Sources of error

Interferences due to overlapping masses ("polyatomic interferences") were not observed for the mass/charge ratios (m/e) measured in this case. Depending on the origin of the lead standard and the exogenous lead exposure, erroneous results of up to

±15% can occur when determining the metal because of local variations in the isotope frequency of lead (final product of a chain of radioactive decay).

When human biological samples are to be determined, sample digestion to destroy the organic matrix is strongly recommended. Thus spectral and non-spectral interferences are distinctly reduced and the long-term stability of the ICP-MS is considerably improved.

The demands placed on sample digestion are normally not very stringent, as there is sufficient thermal energy in the ICP to ensure complete destruction of the organic matrix if this has not already been achieved by sample digestion. UV digestion of the urine samples has proved very practicable for this purpose [42, 63]. In this method, relatively small quantities of acid are added, as H_2O_2 (or rather the OH radicals it generates) represents the real digestion agent. This results in distinctly lower blank values and dispenses with the necessity of sample dilution.

Alternatively, an oxidative digestion with acid, generally with HNO_3 in a closed system (pressure digestion) [64, 65], can achieve satisfactory digestion for analysis. However, limitations are imposed by the relatively high acid concentrations present after an acidic digestion, as they must be lowered by dilution.

The use of two different internal standards (rhodium and iridium) is advisable because of the large range of masses of the elements to be analyzed. However, the precondition for the use of rhodium and iridium is that these elements are not to be subsequently analyzed in other samples in the ultratrace range (risk of memory effects).

It is strongly advisable to shake the samples vigorously after they have been stored, as sedimentation occurring in the urine samples can lead to absorption of the analytes on the surface of the sediment, and thus cause erroneous results [66].

9 Discussion of the method

The method presented here permits the simultaneous analysis of antimony, lead, cadmium, platinum, mercury, tellurium, thallium, bismuth, tungsten and tin in urine.

Apart from bismuth and platinum, the ecological concentration range can be detected. In order to achieve the detection of bismuth and platinum at ecological concentrations, a sector field ICP-MS instrument must be employed.

In this method standard addition was used to evaluate the results. While checking the method it was shown that external aqueous standards can be used for calibration when the metal concentrations of the urine samples are in the concentration range of interest to occupational medicine. However, in this case the accuracy of the calibration must always be checked by means of the standard addition procedure (cf. Appendix).

The examiners of the method carried out the analysis using urine samples which had not been digested. This resulted in a large proportion of the matrix reaching the ICP-MS. This type of test represents the "worst-case scenario". UV digestion of the urine samples minimizes the interfering constituents of the matrix, thus improving the analytical reliability criteria. This was confirmed by comparison with the results for undigested samples obtained by the examiners.

UV digestion of the urine samples is strongly recommended for the investigation of larger series of samples, otherwise clogging of the sample introduction system and the cones is caused by the high salt concentration in the samples, and the removal of such blockages is extremely time-consuming.

According to the examiners, the amount of acid used for digestion can be further reduced if necessary, without reducing the effectiveness of the digestion.

In addition to the isotopes mass spectrometrically evaluated by the author, i.e. ^{111}Cd, ^{202}Hg, ^{208}Pb, ^{121}Sb, ^{118}Sn, ^{205}Tl and ^{182}W, one of the examiners evaluated ^{114}Cd, ^{200}Hg, ^{207}Pb, ^{120}Sn, ^{203}Tl and ^{184}W for comparison. The results of the author were confirmed.

Besides great sensitivity, the use of mass spectrometry for the analysis described here ensures a high degree of specificity with relatively little effort. Thus, this method is superior to the procedures for the analysis of metals such as AAS or ICP-OES which have been available until now.

Instruments used:
ICP mass spectrometer ELAN 5000 (from Perkin Elmer Sciex, Canada)
Autosampler AS-90 (from Perkin Elmer, Germany), PC and printer

UV digestion device UV 1000 (from Kürner)

10 References

[1] *E. Merian* (ed.): Metalle in der Umwelt, Verlag Chemie, Weinheim 1984.
[2] *Deutsche Forschungsgemeinschaft*: MAK- und BAT-Werte-Liste 1998. Maximale Arbeitsplatzkonzentrationen und Biologische Arbeitsstofftoleranzwerte. Mitteilung 34 der Senatskommission zur Prüfung gesundheitsschädlicher Arbeitsstoffe. WILEY-VCH Verlag, Weinheim 1998.
[3] *H. Greim* (ed.): Gesundheitsschädliche Arbeitsstoffe. Toxikologisch-arbeitsmedizinische Begründung von MAK-Werten. Deutsche Forschungsgemeinschaft, WILEY-VCH Verlag, Weinheim, 1st–27th issue, 1998.
[4] *C.G. Elinder* and *L. Friberg*: Antimony. In: *L. Friberg, G.F. Nordberg* and *V.B. Vouk* (eds.): Handbook on the Toxicology of Metals, Vol. II: Specific Metals. Elsevier, Amsterdam 1986.
[5] *E. Merian* and *K.L. Stemmer*: Antimon. In: *E. Merian* (ed.): Metalle in der Umwelt. Verlag Chemie, Weinheim 1984.
[6] *V. Potkonjak* and *M. Pavlovick*: Antimoniosis: A particular form of pneumoconiosis: Etiology, clinical and x-ray findings. Int. Arch. Occup. Environ. Health *51*, 199–207 (1983).
[7] *H. Greim* and *G. Lehnert* (eds.): Biologische Arbeitsstoff-Toleranzwerte (BAT-Werte). Arbeitsmedizinisch-toxikologische Begründungen. Deutsche Forschungsgemeinschaft. WILEY-VCH Verlag, Weinheim, 8th issue, 1996.
[8] *B. Welz* and *M. Sperling*: Atomabsorptionsspektrometrie, 4th revised edition, Wiley-VCH Verlag, Weinheim 1997.
[9] *P. Schramel, I. Wendler* and *J. Angerer*: The determination of metals (antimony, bismuth, lead, cadmium, mercury, palladium, platinum, tellurium, thallium, tin and tungsten) in urine samples by inductively coupled plasma-mass spectrometry. Int. Arch. Occup. Environ. Health *69*, 219–223 (1997).
[10] *J. Konietzko* and *W. Broghammer*: Blei und seine anorganischen Verbindungen. In: *J. Konietzko* and *H. Dupuis* (eds.): Handbuch der Arbeitsmedizin Kapitel IV, ecomed, Mainz 1995.
[11] *H. Seiler*: Lead in blood and urine. In: *J. Angerer* and *K.H. Schaller* (eds.): Analyses of hazardous substances in biological materials, Volume 2. Deutsche Forschungsgemeinschaft, WILEY-VCH Verlag, Weinheim 1988.

[12] *Römpp Chemie Lexikon.* 9th extended and revised edition, Georg Thieme Verlag, Stuttgart, New York 1989.

[13] *L. Friberg, C.G. Elinder, T. Kjellström* and *G. Nordberg* (eds.): Cadmium and Health: A toxicological and epidemiological appraisal, Vol. I and II. CRC Press, Boca Raton, Florida 1985 and 1986.

[14] *R.R. Lauwerys, A.M. Bernard, J.P. Buchet* and *H.A. Roels*: Assessment of the health impact of environmental exposure to cadmium: Contribution of the epidemiologic studies carried out in Belgium. Environ. Res. *62*, 200–206 (1993).

[15] *R.R. Lauwerys, A.M. Bernard, H.H. Roels* and *J.P. Buchet*: Cadmium: Exposure markers as predictors of nephrotoxic effects. Clin. Chem. *40*, 1391–1394 (1994).

[16] *WHO (World Health Organization)*: Environmental Health Criteria 134 – Cadmium. World Health Organization, Geneva, 1992.

[17] *WHO (World Health Organization)*: Recommended health-based biological exposure limits in occupational exposure to heavy metals. WHO Technical Report, Series 543. Geneva 1980.

[18] *P.R. Flanagan, J.S. McLellan, J. Haist, G. Cherian, H.J. Chamberlain* and *L.S. Valberg*: Increased dietary cadmium absorption in mice and human subjects with iron deficiency. Gastroenterology *74*, 841–846 (1978).

[19] *Z.A. Shaikh* and *J.C. Smith*: Metabolism of orally ingested cadmium in humans. Dev. Toxicol. Environ. Sci. *8*, 569–574 (1980).

[20] *D. Szadkowski*: Cadmium – eine ökologische Noxe am Arbeitsplatz? Med. Monatsschr. *26*, 553–556 (1972).

[21] *G. Eisenbrand* and *M. Metzler*: Toxikologie für Chemiker. Georg Thieme Verlag, Stuttgart 1994.

[22] *Kommission „Humanbiomonitoring" des Umweltbundesamtes*: Stoffmonographie Cadmium – Referenz- und Human-Biomonitoring-Werte (HBM). Bundesgesundhbl. *5*, 218–226 (1998).

[23] *J. Begerow, M. Schauer* and *L. Dunemann*: Werkstoffe der konservierenden Zahnheilkunde. In: *H.E. Wichmann, H.-W. Schlipköter* and *G. Fülgraff* (eds.): Handbuch der Umweltmedizin, 11. erg. Lfg. 7/97, ecomed Verlagsgesellschaft, Landsberg/Lech 1997.

[24] *Gesellschaft für Strahlen- und Umweltforschung mbH Neuherberg/Forschungszentrum Jülich GmbH*: Zwischenbericht „Edelmetallemissionen". GSF München 1990.

[25] *F. Zereini, F. Alt, K. Rankenburg, J.-M. Beyer* and *S. Artelt*: Verteilung von Platingruppenelementen (PGE) in den Umweltkompartimenten Boden, Schlamm, Straßenstaub, Straßenkehrgut und Wasser. USWF – Z. Umweltchem. Ökotox. *9*, 193–200 (1997).

[26] *E. Helmers, N. Mergel* and *R. Barchet*: Platin in Klärschlammasche und an Gräsern. USWF – Z. Umweltchem. Ökotox. *6*, 130–134 (1994).

[27] *National Academy of Sciences*: Medical and biological effects of enviromental pollutants: platinum group metals. Washington 1997.

[28] *H.P. König, R.F. Hertel, W. Koch* and *G. Rosner*: Determination of platinum emissions from three-way catalyst-equipped gasoline engines. Atmos. Environ. *26A*, 741–745 (1992).

[29] *O. Innacker* and *R. Malessa*: Abschlußbericht zum Forschungsvorhaben „Experimentalstudie zum Austrag von Platin aus Automobilkatalysatoren". Naturwissenschaftliches und Medizinisches Institut an der Universität Tübingen, Reutlingen 1992.

[30] *F. Alt, A. Bambauer, K. Hoppstock, B. Mergler* and *G. Tölg*: Trace platinum in airborne particulate matter. Determination of whole content, particle size distribution and soluble platinum. Fresenius J. Anal. Chem. *346*, 693–696 (1993).

[31] *P. Goering*: Platinum and related metals: palladium, iridium, osmium, rhodium and ruthenium. In: *J.B. Sullivan* and *G.R. Grieger* (eds.): Hazardous Materials Toxicology. Clinical Principles of Environmental Health. Williams & Wilkens, Baltimore 1992.

[32] *C. Cavelier* and *J. Foussereau*: Kontaktallergie gegen Metalle und deren Salze. Teil IV: Differentialdiagnose und Anhang. Dermatosen *43*, 250–256 (1995).

[33] *G. Rosner* and *R. Merget*: Allergic potential of platinum compounds. In: *A.D. Dayan*: Immunotoxicity of Metals and Immunotoxicology. Plenum Press, New York 1990.

[34] *J. Begerow, U. Sensen, G.A. Wiesmüller* and *L. Dunemann*: Internal platinum, palladium, and gold exposure of environmentally and occupationally exposed persons. Zentralbl. Hygiene, for publication.

[35] *R. Schierl, A.S. Ennslin* and *G. Fruhmann*: Führt der Straßenverkehr zu erhöhten Platinkonzentrationen im Urin beruflich Exponierter? Verhandl. Dt. Ges. Arbeitsmed. *33*, 291–293 (1994).

[36] *K.-H. Schaller, A. Weber, F. Alt, V. Heine, D. Weltle* and *J. Angerer*: Untersuchungen zur Schadstoffbelastung von Beschäftigten einer Autobahnmeisterei. Beitrag anläßlich des 4. Kongresses der Gesellschaft für Hygiene und Umweltmedizin, Graz, 17.–20.4. 1996.

[37] *J. Angerer* and *K.-H. Schaller*: Belastung durch Platin beim Herstellen und Recycling von Katalysatoren. In: *K. Dörner* (ed.) Akute und chronische Toxizität von Spurenelementen. Schriftenreihe der Gesellschaft für Mineralstoffe und Spurenelemente (GMS), Wissenschaftliche Verlagsgesellschaft mbH, Stuttgart 1993.

[38] *R. Schierl, H.-G. Fries, C. van de Weyer* and *G. Fruhmann*: Urinary excretion of platinum from platinum industry workers. Occup. Environ. Med. *55*, 138–140 (1998).

[39] *A.S. Ennslin, A. Pethran, R. Schierl* and *G. Fruhmann*: Urinary excretion in hospital personnel occupationally exposed to platinum-containing antineoplastic drugs. Int. Arch. Occup. Environ. Health *65*, 339–342 (1994).

[40] *J. Messerschmidt, F. Alt, G. Tölg, J. Angerer* and *K.-H. Schaller*: Adsorptive voltammetric procedure for the determination of platinum baseline levels in human body fluids. Fresenius J. Anal. Chem. *343*, 391–394 (1992).

[41] *J. Begerow, M. Turfeld* and *L. Dunemann*: Determination of physiological noble metal levels in human urine applying preconcentration by solvent extraction and Zeeman-ET-AAS. Anal. Chim. Acta *340*, 277–283 (1997).

[42] *J. Begerow, M. Turfeld* and *L. Dunemann*: Determination of physiological platinum levels in human urine using double focusing magnetic sector field inductively coupled plasma mass spectrometry in combination with ultraviolet photolysis. J. Anal. At. Spectrom. *1*, 913–916 (1997).

[43] *J. Begerow, J. Neuendorf, W. Raab, M. Turfeld* and *L. Dunemann*: Der Beitrag von edelmetallhaltigem Zahnersatz zur Gesamtbelastung der Bevölkerung mit Platin, Palladium und Gold. Schriftenreihe der Gesellschaft für Mineralstoffe und Spurenelemente zur 13. Jahrestagung am 24. und 25. 09. 1997 in Dresden.

[44] *P. Schramel, I. Wendler* and *S. Lustig*: Capability of ICP-MS (pneumatic nebulization and ETV) for Pt-analysis in different matrices at ecologically relevant concentrations. Fresenius J. Anal. Chem. *353*, 115–118 (1995).

[45] *O. Nygren* and *C. Lundgren*: Determination of platinum in workroom air and in blood and urine from nursing staff attending patients receiving cis-platinum chemotherapy. Int. Arch. Occup. Environ. Health *70*, 209–214 (1997).

[46] *G. Philippeit* and *J. Angerer*: Bestimmung von Platin, Palladium und Gold in menschlichen Körperflüssigkeiten – Stand der Technik. Vortrag auf der 2. Jahrestagung der International Society of Environmental Medicine 28.–30. 8. 1998, Giessen.

[47] *J. Begerow, J. Neuendorf, M. Turfeld, W. Raab* and *L. Dunemann*: Long-term urinary platinum, palladium, and gold excretion of patients after insertion of noble metal dental alloys. Biomarkers 1998 (in print).

[48] *WHO (World Health Organization)*: Environmental Health Criteria 118 – Inorganic mercury. World Health Organization, Geneva 1991.

[49] *Umweltbundesamt*: Metalle und Metalloide: Quecksilber. Basisdaten Toxikologie (1995).

[50] *NIOSH*: Criteria for a recommended Standard – occupational exposure to inorganic mercury, U.S. Dep. of Health, Education and Welfare, Public Health Service 1973.

[51] *J. Angerer*: Personal communication, 1998.

[52] *J.L. Allsop*: Thallium poisoning. Austr. Ann. Med. *11*, 144–153 (1953).

[53] *O. Grunfeld* and *G. Hinostrozo*: Thallium poisoning. Arch. Int. Med. *114*, 132–140 (1964).

[54] *M. Fleischer* and *K.H. Schaller*: Thallium in urine. In: *J. Angerer* and *K.H. Schaller* (eds.): Analyses of hazardous substances in biological materials, vol. 5. Deutsche Forschungsgemeinschaft, WILEY-VCH Verlag, Weinheim 1996.

[55] *Merck Index* (9th Ed.): Merck & Co. Inc., Rohway, N.Y., USA (1976).

[56] *P. Schramel*: ICP collective method. In: *J. Angerer* and *K.H. Schaller* (eds.): Analyses of hazardous substances in biological materials, vol. 5. Deutsche Forschungsgemeinschaft, WILEY-VCH Verlag, Weinheim 1996.

[57] *E. Berman*: Toxic metals and their analysis. Heyden, London 1980

[58] *Bundesärztekammer*: Qualitätssicherung der quantitativen Bestimmungen im Laboratorium. Neue Richtlinien der Bundesärztekammer. Dt. Ärztebl. *85*, A699–A712 (1988).

[59] *Bundesärztekammer*: Ergänzung der „Richtlinien der Bundesärztekammer zur Qualitätssicherung in medizinischen Laboratorien". Dt. Ärztebl. *91*, C159–C161 (1994).

[60] *J. Angerer* and *K.H. Schaller*: Erfahrungen mit der statistischen Qualitätskontrolle im arbeitsmedizinisch-toxikologischen Laboratorium. Arbeitsmed. Sozialmed. Präventivmed. *1*, 33–35 (1977).

[61] *J. Angerer* and *G. Lehnert*: Anforderungen an arbeitsmedizinisch-toxikologische Analysen. Dt. Ärztebl. *37*, C1753–C1760 (1997).

[62] *G. Lehnert, J. Angerer* and *K.H. Schaller*: Statusbericht über die externe Qualitätssicherung arbeits- und umweltmedizinisch-toxikologischer Analysen in biologischen Materialien. Arbeitsmed. Sozialmed. Umweltmed. *33 (1)*, 21–26 (1998).

[63] *J. Begerow, M. Turfeld* and *L. Dunemann*: Determination of physiological palladium, platinum, iridium and gold levels in human blood using double focusing magnetic sector field inductively coupled plasma mass spectrometry. J. Anal. At. Spectrom. *12*, 1095–1098 (1997).

[64] *P. Schramel* and *S. Hasse*: Destruction of organic materials by pressurized microwave digestion. Fresenius J. Anal. Chem. *346*, 794–799 (1993).

[65] *P. Schramel, A. Wolf, R. Seif* and *B.-J. Klose*: Eine neue Apparatur zur Druckveraschung von biologischem Material. Fresenius Z. Anal. Chem. *302*, 62–64 (1980).

[66] *P. Schramel, I. Wendler, P. Roth, E. Werner*: Method for the Determination of Thorium and Uranium in Urine. Mikrochim. Acta *126*, 263–266 (1997).

Authors: *P. Schramel, I. Wendler*

Examiners: *L. Dunemann, M. Fleischer, H. Emons*

11 Appendix

Comparison of the results achieved by the standard addition method with those of an aqueous calibration curve.

Element	Urine A Measurement 1 Cal. curve [µg/L]	Urine A Measurement 2 Cal. curve [µg/L]	Standard addition [µg/L]	Urine B Measurement 1 Cal. curve [µg/L]	Urine B Measurement 2 Cal. curve [µg/L]	Standard addition [µg/L]
Sb	0.023	0.025	0.017	0.058	0.060	0.047
Pb	0.56	0.58	0.54	0.83	0.81	0.86
Cd	0.067	0.067	0.054	0.316	0.366	0.316
Pt	0.011	0.028	0.0004	0.013	0.028	0.001
Hg	0.030	0.048	0.020	0.51	0.63	0.44
Te	0.023	0.028	0.012	0.018	0.035	0.011
Tl	0.29	0.28	0.28	0.43	0.38	0.41
Bi	<0.010	<0.010	<0.010	<0.010	<0.010	<0.010
W	0.042	0.063	0.020	0.065	0.12	0.040
Sn	0.33	0.36	0.30	2.8	2.7	3.3

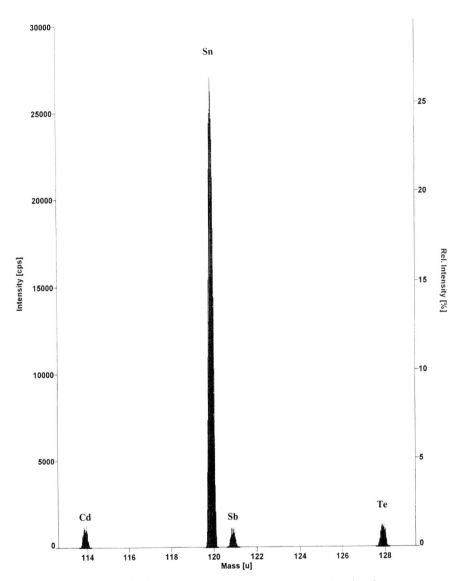

Fig. 1. Q-ICP-MS determination of cadmium, tin, antimony and tellurium in urine.

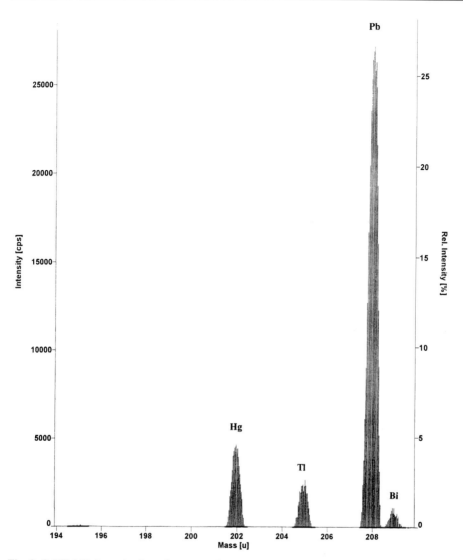

Fig. 2. Q-ICP-MS determination of mercury, thallium, lead and bismuth in urine.

Aromatic Carboxylic Acids (Phenylglyoxylic acid; Mandelic acid; Hippuric acid; o-, m-/p-Methylhippuric acids; Benzoic acid)

Application	Determination in urine
Analytical principle	High Pressure Liquid Chromatography (HPLC)
Completed in	February 1984

Summary

The method described here using high pressure liquid chromatography (HPLC) is the best currently available for the determination of the principal metabolites of styrene, toluene, xylenes and ethylbenzene. A derivatization of the aromatic carboxylic acids is not necessary in this procedure as opposed to gas chromatographic methods. This simplifies the procedure considerably while retaining comparable reliability. This method is thus particularly well-suited to the requirements of medical screening examinations.

The aromatic carboxylic acids are extracted from the acidified urine with diethyl ether. Either 3-hydroxybenzoic acid or 3-chloro-4-hydroxybenzoic acid may be used as internal standard. The dried residue of the ether extract is dissolved in methanol/water (1:5) and the metabolites separated by HPLC and determined with a UV detector at 215 nm.

Aqueous calibration standards are processed and analysed in the same way as the samples. The peak areas determined for the calibration standards are divided by those for the internal standards. The resulting quotients are plotted as a function of the employed concentrations of the aromatic carboxylic acids to produce the calibration curve.

Essential Biomonitoring Methods. DFG, Deutsche Forschungsgemeinschaft
Copyright © 2006 WILEY-VCH Verlag GmbH & Co. KGaA, Weinheim
ISBN: 3-527-31478-4

Phenylglyoxylic acid

Between-day imprecision: Standard deviation (rel.) $s = 2.7 - 5.4\%$
                          Prognostic range $u = 6.0 - 12.3\%$
                          At concentrations ranging from $15 - 100$ mg phenylgly-
                          oxylic acid per litre urine and where $n = 10$ days

Inaccuracy: Recovery rate $r = 86 - 108\%$

Detection limit: 15 mg Phenylglyoxylic acid per litre urine

Mandelic acid

Between-day imprecision: Standard deviation (rel.) $s = 5.8 - 1.6\%$
                          Prognostic range $u = 13.2 - 3.5\%$
                          At concentrations ranging from $25 - 100$ mg mandelic acid
                          per litre urine and where $n = 10$ days

Inaccuracy: Recovery rate $r = 91 - 102\%$

Detection limit: 25 mg Mandelic acid per litre urine

Hippuric acid

Between-day imprecision: Standard deviation (rel.) $s = 3.4 - 12.8\%$
                          Prognostic range $u = 7.6 - 28.9\%$
                          At spiked concentrations ranging from $250 - 1000$ mg hipp-
                          uric acid per litre urine and where $n = 10$ days

Inaccuracy: Recovery rate $r = 99\%$

Detection limit: 25 mg Hippuric acid per litre urine

o-Methylhippuric acid

Between-day imprecision: Standard deviation (rel.) $s = 9.2 - 1.6\%$
                          Prognostic range $u = 20.7 - 3.6\%$
                          At concentrations ranging from $25 - 100$ mg o-methyl-
                          hippuric acid per litre urine and where $n = 10$ days

Inaccuracy: Recovery rate $r = 100 - 116\%$

Detection limit: 15 mg o-Methylhippuric acid per litre urine

m-Methylhippuric acid

Between-day imprecision: Standard deviation (rel.) $s = 4.1 - 1.9\%$
                          Prognostic range $u = 4.4 - 9.3\%$
                          At concentrations ranging from $25 - 100$ mg m-methylhipp-
                          uric acid per litre urine and where $n = 10$ days

Inaccuracy: Recovery rate $r = 98 - 114\%$

Detection limit: 15 mg m-Methylhippuric acid per litre urine

Benzoic acid

Between-day imprecision:	Standard deviation (rel.)	$s = 2.9 - 0.9\%$

Between-day imprecision: Standard deviation (rel.) $s = 2.9 - 0.9\%$
                          Prognostic range $u = 6.6 - 2.0\%$
                          At concentrations ranging from $25 - 100$ mg benzoic acid
                          per litre urine and where $n = 10$ days

Inaccuracy: Recovery rate $r = 90 - 103\%$
Detection limit: 15 mg Benzoic acid per litre urine

Phenylglyoxylic acid, mandelic acid, hippuric acid, o-, m-, p-methylhippuric acids, benzoic acid

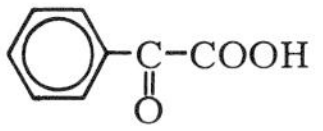

◯–C–COOH
‖
O

Phenylglyoxylic acid

◯–CH–COOH
|
OH

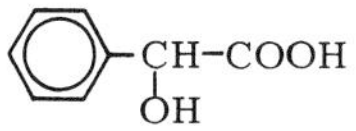

DL-Mandelic acid
(DL-Phenylhydroxyacetic acid)

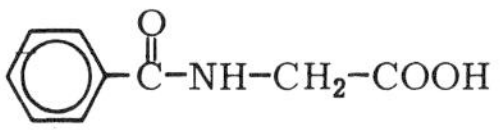

O
‖
◯–C–NH–CH$_2$–COOH

Hippuric acid
(N-Benzoylaminoacetic acid)

O
‖
◯–C–NH–CH$_2$–COOH
|
CH$_3$

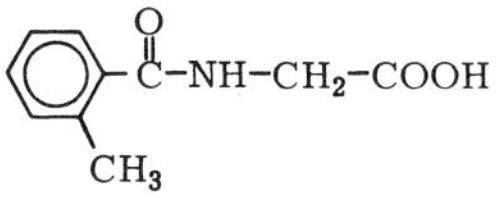

o-Methylhippuric acid
(o-Toluric acid, N-(2-Methylbenzoyl)aminoacetic acid)

◯–COOH

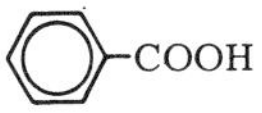

Benzoic acid

These aromatic carboxylic acids are metabolites of the aromatic hydrocarbons styrene, toluene, the xylenes and ethylbenzene which are important in industry and are widely used as starting materials, intermediates and end products of large scale industrial processes. Toluene, the xylenes and ethylbenzene are also applied frequently as solvents in the production and processing of synthetic resins, pigments, paints, etc. The principal application of styrene is in the production of polymers (polystyrene).

These aromatic carbohydrates are taken up mainly by inhalation. However, not insignificant amounts can be absorbed through the skin as well. In humans, the metabolic oxidation of these aromatics takes place preferentially on the side chain, and to a markedly lesser extent also on the aromatic ring [1 – 3]. The details of the various mechanisms are largely known and are described extensively in the "Begründungen der BAT-Werte" (BAT value justification reports) for these substances [4 – 6]. For this reason only a few characteristics are presented here, the understanding of which facilitates the choice of the most appropriate analytical monitoring parameter in each case.

Styrene is converted in the human body to mandelic acid and phenylglyoxylic acid and, according to older studies, is excreted in urine in the form of these metabolites, up to 85 % as mandelic acid and up to 10 % as phenylglyoxylic acid [7]. The excretion pattern is now known to be more complex. It is apparently subject to a two-phase kinetic [8, 9]. Only 1 – 2 % of the internal styrene load is exhaled unchanged [7]. There are conflicting reports as to the relative amounts of mandelic and phenylglyoxylic acids excreted in urine with relative proportions ranging from 1.1 to 3.2 [7, 10, 11]. Our own studies on styrene-exposed individuals with mandelic and phenylglyoxylic acid excretion concentrations up to approximately 200 mg/L indicate that the acids are excreted in equimolar quantities [12]. With increasing styrene exposure and mandelic acid excretion levels above 500 or 1000 mg/L, the mandelic acid/phenylglyoxylic acid ratio seems to increase to 1.5 to 2.0. These ratios can usually be reproduced surprisingly well in each case. Styrene taken up by humans is metabolized to hippuric acid only in trace quantities [13]. The determination of these metabolites seems thus to be suitable for the biological monitoring of persons exposed to styrene [14, 15].

The BAT value [16] for the excretion of mandelic acid in urine is 2 g/L and for the sum of the concentrations of mandelic and phenylglyoxylic acid in urine is 2.5 g/L. The urine specimens should be collected at the end of the exposure period. A detailed justification of these values based on the currently available relevant literature has been published [4].

Hippuric acid is the main metabolite of toluene. However, it is excreted physiologically in inter- and intra-individually markedly variable quantities in the urine of unexposed persons [17]. Since hippuric acid excretion is also strongly influenced by nutritional factors [18] it is unsuitable for estimating individual exposure. Nevertheless it is of practical use in the screening of toluene-exposed groups of individuals. Average night shift urine concentrations of hippuric acid up to 8.5 g/L were found after 8 h exposures to 200 ml/m³ toluene [19, 20].

The isomers of xylene are metabolized for the most part to methylhippuric acids (toluric acids) which are excreted in the urine. Methylhippuric acids are thus suitable for the biological monitoring of persons exposed to xylene, especially as, unlike hippuric acids, they do not occur physiologically in urine. The biological tolerance value for a working material (BAT value) for the level of toluric acids in the urine specimen collected at the end of the shift was defined as 2 g/L [6, 16] after reviewing the currently available literature.

Mandelic and phenylglyoxylic acids, the principal metabolites of ethylbenzene, are suitable for the biological monitoring of this important aromatic hydrocarbon. The American Conference of Governmental Industrial Hygienists (ACGIH) has established a Biological Exposure Index (BEI) for ethylbenzene exposure of 2 g/L mandelic acid in urine. The specimen must be collected at the end of the last shift of a working week [21]. A BAT value for the urine level of mandelic and phenylglyoxylic acids as a result of ethylbenzene exposure is currently in preparation by the working group, Aufstellung von

Grenzwerten in biologischem Material der Senatskommission zur Prüfung gesundheits-schädlicher Arbeitsstoffe (Establishment of threshold values in biological material of the Commission for the Investigation of Health Hazards of Chemical Compounds in the Work Area [Deutsche Forschungsgemeinschaft]) [22].

Benzoic acid, which has frequently been put forward as a metabolite of toluene and styrene [23] seems, according to our own investigations [12], to occur in urine only as a bacterial breakdown product of the main metabolite of these aromatic hydrocarbons. Even in cases of marked exposure only very low levels (< 1000 μg benzoic acid per litre) are detectable in fresh urine samples and may be ignored.

Authors: *J. Lewalter, Th. Schucht*
Examiners: *K. H. Schaller, J. Angerer*

Aromatic Carboxylic Acids (Phenylglyoxylic acid; Mandelic acid; Hippuric acid; o-, m-/p-Methylhippuric acids; Benzoic acid)

Application   Determination in urine

Analytical principle   High Pressure Liquid Chromatography (HPLC)

Completed in   February 1984

Contents

Essential Biomonitoring Methods. DFG, Deutsche Forschungsgemeinschaft
Copyright © 2006 WILEY-VCH Verlag GmbH & Co. KGaA, Weinheim
ISBN: 3-527-31478-4

1 General principles

The aromatic carboxylic acids are extracted from the acidified urine with diethyl ether. Either 3-hydroxybenzoic acid or 3-chloro-4-hydroxybenzoic acid may be used as internal standard. After evaporation of the ether extract to dryness the residue is dissolved in methanol/water (1:5) and the metabolites are separated by HPLC and determined using a UV detector at 215 nm.

Calibration is carried out using aqueous standards which are processed and analysed as described for the samples. The peak areas obtained for the calibration standards are expressed in terms of those for the internal standards. The resulting quotients are plotted as a function of the concentration of the aromatic carboxylic acids employed.

2 Equipment, chemicals and solutions

2.1 Equipment

High pressure liquid chromatograph with gradient former, UV-detector with measurement facility at 215 nm and recorder, preferably with integrator

Steel column: Length: 25 cm; Inner diameter: 4.0 mm
Column packing: Li-Chrosorb RP 18; 5 µm (e.g. Hibar from Merck)

10 and 20 µL Syringes for HPLC
Vacuum centrifuge for evaporation of the solvent from the ether extract or equipment for evaporation under a stream of nitrogen gas
Vortex mixer (e.g. Cenco, Netherlands)
10 mL Test tubes
5 mL Disposable polyethylene tubes
Dispenser (manual) for various fixed volumes between 10 µL and 5 mL (e.g. Multi-pipette 4780 from Eppendorf)
0.5 mL Transfer pipettes
10 and 100 mL Volumetric flasks

2.2 Chemicals

Benzoic acid, p.a. (e.g. Merck)
Hippuric acid (N-benzoylaminoacetic acid), LAB (e.g. Merck) (the commercially available hippuric acid must be purified by recrystallization from water and dried to constant weight at 1 hPa)
o-Methylhippuric acid (o-toluric acid, N-(2-methylbenzoyl)aminoacetic acid), p.a. (e.g. Tokyo Kasei, Japan)

m-Methylhippuric acid (m-toluric acid, N-(3-methylbenzoyl)aminoacetic acid), p.a. (e.g. Tokyo Kasei, Japan)

DL-Mandelic acid (DL-phenylhydroxyacetic acid), p.a. (e.g. Fluka)

Phenylglyoxylic acid, p.a. (e.g. Fluka)

3-Hydroxybenzoic acid, approx. 99 % (e.g. EGA)

3-Chloro-4-hydroxybenzoic acid hemihydrate, at least 99 % (e.g. EGA)

Diethyl ether, p.a.

37 % Hydrochloric acid

Tripotassium phosphate trihydrate, p.a.

Orthophosphoric acid, at least 85 %, density approx. 1.71 g/cm^3, p.a.

Methanol for HPLC (e.g. Lichrosolv, Merck)

Ultrapure water (ASTM type 1) or double-distilled water

Nitrogen (99.99 %)

2.3 Solutions

Eluent A (phosphate buffer):
2.66 g tripotassium phosphate trihydrate is made up to 1 L with ultrapure water and the pH is adjusted to 3.5 with concentrated phosphoric acid.

Eluent B: methanol

Both eluents are degassed, e.g. by stirring under vacuum at 65 °C for the buffer and 45 °C for methanol. They are then mixed automatically in the HPLC system.

Internal standard solution:
100 mg 3-hydroxybenzoic acid or 3-chloro-4-hydroxybenzoic acid is weighed into a 100 mL volumetric flask which is then filled to the mark with a mixture of methanol and ultrapure water (1 + 1). The solution must be freshly prepared at least every 7 days.

Methanol/ultrapure water mixture (1:5)

2.4 Calibration standards

Stock solution:
500 mg hippuric acid and 100 mg each of mandelic, o- and m-methylhippuric and benzoic acids together with 60 mg phenylglyoxylic acid are dissolved in a mixture of methanol and ultrapure water (1 + 1) in a 100 mL volumetric flask. The flask contents are mixed with the solvent and then filled to the mark (5 g/L, 1 g/L and 0.6 g/L, respectively).

Dilution of this stock solution with a mixture of methanol and ultrapure water (1 + 1) yields calibration standards in the concentration range from 15 to 1000 mg/L as shown in the following table:

Volume of stock solution	Final volume of calibration standard	Concentration of calibration standard		
		mandelic acid, o-, m-methyl- hippuric acids, benzoic acid	phenyl- glyoxylic acid	hippuric acid
μL	mL	mg/L	mg/L	g/L
250	10	25	15	0.125
500	10	50	30	0.250
1000	10	100	60	0.500
2000	10	200	120	1.000

The stock solution and the calibration standards should be freshly prepared at least every 7 days.

3 Specimen collection and sample preparation

Urine is collected in plastic bottles at the end of the shift and if possible at the end of three consecutive working days. It can be advantageous to determine the total volume, creatinine level, density and pH even of spontaneous urine specimens. If the specimens cannot be processed immediately they shoud be acidified after collection. The addition of 1 mL acetic acid to 100 mL urine has proved to be appropriate. The specimens can be stored for 1–2 days in this condition without influencing the analytical results. The urine specimens are deep frozen in aliquots in the laboratory until analysis. After storage in the deep freeze ($-$ 18 °C) the specimens may be thawed in a water bath at 60 °C. After cooling to room temperature and thorough shaking of the specimens, aliquots are taken for analysis.

0.5 mL urine, 50 μL internal standard solution and 25 μL concentrated hydrochloric acid are thoroughly mixed in a 10 mL test tube at room temperature. The sample is then extracted with 3 mL diethyl ether for 1 min on the vortex mixer. 2 mL of the ether phase is transferred to a 5 mL disposable polyethylene tube. The ether is then evaporated carefully at room temperature in a vacuum centrifuge or, according to the examiners, by passing nitrogen gas over it until the residue is dry. The residue is then taken up in 0.5 mL methanol/ultrapure water (1:5).

4 Operational parameters for HPLC

Column:	Material:	steel
	Length:	25 cm
	Inner diameter:	4.0 mm
Column packing:	Li-Chrosorb RP 18; 5 μm	
Separation mode:	Reversed phase	
Detector:	UV-detector with measurement facility at 215 nm (see Section 8.4 and Fig. 4)	
Column temperature:	35 °C	
Mobile phase:	0.01 M Phosphate buffer (75 %)	
	Methanol (25 %)	
	The solutions are mixed in the HPLC apparatus.	
Pressure:	c. 163 bar (16.3 MPa)	
Flow rate:	0.8 mL/min	

5 Analytical determination

10 μL of the extraction residue solution in methanol/ultrapure water (1:5) is injected into the HPLC apparatus by means of a syringe. If the analytical results lie outside the linear portion of the calibration curve, the specimens are diluted and reprocessed.

6 Calibration

The aqueous calibration standards (see Section 2.4) are processed and analysed as described for the samples. The calibration curve is established by plotting the ratios of the peak areas for each of the carboxylic acids to the areas for the internal standards against the concentrations employed. The internal standard whose retention time is closest to that of the carboxylic acid in question should be chosen (see Fig. 1 and 2). The linear range of the method was tested up to a concentration of 100 mg/L for benzoic acid, 200 mg/L for mandelic, phenylglyoxylic, o- and m-methylhippuric acids and 1 g/L for hippuric acid. Examples of calibration curves for the determination of the aromatic carboxylic acids are shown in Fig. 3.

7 Calculation of the analytical result

The peak areas for the aromatic carboxylic acids are divided by the peak areas for the internal standards. Using this quotient, the urine concentration of the acid in mg/L or g/L (for hippuric acid) is read off the calibration curve.

8 Reliability of the method

The reliability of the method was examined within the following concentration ranges: 15 – 100 mg/L for phenylglyoxylic acid, 25 – 100 mg/L for mandelic acid, o- and m-methylhippuric acids and benzoic acid and 0.25 – 1 g/L for hippuric acid.

8.1 Precision

The between-day imprecision was determined for each of 33 individual urine specimens which were spiked with the aromatic carboxylic acids in the concentration ranges listed above. These specimens were analysed on $n = 10$ days. The relative standard deviations and the corresponding prognostic ranges are given in Tab. 1.

8.2 Accuracy

Recovery experiments were carried out to check the accuracy of the method. Urine was mixed with three levels of each of the aromatic carboxylic acids. For the concentrations tested, recovery rates between $r = 86\%$ and 116% were obtained. The values are shown in Tab. 2.
In addition, the losses during the sample processing were investigated. Approximately 33 individual urine specimens were spiked as above, processed and analysed. The results were compared with those from calibration standards which had not been subjected to the experimental procedure. Losses ranged from 8 to 61 %. The individual values are presented in Tab. 1.

8.3 Detection limit

Under the described conditions of sample processing and HPLC analysis, results indicating urine levels of more than 15 mg/L (phenylglyoxylic acid, o-, m-methylhippuric acids, benzoic acid) or 25 mg/L (mandelic acid, hippuric acid) were, with a probability of error of less than 10 %, due to the aromatic carboxylic acid content of the urine.
The detection limit can be lowered by increasing the urine volume analysed and/or taking up the dry residue from the ether extract in a smaller volume.

8.4 Sources of error

The sensitivity of the method to disturbance from the following physiological urine components was tested and excluded:

Substance	Retention time min	Substance	Retention time min
α-Ketoglutarate	2.93	DL-2-Aminobutyric acid	10.13
Barbituric acid	3.04	Salicylic acid	12.36
L(+)-Ascorbic acid	3.17	L(+)-Malic acid	18.88
Nicotinuric acid	4.04	Phenylsuccinic acid	22.27
Creatinine	4.08	Phenylacetic acid	25.91
Trichloroacetic acid	5.03	L-Aspartic acid	./.
2,4-Dinitrobenzoic acid	5.50	Citric acid	./.
Phthalic acid	6.30	L-Glutamic acid	./.
L-Tryptophan	6.93	Urea	./.
Tobias acid	7.24	Lactic acid	./.
4-Hydroxybenzoic acid	9.61	5-Aminolaevulinic acid	./.

The individual aromatic carboxylic acids have individually different absorption maxima. The detection wavelength of 215 nm chosen here is a compromise. At lower wavelengths mandelic acid in particular may be determined with greater sensitivity (see Fig. 4). This increased sensitivity, however, leads to increased background levels in the analysis. Inappropriate storage of samples may result in losses of the aromatic carboxylic acids. For details of the proper storage conditions, see Section 3.

9 Discussion of the method

With the method described here, the principle metabolites of styrene, toluene, xylene, and ethylbenzene may be reliably determined in urine in the concentration ranges relevant to occupational medicine. The method has the advantage over gas chromatographic procedures that derivatization is not necessary [24 – 27]. Other methods of determination (e.g. photometry [11, 28, 29]) no longer fulfill the requirements of present day analysis, in particular because of their inadequate specificity. HPLC methods for quantification of hippuric acid, mandelic acid and phenylglyoxylic acid have been described previously [30, 31] but the analytical conditions chosen did not allow a reliable separation of the metabolites in human body fluids.

The separation of m- from p-toluric acid was not possible with the present method. However, in view of the similar toxicity of the two initial substances, such a separation should prove unnecessary.

With the liquid-liquid extraction method used here, not only the metabolites of interest but also interfering substances in the urine are transferred to the organic phase. The liquid chromatographic procedure was therefore optimized so that a reliable determination of the required metabolites was possible even when they were extracted together with large amounts of other urine components.

The particular advantage of the method is the simultaneous determination of the most important metabolites of the six industrially common aromatic hydrocarbons. Thus the method fulfills the requirements of occupational medical screening examinations (biological monitoring) particularly well, especially as the occurrence of mixtures of substances, including those dealt with here, at workplaces is the rule rather than the exception. The possibility of detecting the metabolites of these workplace substances in a single analytical process seems particularly practical and efficient.

As a variation of the method described here, the eluents can be mixed volumetrically, i.e. not in the apparatus. In this case, slight variations in the retention times can occur.

Using packed steel cartridges instead of the classical columns, one of the examiners was able to produce results which agree with those given here.

Instruments used:
Liquid chromatograph, model 1084 B, UV detector with adjustable wavelength setting, recorder and integrator from Hewlett Packard

10 References

[1] *V. Šedivek* and *J. Flek:* The absorption, metabolism, and excretion of xylenes in man. Int. Arch. Occup. Environ. Health *37,* 205 – 217 (1976).

[2] *J. Angerer:* Occupational chronic exposure to organic solvents. VII. Metabolism of toluene in man. Int. Arch. Occup. Environ. Health *43,* 63 – 67 (1979).

[3] *J. Angerer* and *G. Lehnert:* Occupational chronic exposure to organic solvents. VIII. Phenolic compounds – metabolites of alkylbenzenes in man. Int. Arch. Occup. Environ. Health *43,* 145 – 150 (1979).

[4] *K. H. Schaller:* Styrol. In: *D. Henschler* and *G. Lehnert* (eds.): Biologische Arbeitsstoff-Toleranz-Werte (BAT-Werte). Arbeitsmedizinisch-toxikologische Begründungen. Deutsche Forschungsgemeinschaft, Verlag Chemie, Weinheim, 2nd issue 1985.

[5] *J. Angerer, K. Behling,* and *G. Lehnert:* Toluol. In: *D. Henschler* and *G. Lehnert* (eds.): Biologische Arbeitsstoff-Toleranz-Werte (BAT-Werte). Arbeitsmedizinisch-toxikologische Begründungen. Deutsche Forschungsgemeinschaft, Verlag Chemie, Weinheim, 1st issue 1983.

[6] *J. Angerer:* Xylole. In: *D. Henschler* and *G. Lehnert* (eds.): Biologische Arbeitsstoff-Toleranz-Werte (BAT-Werte). Arbeitsmedizinisch-toxikologische Begründungen. Deutsche Forschungsgemeinschaft, Verlag Chemie, Weinheim, 2nd issue 1985.

[7] *Z. Bardodej* and *E. Bardodejova:* Biotransformation of ethyl benzene, styrene, and α-methylstyrene in man. Am. Ind. Hyg. Ass. J. *31,* 206 – 209 (1970).

[8] *M. P. Guillemin* and *D. Bauer:* Human exposure to styrene. III. Elimination kinetics of urinary mandelic and phenylglyoxylic acids after single experimental exposure. Int. Arch. Occup. Environ. Health *44,* 249 – 263 (1979).

[9] *J. Engström, R. Bjurström, I. Åstrand,* and *P. Övrum:* Uptake, distributions and elimination of styrene in man: Concentration in subcutaneous adipose tissue. Scand. J. Work Environ. Health *4,* 315 – 323 (1978).

[10] *J. K. Piotrowski:* Exposure tests for organic compounds in industrial toxicology. US. Government Printing Office, Washington, D.C. 1977, p. 60 – 65.

[11] *H. Ohtsuji* and *M. Ikeda:* A rapid colorimetric method for the determination of phenyl-glyoxylic and mandelic acids: its application to the urinary analysis of workers exposed to styrene vapour. Br. J. Ind. Med. *27,* 150 – 154 (1970).

[12] *J. Lewalter:* Unpublished results from persons exposed to styrene (1982).

[13] *M. Ikeda, T. Imamura, M. Hayashi, T. Tabuchi,* and *I. Hara:* Evaluation of hippuric, phenylglyoxylic and mandelic acids in urine as indices of styrene exposure. Int. Arch. Arbeitsmed. *32,* 93 – 101 (1974).

[14] *Z. Bardodej:* Styrene, its metabolism and the evaluation of hazards in industry. Scan. J. Work. Environ. Health *4,* 85 – 103 (1978).

[15] *M. Ikeda* and *H. Ohtsuji:* Significance of urinary hippuric acid determination as an index of toluene exposure. Br. J. Ind. Med. *26,* 244 – 246 (1969).

[16] *Deutsche Forschungsgemeinschaft:* Maximale Arbeitsplatzkonzentrationen und Biologische Arbeitsstofftoleranzwerte. Mitteilung XXII der Senatskommission zur Prüfung gesundheitsschädlicher Arbeitsstoffe, Verlag Chemie, Weinheim 1986.

[17] *L. D. Pagnotto, H. B. Elkins,* and *H. G. Brugsch:* Benzene exposure in the rubber coating industry – a follow up. Am. Ind. Hyg. Ass. J. *40,* 137 – 146 (1979).

[18] *D. Szadkowski, A. Borkamp,* and *G. Lehnert:* Hippursäureausscheidung im Harn in Abhängigkeit von Tagesrhythmik und alimentären Einflüssen. Int. Arch. Occup. Environ. Health *45,* 141 – 152 (1980).

[19] *T. Gutewort, J. Gartzke,* and *R. Pannier:* Die Hippursäureausscheidungsrate im Harn zur Bewertung einer beruflichen Toluolexposition. Z. ges. Hyg. *27,* 57 – 63 (1981).

[20] *J. Angerer:* Prävention beruflich bedingter Gesundheitsschäden durch Benzol, Toluol, Xylole und Ethylbenzol. Vol. 71 of the series Arbeitsmedizin, Sozialmedizin, Präventivmedizin. A.W. Gentner Verlag, Stuttgart 1983.

[21] *American Conference of Governmental Industrial Hygienists:* Threshold limit values for chemical substances in the work environment. Cincinnati 1984.

[22] *J. Angerer:* Ethylbenzol. In: *D. Henschler* and *G. Lehnert* (eds.): Biologische Arbeitsstoff-Toleranz-Werte (BAT-Werte). Arbeitsmedizinisch-toxikologische Begründungen. Deutsche Forschungsgemeinschaft, Verlag Chemie, Weinheim (in preparation).

[23] *Z. Bardodej:* Beurteilung der Gefährdung durch Toluol in der Industrie mittels der Hippursäurebestimmung im Harn. Arbeitsmed. Sozialmed. Arbeitshyg. *3,* 254 (1968).

[24] *M. Guillemin* and *D. Bauer:* Human exposure to styrene. II. Quantitative and specific GC analysis of urinary MA and PhGA as an index of styrene exposure. Int. Arch. Occup. Environ. Health *37,* 57 – 64 (1976).

[25] *K. Engström, K. Husman,* and *J. Ratanen:* Measurement of toluene and xylene metabolites by gas chromatography. Int. Arch. Occup. Environ. Health *36,* 153 – 160 (1976).

[26] *J. R. Caparos* and *J. G. Fernandez:* Simultaneous determination of toluene and xylene metabolites in urine by gas chromatography. Br. J. Ind. Med. *34,* 229 – 233 (1977).

[27] *P. van Roosmalen* and *I. Drummond:* Simultaneous determination by GC of the major metabolites in urine of toluene, xylene, and styrene. Br. J. Ind. Med. *35,* 56 – 60 (1978).

[28] *K. Tomokum* and *M. Ogata:* Direct colorimetric determination of hippuric acid in urine. Clin. Chem. *18,* 349 – 351 (1972).

[29] *P. L. Paggiaro, D. Taddeo, A. M. Loi, G. Pagano, N. Serretti, G. Toma,* and *G. Simonini:* Controllo dell' esponzione a solventi con il dosaggio dell'acido ippurico urinario: discussione sul metodo e risultati preliminari. Atti 41 Cong. Soc. It. Med. Lav. Ig. Ind., S. Margherita Ligure 1978, 243 – 246.

[30] *M. Ogata* and *R. Sugihara:* HPLC procedures for quantitative determination of urinary PhGA, MA and HA as indices of styrene exposure. Int. Arch. Occup. Environ. Health *42,* 11 – 20 (1978).

[31] *G. Poggi, M. Giussiani, M. Palagi, P. L. Paggiaro, A. M. Loi, F. Dazzi, C. Siclari,* and *L. Baschieri:* HPLC for the quantitative determination of the urinary metabolites of toluene, xylene and styrene. Int. Arch. Occup. Environ. Health *50,* 25 – 31 (1982).

Authors: *J. Lewalter, Th. Schucht*
Examiners: *K. H. Schaller, J. Angerer*

Tab. 1. Between-day imprecision and processing losses for the determination of aromatic carboxylic acids in urine.

Substance	Spiked concentration	Between-day imprecision		Losses during processing
		s	u	
	mg/L	%	%	%
Phenyl-glyoxylic acid	15	2.7	6.0	18
	30	5.4	12.3	13
	100	2.9	6.5	13
Mandelic acid	25	5.8	13.2	41
	50	1.6	3.5	48
	100	1.6	3.6	61
Hippuric cid	250	5.5	12.5	34
	500	12.8	28.9	8
	1000	3.4	7.6	33
o-Methyl-hippuric acid	25	9.2	20.7	20
	50	4,9	11.1	22
	100	1.6	3.6	25
m-Methyl-hippuric acid	25	4.1	9.3	19
	50	3.3	7.5	19
	100	1.9	4.4	20
Benzoic acid	25	2.9	6.6	58
	50	1.6	3.7	59
	100	0.9	2.0	59

Tab. 2. Recovery experiments for the HPLC analysis of aromatic carboxylic acids in urine.

Substance	n	Calculated value mg/L	Measured value mg/L	Recovery rate %
Phenyl-	4	25	27.1	108
glyoxylic	4	50	50.9	102
acid	4	100	86.4	86
Mandelic	4	25	25.5	102
acid	4	50	45.3	91
	4	100	94.1	94
Hippuric acid		6000	5930	99
o-Methyl-	4	25	29.1	116
hippuric acid	4	50	51.0	102
	4	100	99.6	100
m-Methyl-	4	25	28.4	114
hippuric acid	4	50	49.2	98
	4	100	104.3	104
Benzoic acid	4	25	25.8	103
	4	50	45.5	91
	4	100	90.3	90

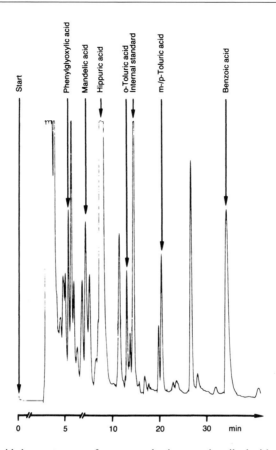

Fig. 1. High pressure liquid chromatogram of a processed urine sample spiked with aromatic carboxylic acids in the following concentrations:

Phenylglyoxylic acid	50 mg/L
Mandelic acid	50 mg/L
Hippuric acid	1060 mg/L
o-Methylhippuric acid (o-toluric acid)	50 mg/L
m-, p-Methylhippuric acid (m-, p-toluric acid)	50 mg/L
Benzoic acid	383 mg/L

Internal standard: 3-hydroxybenzoic acid
Operational parameters for HPLC, see Section 4.

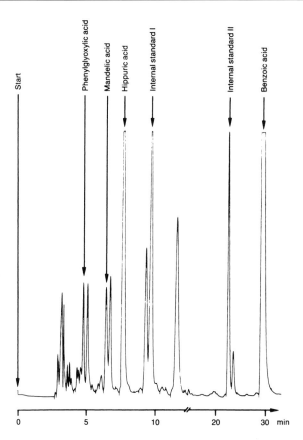

Fig. 2. High pressure liquid chromatogram of a urine sample from a person exposed to styrene.
Internal standard I: 3-hydroxybenzoic acid
Internal standard II: 3-chloro-4-hydroxybenzoic acid
Operational parameters for HPLC, see Section 4.

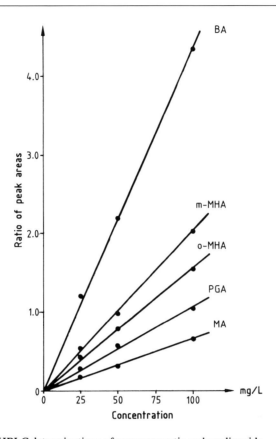

Fig. 3. Examples of HPLC determinations of some aromatic carboxylic acids

MA	Mandelic acid
PGA	Phenylglyoxylic acid
o-MHA	o-Methylhippuric acid
m-MHA	m-Methylhippuric acid
BA	Benzoic acid

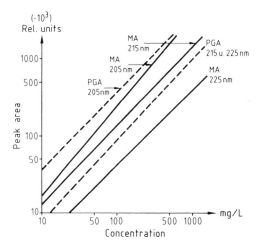

Fig. 4. Dependence of the UV detection of mandelic acid (MA) and phenylglyoxylic acid (PGA) on the wavelength.

Arsenic

Application	Determination in urine
Analytical principle	Atomic absorption spectrometry – hydride technique
Completed in	June 1990

Summary

The method described here is used to determine only the toxicologically significant arsenic compounds, in particular, inorganic compounds as well as the metabolic products monomethylarsonic and dimethylarsinic acids. These substances react with sodium borohydride to form volatile arsenic hydrides which can be assayed by means of atomic absorption spectrometry using the so-called hydride technique. In contrast, aromatic arsenic compounds do not form volatile hydrides with sodium borohydride and thus evade determination. As aromatic arsenic compounds pass through the body unchanged and are therefore of low toxicity, the hydride technique offers a welcome opportunity of differentiating between toxic and less toxic arsenic compounds. Using this method arsenic excreted in urine as a result of environmental exposure can be determined.

Arsenic is determined in urine by means of hydride atomic absorption spectrometry. After adding sodium borohydride, the volatile arsenic hydrides are flushed out of the reaction solution, thermally decomposed in a quartz cuvette and measured by means of atomic absorption spectrometry.

The quantitative evaluation is carried out using the standard addition procedure.

Within-series imprecision: Standard deviation (rel.) $s_w = 0.6 - 5.6\%$
Prognostic range $u = 2.6 - 24.1\%$
At concentrations ranging from 26–754 µg arsenic per litre urine and where $n = 3-5$ determinations

Between-day imprecision: Standard deviation (rel.) $s = 5.1$ and 6.0%
Prognostic range $u = 10.5$ and 13.1%
At concentrations of 19.1 and 48.8 µg arsenic per litre urine and where $n = 25$ or 13 days

Inaccuracy: Recovery rate $r = 88.3-104.6\%$

Detection limit: 2 µg Arsenic per litre urine (evaluation of peak areas)
1 µg Arsenic per litre urine (evaluation of peak heights)

Essential Biomonitoring Methods. DFG, Deutsche Forschungsgemeinschaft
Copyright © 2006 WILEY-VCH Verlag GmbH & Co. KGaA, Weinheim
ISBN: 3-527-31478-4

Arsenic

Arsenic (As, relative atomic mass 74.92; atomic number 33) is an element which shows both metallic and non-metallic properties. It belongs to group V of the 4th period of the periodic table and occurs in the oxidation states $+V$, $+III$, 0 and $-III$.

Arsenic occurs ubiquitously in animate and inanimate nature. The earth's crust contains 1.5–2.0 mg/kg of arsenic. It is the 20th most common element. Its sulfides and oxides are found in association with various metals. The best known metal arsenides are arsenopyrite (FeAsS), cobaltite (CoAsS), proustite (Ag_3AsS_3), arsenical copper (Cu_3AS), arsenical pyrite ($FeAs_2$), white nickel ($NiAs_2$), arsenical nickel (NiAs), cobalt arsenide ($CoAs_2$). Realgar (As_4S_4) and orpiment (As_2S_3) are naturally occurring sulfides of arsenic. The most important compound of arsenic with oxygen is arsenic trioxide or arsenic which is formed as a natural product when arsenic sulfides are oxidized.

The chief occupational sources of danger to health are:
- Extraction of the metal from ores containing arsenic
- Arsenic extraction by roasting sulfur ores
- Residues in the lead chambers in which sulfuric acid is manufactured
- Use of raw materials containing arsenic in the glass industry
- Manufacture of dyes, paints, pigments, fusion glazes, varnishes containing arsenic as well as their use, e.g. in the dyeing of wallpaper or fabrics
- Use in the fur-processing industry, in the manufacture of insecticides and pesticides, in pyrotechnics and for the treatment of metal surfaces with As(III) chloride
- Production of pharmaceuticals containing arsenic
- Repair and cleaning of flue dust installations

In recent years several reviews have been published on the subject of arsenic in the environment and at the workplace [1–4]. The metabolism, toxicology of arsenic as well as the clinical symptoms of arsenic poisoning are comprehensively described in these publications. In the case of acute poisoning one must differentiate between the paralytic and the gastro-intestinal form of poisoning. Paralytic symptoms occur when a large amount of poison is rapidly resorbed. General paralysis of the capillaries leads to a severe state of shock with acute stimulus of the CNS, partly accompanied by delirium, within 1 to 2 hours. Death occurs as a result of total paralysis. There are a whole range of possible transition forms between the symptoms of paralytic and gastro-intestinal poisoning. Gastro-intestinal poisoning is characterized by failure of the central regulatory mechanisms, particularly of the capillaries. Accompanied by intense colic-like pains, blood accumulates in the intestinum with the participation of the liver, kidneys and spleen. Furthermore, the stool resembles the water in which rice has been cooked and the victim suffers uncontrollable vomiting. The victim falls into a coma and death is caused by heart and circulatory failure.

Chronic arsenic poisoning is generally the result of long-term intake of subtoxic doses of arsenic. A protracted cachexia is noticeable when poisoning occurs over a longer period. The uncovered parts of the skin exhibit brown pigmentation. Ketatosis develops in the areas of pigmentation. Cross stripes appear on the nails (Mees' stripes) due to disturbance of their growth. The blood is frequently hyperchromic; sometimes pernicious anaemia with a reduction in the number of leucocytes occurs. The chronic arsenic poisoning suffered by winegrowers with accompanying cirrhosis of the liver is regarded as a combination of damage caused by arsenic and a high consumption of alcohol (review of the symptoms in [4]).

Arsenic and its inorganic compounds (arsenic trioxide, arsenic pentoxide, arsenious acids, arsenic acid and its salts) are "working materials which have been unequivocally proven carcinogenic" for humans (so-called A 1 substances) [5]. Therefore no maximum exposure limit for the concentration of arsenic in the air can be assigned without reservations about risks to health. The TRK value (Technical Guiding Concentration) for this group of substances is 0.1 mg/m^3, calculated as arsenic in total dust [6]. Above all, skin cancer in the form of the malignant degeneration of hyperkeratoses as well as carcinomas of the respiratory organs must be pointed out.

The relationship between the arsenic concentration in the air at the workplace and arsenic excretion in urine shown in Table 1 is included in Chapter VIII "Carcinogenic Working Materials" of the 1990 MAK/BAT value list [5].

Table 1: Exposure equivalents for arsenic trioxide.

Air Arsenic (μg/m^3)	Sampling time: end of exposure or end of shift Urine Arsenic (μg/L)
10	50
20	80
50	175
100	330
200	640

The values given for urinary excretion of arsenic refer to "certain volatile arsenic compounds produced by direct hydrogenation" in urine. The analytical method described here, with which only the hydrogenable arsenic is detected, takes this into account.

Arsenic can occur in human biological material as part of different bonds and in different oxidation states. Table 2 gives an overview of the arsenic compounds which can occur in biological material. Arsenic occurs in both inorganic and organic forms.

Following oral intake of inorganic arsenic compounds, e.g. in the form of arsenic trioxide or arsenic acid, only 10 to 20 % of the arsenic is excreted in inorganic form. The rest is methylated in vivo. The two main metabolites are monomethylarsonic acid (CH_3-$AsO(OH)_2$) and dimethylarsinic acid (($CH_3)_2AsO(OH)$), which are excreted in a ratio of approximately 1:3. Experiments by Mappes [7] and the results obtained by Buchet et al. [8–11] show that 60 to 70 % of the ingested arsenic is excreted in the urine. The biological half-life varies between 30 and 37 hours. It depends on the dose.

In addition to the above-mentioned organic arsenic compounds and the metabolites of inorganic arsenic compounds, other organic, mostly aromatic compounds, are found in the human body. The latter arsenic compounds are primarily found in food of "maritime

Table 2: Arsenic compounds occurring in biological material (cf. [20]).

1 Trivalent arsenic compounds

Arsenic trioxide (arsenic)	As_2O_3
Sodium arsenate (meta form)	$NaAsO_2$

Arsenic-thiol complexes
e.g. arsenic British anti-Lewisite
(British anti-Lewisite = BAL, dimercaprol,
2,3-dimercapto-1-propanol)
BAL was developed as a highly effective
antidote against the arsenic-containing Lewisite
series

$$
\begin{array}{ccc}
H_2C-SH & Cl & \\
| & \diagdown & \\
HC-SH & + & As-CH{=}CHCl \\
| & \diagup & \\
H_2C-OH & Cl & \\
BAL & & Lewisite\ I
\end{array}
$$

$$\downarrow -2HCl$$

$$
\begin{array}{l}
H_2C-S\diagdown \\
\quad|\qquad As-CH{=}CHCl \\
HC-S\diagup \\
\quad| \\
H_2C-OH
\end{array}
$$

As-Thiol-Complex

Trimethylarsine	$(CH_3)_3As$

2 Pentavalent arsenic compounds

Arsenic pentaoxide	As_2O_5
Phenylarsonic acid	$\bigcirc\!\!-AsO(OH)_2$
Arsenobetaine	$(CH_3)_3\overset{\oplus}{As}CH_2COO^{\ominus}$
Arsenocholine	$\left[\begin{array}{cc} CH_2-CH_2 \\ |\qquad| \\ OH\quad \overset{\oplus}{As}(CH_3)_3 \end{array}\right]^{\ominus} OH$
Monomethylarsonic acid	$CH_3AsO(OH)_2$
Dimethylarsinic acid (cacodylic acid)	$(CH_3)_2AsO(OH)$

origin". One of these compounds o-phosphatidyl-trimethylarsonium-lactic acid is hardly toxic and evidently prevalent in fish. It is converted to trimethylarsonium-betaine. Seawater fish and molluscs can contain up to 122 mg arsenic per kg [3].

It is evident from this information that occupational medicine must make a differentiated toxicological assessment of the urinary excretion of arsenic. Only the "toxicologically relevant arsenic" is of importance for this assessment.

The chemical properties of the arsenic excreted in urine offer the opportunity to differentiate between the biologically effective arsenic compounds and those of low bioavailability. Thus, in an acid milieu, the "toxicologically relevant" arsenic compounds are converted by sodium borohydride to volatile hydrides and separated from the biological matrix so that they can be quantitatively determined. This determination can be carried out practically and reliably [12–15]. The compounds of low bioavailability, e.g. those which occur in seafood, are not detected by the hydride method. The results obtained with this method are the same those from the determination of the total arsenic in the urine of a person who has not eaten seafood for a longer period of time.

Investigation of urine samples of 25 people who were not exposed to arsenic at the workplace showed a mean value of 7.4 µg arsenic per litre. The upper normal limit is 25 µg/L. The content of the toxicologically relevant arsenic in urine does not exhibit large intraindividual fluctuations. This was shown by examination of urine samples over a period of 14 days [12].

Vahter [16] found a mean value of 8.2 µg arsenic per g creatinine in unexposed persons. Four urine samples contained more than 30 µg arsenic per g creatinine, the highest value was 52 µg arsenic per g creatinine. Similar renal excretion of arsenic was reported by working groups from Finland [17], the Federal Republic of Germany [18] and Italy [19], somewhat higher values were found in Belgium [19]. It is assumed that the daily renal excretion is approximately 60 % of the ingested dose [9 or 10]. Thus, it can be concluded that the daily intake of bioavailable arsenic is about 15 to 25 µg in the European countries [16]. The arsenic concentrations in the urine of occupationally unexposed persons are summarized in Table 3.

Table 3: Arsenic concentration in urine samples of persons not occupationally exposed to arsenic.

Toxicologically relevant arsenic µg/L	Total arsenic µg/L	No. of test persons	Origin of test persons	Authors
22.6	22.5	4		*Braman* and *Foreback*, 1973 [21]
17.5	21.2 ± 2[1]	41		*Smith* et al., 1977 [22]
15.5–113.1	28–170	10	Belgium	*Buchet* et al., 1980 [8]
6.9–25.6	36–1180	10	Mediterranean region	*Buchet* et al., 1980 [8]
7.4 (up to 25)		25		*Schierling* et al., 1982 [12]
5.9	17.2 ± 11	148	Italy	*Foa* et al., 1984 [19]
8.2[2] (up to 52[2])				*Vahter,* 1986 [16]
12.4 ± 11[3]	72.9 ± 90[3]	49	Stockholm	*Vahter* and *Lind,* 1986 [23]
9.7 ± 7[3]	41.1 ± 46[3]	50	Västeras	
5.5 (up to 29.0)		70		
6.2 (up to 29.0)		50	Men	
3.9 (up to 8.3)		20	Women	*Angerer,* 1987 [24]
6.8 (up to 29.0)		29	Smokers	
4.6 (up to 17.7)		41	Non-smokers	

1) Geometric mean
2) Arsenic per g creatinine
3) Values corrected to a specific density of 1.016

Author: *K.H. Schaller*
Examiners: *M. Fleischer, J. Angerer, J. Lewalter*

Arsenic

Application   Determination in urine

Analytical principle Atomic absorption spectrometry
                          – hydride technique

Completed in   June 1990

Contents

1 General principles

This method permits the determination of those arsenic compounds which form volatile arsenic hydrides when they react with sodium borohydride in hydrochloric acid. The arsenic compounds assayed in this way are, at the same time, those compounds of toxicological significance. Other harmless organo-arsenic compounds are not detected by this method.

Essential Biomonitoring Methods. DFG, Deutsche Forschungsgemeinschaft
Copyright © 2006 WILEY-VCH Verlag GmbH & Co. KGaA, Weinheim
ISBN: 3-527-31478-4

Arsenic is determined in urine by means of hydride atomic absorption spectrometry. After adding sodium borohydride, the volatile arsenic hydrides are flushed out of the reaction solution, thermally decomposed in a quartz cuvette and measured by means of atomic absorption spectrometry.

The quantitative evaluation is carried out using the standard addition procedure.

2 Equipment, chemicals and solutions

2.1 Equipment

Atomic absorption spectrometer which enables measurement and background correction at 193.7 nm and a chart recorder

Hydride system with a quartz cuvette (Either a system constructed in the laboratory or a commercially available system can be used. The settings given here are for the commercially available system MHS-20 from Perkin Elmer. The settings must be modified if other systems are used.)

Arsenic ED lamp with power supply

Hand dispensers, adjustable for various dosages between 100 µL and 5 mL (e.g. Multipette from Eppendorf)

50, 100, 500 and 1000 mL Volumetric flasks

2.2 Chemicals

All the chemicals used must be of the highest available purity.

Arsenic standard (e.g. Titrisol from Merck) containing 1 g of arsenic (in the form of As_2O_5)

65 % Nitric acid (e.g. Suprapur from Merck)

96 % Acetic acid (e.g. Suprapur from Merck)

30 % Hydrochloric acid (e.g. Suprapur from Merck)

Sodium borohydride, p.a. (e.g. from Merck)

Sodium hydroxide monohydrate (e.g. Suprapur from Merck)

Silicon anti-foaming emulsion (e.g. Antifoam 110 A from Dow Corning)

Ultrapure water (ASTM type 1) or double-distilled water

Purified nitrogen (99.999 %) or argon

2.3 Solutions

1.5 % Hydrochloric acid:

25 mL 30 % hydrochloric acid is pipetted into a 500 mL volumetric flask containing about 200 mL ultrapure water. After thorough mixing the flask is filled to the mark with ultrapure water.

1% Silicon anti-foaming solution:
1 g anti-foaming emulsion is dissolved in ultrapure water. This solution is transferred to a 100 mL volumetric flask which is filled to the mark with ultrapure water.

Reduction solution (3 % sodium borohydride solution in 1 % sodium hydroxide solution):
30 g of sodium borohydride are dissolved with 14.5 g sodium hydroxide in 1 litre ultrapure water. This solution must be freshly prepared at least every second day.
Unused sodium borohydride solution as well as the assay solutions containing any remaining sodium borohydride after atomic absorption spectrometric determination must be broken down by dropwise addition of dilute sulfuric acid in the fume cupboard.

1 M Nitric acid (for cleaning tubes and glassware):
72 mL 65 % nitric acid is pipetted into about 400 mL ultrapure water in a 1000 mL volumetric flask. After thorough mixing the flask is filled to the mark with ultrapure water.

2.4 Calibration standards

Starting solution:
The standard solution containing 1 g arsenic is placed in a 1000 mL volumetric flask and ultrapure water is filled to the mark (1 g/L).

Stock solution:
100 µL of the starting solution is pipetted into a 100 mL volumetric flask and filled to the mark with ultrapure water (1 mg/L).
The calibration standards are prepared from this stock solution by diluting with ultrapure water in 50 mL volumetric flasks according to Table 4.

Table 4: Pipetting scheme for the preparation of the calibration standards.

Volume of the stock solution mL	Final volume of calibration standard mL	Concentration of calibration standard µg/L	Designation of calibration standard
0.5	50	10	I
1.0	50	20	II
2.5	50	50	III
4.0	50	80	IV
5.0	50	100	V

The calibration standards must be freshly prepared daily.

3 Specimen collection and sample preparation

As in all trace element analyses, reagents must be of the highest purity, and glassware and tubes scrupulously clean.

Urine is collected in plastic bottles, acidified with acetic acid (1 mL per 100 mL urine) and stored in the refrigerator or deep frozen until the assay samples are prepared. Before aliquotation the urine specimen is homogenized. Any sediment which may have formed is dissolved by warming. If this is not completely successful the remaining sediment should be distributed as homogeneously as possible by careful mixing before withdrawing the aliquots. Urine specimens which have been cooled should be warmed in a shaker water bath before aliquotation. After homogenization the specimen is allowed to cool to room temperature.

For the arsenic determination the given volumes of 1.5 % hydrochloric acid, anti-foaming emulsion, urine and calibration standards (see Table 5) are pipetted into each of the five reaction vessels of the hydride system.

Each analytical series includes a reagent blank, in which ultrapure water is used instead of urine.

Table 5: Preparation of the assay solutions.

| Sample no. | 1.5 % HCl | Anti-foaming emulsion | Urine | Ultra-pure water | Calibration standards | | | | | Spiked arsenic conc. in terms of urine sample volume |
| | | | | | I 10 µg/L | II 20 µg/L | III 50 µg/L | IV 80 µg/L | V 100 µg/L | |
	mL	mL	mL	mL	mL	mL	mL	mL	mL	µg/L
1	10	1	1	1						0
2	10	1	1		1					10
3	10	1	1			1				20
4	10	1	1				1			50
5	10	1	1					1		80
6	10	1	1						1	100

4 Operational parameters for atomic absorption spectrometry

Atomic absorption spectrometer:
Wavelength: 193.7 nm
Background correction: Deuterium lamp
Spectral slit width: 0.7 nm
Lamp current: According to the manufacturer's instructions
Analytical determination: Determination of peak heights or peak areas
                                     (see Section 8.4)

Hydride system:
In principle, the atomic absorption spectrometric determination of arsenic by the hydride technique may be carried out with apparatus constructed in the laboratory. In general, however, systems are used which are commercially available from various manufacturers as accessories for atomic absorption spectrometers.
The settings of the hydride system depend on the equipment being used. The duration of the various purge and transfer steps must be optimized by each operator for his own equipment. The program shown in Table 6 has proved successful for the author and examiner and is intended as a guide.

Table 6: Program for arsenic determination in the hydride system.

Analytical step	Step duration
Purging the whole system with the inert gas (approx. 1500 mL/min)	30 s
Reducing the inert gas flow rate (to about 600 mL/min)	
Flushing the reduction solution (0.5 mL/s) into the reaction vessel by applying a pressure of 150 hPa on the reduction solution vessel	12 s
Flushing the arsenic hydride into the quartz cuvette with hydrogen which is simultaneously generated	
Purging the whole system with the inert gas (approx. 1500 mL/min)	30 s

Purified nitrogen or argon is used as an inert gas. The temperature of the quartz cuvette should be 1000 °C.
Further details can be found in the instructions given by the manufacturers of hydride equipment. The duration of the steps given here apply to the commercially available system MHS-20.

5 Analytical determination

The peak height or peak area of the signal is recorded. A chart recorder for registering the signals is recommended, as the shape of the peak indicates the quality of the reduction process.

When 1 mL of urine is used the calibration curve is linear up to 150 µg/L. If the concentration of arsenic in the sample exceeds 100 µg/L of urine, a smaller volume of biological material must be used and the analysis must be carried out anew.

6 Calibration and calculation of the analytical result

The standard addition procedure is used. The extinctions of the unspiked and spiked assay solutions are plotted as a function of the arsenic concentration added to the urine. The intersection of the resulting straight line with the concentration axis gives the arsenic concentration of the urine sample.

The calculation can also be carried out using a computer. Then the correlation coefficient provides a useful check on the linearity as well as on the quality of the analysis.

If necessary, the reagent blank value is subtracted from the results.

7 Standardization and quality control

Quality control of the analytical results is carried out as stipulated in TRgA 410 of the German Arbeitsstoffverordnung (Regulation 410 of the German Code on Hazardous Working Materials) [25]. Lyophylized urine samples with a specified arsenic content (inorganic arsenic) are commercially available for internal quality control (control material from Nycomed, Oslo, Norway or from Bio-Rad Laboratories, Munich, FRG).

8 Reliability of the method

8.1 Precision

In order to determine the within-series imprecision native urine specimens as well as control urine were each repeatedly analysed. For 5 determinations of spiked urine samples with arsenic concentrations of 26 µg/L or 46 µg/L, the relative standard deviations were 2.3 or 4.2 % which is equivalent to prognostic ranges of 6.4 or 11.7 % respectively. Analysis of control urine containing 107 µg/L, 332 µg/L and 754 µg/L of arsenic resulted in relative standard deviations ranging from 0.6 to 5.6 %. The corresponding prognostic ranges lay between 2.6 and 24.1 %.

Comprehensive tests were carried out in two laboratories using quality control material to evaluate the between-day imprecision. Measurement of a urine sample with a mean arsenic content of 48.8 µg/L over a period of 10 months ($n = 13$) gave a relative standard deviation of 6.0 %, the corresponding prognostic range was 13.1 %. Analogous tests carried out over a period of 8 months ($n = 25$) using a mean arsenic concentration of 19.1 µg/L showed a relative standard deviation of 5.1 % and the corresponding prognostic range was 10.5 %.

8.2 Accuracy

Recovery experiments, using spiked urine with arsenic concentrations of 26 or 46 µg/L, as well as control urine containing 107 µg/L, 332 µg/L and 754 µg/L of arsenic, were carried out to check the accuracy of the method. The recovery rates lay between 88.3 and 104.6 %.

Table 7: Comparison of the results obtained with this method in 4 different laboratories.

Sample No.	Evaluation of Standard	Laboratory 1 Peak height As(V)	Laboratory 2 Peak height As(V)	Peak area As(V)	Laboratory 3 Peak area As(V)	Laboratory 4 Peak area As(V)
1		36.4	24.00	25.75	36	42.1
2		45.2	34.40	30.00	46	189.0
3		6.9	7.50	6.50	7	7.2
4		15.2	10.60	11.00	20	16.0
5		2.5	< 1.00	2.50	3	1.7
6		50.7	41.13	43.00	45	41.7
7		56.4	58.13	60.00	55	62.2
8		17.3	10.00	13.00	14	15.8
9		44.6	35.00	33.00	46	39.5
10		100.0	71.30	76.00	85	75.1
11		17.9	15.00	15.00	13	14.8
12		11.6	10.30	11.00	12	11.0
13		2.6	< 1.00	< 1.00	2	2.0
14		5.8	2.00	3.30	5	3.8
15		30.9	23.44	25.50	25	26.8
16		2.6	3.25	2.60	5	4.1
17		5.8	4.06	4.50	6	5.4
18		10.8	9.06	10.50	10	9.6
19		5.2	5.00	6.75	7	5.1
20		15.6	12.80	14.00	15	15.2
21		13.4	10.00	12.50	–	13.4
22		10.8	10.63	9.50	10	10.3

The expected values for the control urine were established in three laboratories which were especially experienced in metal determination by means of atomic absorption spectrometry. The expected values were calculated statistically on the basis of the results from those laboratories.

Furthermore, experiments were carried out to recover the physiological metabolite dimethylarsinic acid. After addition of dimethylarsinic acid which resulted in arsenic concentrations of 20 and 50 µg/L urine, between 94 and 98 % of the arsenic was recovered. In contrast, the same concentrations of arsenic added in the form of aromatic compounds (4-aminobenzenearsonic acid) could not be detected.

In order to compare the results obtained by this method 22 urine samples from persons occupationally exposed to arsenic were analysed in four laboratories. The evaluation was carried out using either peak heights or peak areas. Arsenic(V) compounds were used for calibration. The results are presented in Table 7. As a rule the results were in agreement. There were only very few deviations which cannot be explained at present. Table 8 shows the correlations which were calculated from these results. The correlation coefficients, the slopes and the intersection points of the linear regression curves with the ordinate demonstrate that, with a few exceptions, the results are closely correlated.

Table 8: Correlation calculations for the results given in Table 7.

Correlation between	No. of samples	Correlation coefficient	Slope of the linear regression curves	Intersection point on the ordinate
	n	r	b	a
Laboratories 2 and 1	21	0.979	1.23	0.6
Laboratories 2 and 3	21	0.979	1.10	1.7
Laboratories 2 and 4	21	0.981	1.05	1.9
Laboratories 3 and 1	21	0.991	1.11	−1.1
Laboratories 4 and 1	21	0.975	1.15	−1.1
Laboratories 4 and 3	20	0.985	1.02	0.1

8.3 Detection limit

When peak areas are evaluated the detection limit for arsenic determination in 1 mL of urine sample is about 2 µg/L. When peak heights are measured the detection limit is about 1 µg/L.

8.4 Sources of error

Trivalent as well as pentavalent arsenic is excreted in the urine of occupationally exposed and unexposed persons, but the pentavalent form seems to predominate. Arsenic species, present in different oxidation states and in different bonds, are converted to the corresponding volatile compounds with differing reaction kinetics. As a result the atomic absorption spectrometric signal is more or less deformed. For this reason evaluation of the peak areas is preferable. In addition, when the peak areas are evaluated calibration is possible using one of the oxidation states of arsenic. Pentavalent arsenic has proved useful for calibration.

This was illustrated by the investigation of 22 urine samples of persons occupationally exposed to arsenic. The results obtained by evaluation of peak heights and peak areas are compared in Table 9. There is good agreement between the results.

Table 9: Comparative investigations to determine the toxicologically relevant arsenic in urine samples of 22 occupationally exposed persons. Arsenic(V) was used for the calibration and both the peak heights and peak areas were evaluated.

| Sample No. | Arsenic concentration (µg/L) spiked with arsenic(V) in each case | |
	Peak height	Peak area
1	24.00	25.75
2	34.40	30.00
3	7.50	6.50
4	10.60	11.00
5	< 1.00	2.50
6	43.13	43.00
7	58.13	60.00
8	10.00	13.00
9	35.00	33.00
10	71.30	76.00
11	15.00	15.00
12	10.30	11.00
13	< 1.00	< 1.00
14	2.80	3.30
15	23.44	25.50
16	3.25	2.60
17	4.06	4.50
18	9.06	10.50
19	5.00	6.75
20	12.80	14.00
21	10.00	12.50
22	10.63	9.50

In the case of acute poisoning with inorganic arsenic compounds a relatively large proportion is excreted as trivalent arsenic. The relationship between trivalent and pentavalent arsenic seems to exhibit great individual differences [15]. Even in these cases evaluation of the peak areas is preferable to evaluation of the peak heights.

In certain cases when 1 mL of a urine sample was used relatively flat standard addition curves could be observed. This loss of sensitivity cannot be satisfactorily explained at present. It is evidently due to the effect of the urine matrix. In this case the sample should be diluted or a smaller volume of urine (0.2–0.5 mL) should be used.

Interference from other elements has not yet been observed.

Arsenic is measured in the near UV range (193.7 nm). It is especially important to ensure that the quartz windows of the cuvette are sufficiently transparent. After the cuvette has been in use for a longer period considerable changes in the signal occur, evidently as a result of the high thermal strain on the cuvette (approx. 1000 °C) as well as chemical reactions with the hydrides. This error can be remedied by treating the cuvette with 1 % hydrofluoric acid. If necessary, the cuvette must be replaced.

The reagent blank value should be regularly checked. Experience has shown that high blank values can be caused, especially by the sodium borohydride solution, depending on the charge used.

9 Discussion of the method

Reviews on the determination of arsenic can be found in *Braman* [26, 27] as well as in *Ishinishi* et al. [2]. According to present knowledge, the hydride-AAS technique seems to be the most suitable for the determination of arsenic in biological material. The determination of total arsenic and specific arsenic compounds (arsenic(V), arsenic(III), aromatic and aliphatic arsenic derivatives) has been reviewed by *Lauwerys* et al. [20]. Special consideration was given to urine as the biological material.

Methodical instructions for the determination of total arsenic have already been published in the German loose-leaf collection [28, see also 29]. In contrast, the method described here enables determination of the toxicologically relevant arsenic only. Experiments showed that in addition to the inorganic arsenic the methylated arsonic acids are reduced by sodium borohydride under the chosen working conditions [12, 30]. However, aromatic arsenic compounds, which primarily occur in marine animals, are not reduced. This was confirmed by investigations using urine samples spiked with 4-aminobenzenearsonic acid, arsenobetaine, and arsenocholine [14, 20].

With this method the calibration can be carried out without using the standard addition procedure. Calibration using aqueous calibration standards also gives reliable results.

At sensitivity no other simple analytical method is known which permits the quantitative determination of this toxicologically relevant arsenic fraction.

The sensitivity of this analytical procedure allows the determination of arsenic excretion in persons who are not occupationally exposed to the substance. Parallel analyses of urine samples were carried out in different laboratories to check the comparability of the analytical results obtained by this method (cf. Section 8.4). Furthermore the reliability of the method was confirmed in numerous external quality control programs.

The practicability of this procedure is documented by its use in routine occupational medicine surveillance [13, 30].

Instruments used:

Atomic absorption spectrometer model AA 290 with hydride system MHS-20 from Perkin Elmer.

10 References

[1] *J. Savory* and *M. R. Wills:* Arsen. In: *E. Merian* (ed.): Metalle in der Umwelt. Verlag Chemie, Weinheim 1984.

[2] *N. Ishinishi, K. Tsuchiya, M. Vahter,* and *B. A. Fowler:* Arsenic. In: *L. Friberg, G. F. Nordberg,* and *V. B. Vouk* (eds.): Handbook on the Toxicology of Metals, 2nd ed, Volume II: Specific Metals. Elsevier, Amsterdam 1986, pp. 43–83.

[3] *W. Arnold:* Arsenic. In: *H. G. Seiler* and *H. Sigel* (eds.): Handbook on Toxicity of Inorganic Compounds. Marcel Dekker Inc., New York 1987, pp. 79–93.

[4] *W. Arnold:* Arsen, ein allgegenwärtiges Gift in der Umwelt des Menschen. Öff. Gesundh.-Wes. *49*, 240–249 (1987).

[5] *Deutsche Forschungsgemeinschaft:* Maximum Concentrations at the Workplace and Biological Tolerance Values for Working Materials 1990. Report No. XXVI of the Commission for the Investigation of Health Hazards of Chemical Compounds in the Work Area. VCH, Weinheim 1990.

[6] TRGS 102: Technische Richtkonzentrationen (TRK) für gefährliche Stoffe. Ausgabe September 1988. In: *W. Weinmann* and *H.-P. Thomas:* Gefahrstoffverordnung Teil 2: Technische Regeln (TRGS) und ergänzende Bestimmungen zur Verordnung über gefährliche Stoffe. Heymanns, Köln.

[7] *R. Mappes:* Versuche zur Ausscheidung von Arsen im Urin. Int. Arch. Occup. Environ. Health *40*, 267–272 (1977).

[8] *J. P. Buchet, R. Lauwerys,* and *H. Roels:* Comparison of several methods for the determination of arsenic compounds in water and in urine. Int. Arch. Occup. Environ. Health *46*, 11–29 (1980).

[9] *J. P. Buchet, R. Lauwerys,* and *H. Roels:* Comparison of the urinary excretion of arsenic metabolites after a single oral dose of sodium arsenite, monomethylarsonate or dimethylarsinate in man. Int. Arch. Occup. Environ. Health *48*, 71–79 (1981).

[10] *J. P. Buchet, R. Lauwerys,* and *H. Roels:* Urinary excretion of inorganic arsenic and its metabolites after repeated ingestion of sodium metaarsenite by volunteers. Int. Arch. Occup. Environ. Health *48*, 111–118 (1981).

[11] *J. P. Buchet* and *R. Lauwerys:* Study of inorganic arsenic methylation by rat liver in vitro: Relevance for the interpretation of observations in man. Arch. Toxicol. *57*, 125–129 (1985).

[12] *P. Schierling, Ch. Oefele,* and *K. H. Schaller:* Bestimmung von Arsen und Selen in Harnproben mit der Hydrid-AAS-Technik. Ärztl. Lab. *28*, 21–27 (1982).

[13] *K. H. Schaller, P. Schierling,* and *R. Schiele:* Die Bestimmung von Arsen im Harn für arbeitsmedizinische Vorsorgeuntersuchungen. In: *T. M. Fliedner* (ed.): Bericht über die 22. Jahrestagung der Deutschen Gesellschaft für Arbeitsmedizin e.V., Ulm, 27.–30.4.1982. Gentner, Stuttgart, pp. 217–221.

[14] *H. Norin* and *M. Vahter:* A rapid method for the selective analysis of total urinary metabolites of inorganic arsenic. Scand. J. Work Environ. Health *7*, 38–44 (1981).

[15] *M. A. Lovell* and *J. G. Farmer:* Arsenic speciation in urine from humans intoxicated by inorganic arsenic compounds. Human Toxicol. *4*, 203–214 (1985).

[16] *M. Vahter:* Environmental and occupational exposure to inorganic arsenic. Acta Pharmacol. Toxicol. *59*, suppl. 7, 31–34 (1986).

[17] *S. Valkonen, J. Jarvisalo,* and *A. Aitio:* Urinary arsenic in a Finnish population without occupational exposure to arsenic. In: *P. Brätter* and *P. Schramel* (eds.): Trace Element Analytical Chemistry in Medicine and Biology. Vol. 2, de Gruyter, Berlin 1983.

[18] *M. Apel* and *M. Stoeppler:* Speciation of arsenic in urine of occupationally non-exposed persons. 4th Int. Conf. Heavy Metals in the Environment. CEP Consultants Ltd. Vol. 1, Heidelberg 1983, pp. 517–520.

[19] *V. Foa, A. Columbi,* and *M. Maroni:* The speciation of the forms of arsenic in the biological monitoring of exposure to inorganic arsenic. Sci. Total Environ. *34*, 241–259 (1984).

[20] *R. R. Lauwerys, J. P. Buchet,* and *H. Roels:* The determination of trace levels of arsenic in human biological materials. Arch. Toxicol. *41*, 239–247 (1979).

[21] *R. S. Braman* and *C. C. Foreback:* Methylated forms of arsenic in the environment. Science *182*, 1247–1249 (1973).

[22] *T. J. Smith, E. A. Crecelius,* and *J. C. Reading:* Airborne arsenic exposure and excretion of methylated arsenic compounds. Environ. Health Perspect. *19*, 89–93 (1977).

[23] *M. Vahter* and *B. Lind:* Concentrations of arsenic in urine of the general population in Sweden. Sci. Total Environ. *54*, 1–12 (1986).

[24] *J. Angerer,* unpublished results, 1987.

[25] TRgA 410: Statistische Qualitätssicherung. Ausgabe April 1979. In: *W. Weinmann* and *H.-P. Thomas:* Gefahrstoffverordnung Teil 2: Technische Regeln (TRGS) und ergänzende Bestimmungen zur Verordnung über gefährliche Stoffe. Heymanns, Köln.

[26] *R. S. Braman, D. L. Johnson, C. C. Foreback, J. M. Ammons,* and *J. L. Bricker:* Separation and determination of nanogram amounts of inorganic arsenic und methylarsenic compounds. Anal. Chem. *49*, 621 (1977).

[27] *R. S. Braman:* Environmental reaction and analysis methods. In: *B. A. Fowler* (ed.): Biological and Environmental Effects of Arsenic. Elsevier, Amsterdam 1983, pp. 141–154.

[28] *J. Angerer, M. Stoeppler,* and *H. Zorn:* Arsen in Harn. In: *D. Henschler* (ed.): Analytische Methoden zur Prüfung gesundheitsschädlicher Arbeitsstoffe, Vol. 2: Analysen in biologischem Material. VCH, Weinheim, 5th issue 1981.

[29] *W. Mundt, J. Angerer,* and *J. Maassen:* Bestimmung von Arsen im Harn mit Hilfe der flammenlosen Atomabsorptionsspektrometrie. Arbeitsmed. Sozialmed. Präventivmed. *13*, 62–64 (1978).

[30] *M. Vahter, L. Friberg, B. Rahnster, A. Nygren,* and *P. Nolinder:* Airborne arsenic and urinary excretion of metabolites of inorganic arsenic among smelter workers. Int. Arch. Occup. Environ. Health *57*, 79–91 (1986).

Author: *K. H. Schaller*
Examiners: *M. Fleischer, J. Angerer, J. Lewalter*

Benzene and alkylbenzenes (BTX aromatics)

Application   Determination in blood

Analytical principle   Capillary gas chromatography Headspace technique

Completed in   May 1993

Summary

The level of aromatic hydrocarbons in blood due to occupational exposure can be simply, sensitively and specifically determined by the static headspace technique as described in this method. In order to detect the range which results from environmental exposure (traffic, passive smoking) it is necessary to use the so-called dynamic headspace analysis technique.

The highly volatile aromatic hydrocarbons contained in blood are determined by means of capillary gas chromatography using the headspace technique. For this purpose the blood samples are warmed to 40 °C in airtight crimp top vials. After distribution of the aromatic hydrocarbons between the liquid and vapour phases has reached equilibrium, an aliquot of the headspace is withdrawn and analysed by gas chromatography with a flame ionisation detector (FID).

Calibration curves are obtained by analysing blood samples to which known quantities of aromatic hydrocarbons have been added. The peak areas are plotted as a function of the concentrations used.

Benzene

Within-series imprecision: Standard deviation (rel.) $s_w = 2.1\%$
                                   Prognostic range $u = 5.7\%$
At a concentration of 718 μg benzene per litre blood and where $n = 5$ determinations

Essential Biomonitoring Methods. DFG, Deutsche Forschungsgemeinschaft
Copyright © 2006 WILEY-VCH Verlag GmbH & Co. KGaA, Weinheim
ISBN: 3-527-31478-4

Between-day imprecision:	Standard deviation (rel.)	$s = 5.8\%$
	Prognostic range	$u = 13.6\%$

At a concentration of 718 µg benzene per litre blood and where $n = 8$ days

Inaccuracy:	Recovery rate	$r = 91.0–96.5\%$

Detection limit:	3 µg benzene per litre blood

Toluene

Within-series imprecision:	Standard deviation (rel.)	$s_w = 2.0\%$
	Prognostic range	$u = 5.4\%$

At a concentration of 698 µg toluene per litre blood and where $n = 5$ determinations

Between-day imprecision:	Standard deviation (rel.)	$s = 7.5\%$
	Prognostic range	$u = 17.6\%$

At a concentration of 698 µg toluene per litre blood and where $n = 8$ days

Inaccuracy:	Recovery rate	$r = 93.8–97.3\%$

Detection limit:	5 µg toluene per litre blood

Ethylbenzene

Within-series imprecision:	Standard deviation (rel.)	$s_w = 1.5\%$
	Prognostic range	$u = 4.2\%$

At a concentration of 694 µg ethylbenzene per litre blood and where $n = 5$ determinations

Between-day imprecision:	Standard deviation (rel.)	$s = 6.3\%$
	Prognostic range	$u = 14.8\%$

At a concentration of 694 µg ethylbenzene per litre blood and where $n = 8$ days

Inaccuracy:	Recovery rate	$r = 90.7–94.6\%$

Detection limit:	8 µg ethylbenzene per litre blood

m-Xylene

Within-series imprecision:	Standard deviation (rel.)	$s_w = 2.0\%$
	Prognostic range	$u = 5.6\%$

At a concentration of 694 µg m-xylene per litre blood and where $n = 5$ determinations

Between-day imprecision:	Standard deviation (rel.)	$s = 5.9\%$
	Prognostic range	$u = 13.9\%$
	At a concentration of 694 µg m-xylene per litre blood and where $n = 8$ days	

| Inaccuracy: | Recovery rate | $r = 94.2–97.9\%$ |

| Detection limit: | 8 µg m-xylene per litre blood | |

o-Xylene

Within-series imprecision:	Standard deviation (rel.)	$s_w = 1.7\%$
	Prognostic range	$u = 4.8\%$
	At a concentration of 706 µg o-xylene per litre blood and where $n = 5$ determinations	

Between-day imprecision:	Standard deviation (rel.)	$s = 6.0\%$
	Prognostic range	$u = 14.1\%$
	At a concentration of 706 µg o-xylene per litre blood and where $n = 8$ days	

| Inaccuracy: | Recovery rate | $r = 95.7–101.4\%$ |

| Detection limit: | 8 µg o-xylene per litre blood | |

Benzene and alkylbenzenes (BTX aromatic hydrocarbons)

The aromatic hydrocarbons benzene, toluene, ethylbenzene and the three xylene isomers o-, m- and p-xylene are colourless, volatile, flammable liquids with a characteristic odour. They are miscible with organic solvents.

The so-called BTX aromatics (benzene, toluene and the three xylene isomers) used to be obtained mainly by distillation from coal and by washing out of coke gas. As a result of the increase in demand for the BTX aromatics, new production processes were developed which used mineral oil as their source. Thus, during the process of refining oil, the upgrading of petroleum by reforming as well as the cracking process for the manufacture of olefins, fractions occur which are rich in aromatic hydrocarbons. These are reformate, pyrolysis or cracked gasoline which represent valuable sources for the production of BTX aromatics [1, 2].

Benzene is added to motor fuels. It is also used as the starting material for the synthesis of many benzene derivatives, such as aniline, nitrobenzene, styrene, synthetic rubber, plastics, phenol and dyestuffs. However, benzene is no longer used as a solvent because of its carcinogenic effect [1].

Toluene is primarily employed as a solvent for paints, resins, varnishes and plastics as well as for the extraction of natural substances. In addition, toluene is an ingredient of automobile gasoline. It is also the starting material for important chemical syntheses [1].

Technical xylene occurs as a mixture of the three isomers o-, m- and p-xylene which are rarely separated in industry. The xylenes are used as solvents for oils, fats, resins, rubber, varnishes and paints as well as for removing grease from metals. o-Xylene and p-xylene are the starting materials for the manufacture of phthalic anhydride (precursor for plasticizers, polyester) and terephthalic acid (synthetic fibres, PET) [1, 2].

Ethylbenzene is produced by the alkylation of benzene with ethylene in the presence of Friedel-Crafts catalysts, such as $AlCl_3$. It is used as a solvent and diluting agent, as well as a starting material for the synthesis of styrene [1, 2].

Inhalation is the main intake route of the highly volatile BTX aromatics at the workplace. Considerable amounts of these organic solvents can be absorbed through the intact skin. Cutaneous absorption is indicated by „H" in the list of MAK values [3]. Like all organic solvents, inhaled BTX aromatics cause narcotic effects. Depending on the absorbed dose all stages of narcosis can be observed, including death due to cardiac arrest. At worst, under present day working conditions, their perceptible effect on the CNS takes the form of headaches, nausea, lack of appetite, poor concentration, impaired memory, etc. In particular, irritation of the mucous membranes can be observed when workers are exposed to small concentration peaks of the alkylbenzenes. Thus, the MAK value for styrene mainly aims to prevent irritation of the mucous membranes.

Benzene represents an exception within the BTX aromatics, as it has proved carcinogenic in humans. Therefore, it has been assigned to category III A1 of the carcinogenic substances by the Commission for the Investigation of Health Hazards of Chemical Compounds in the Work Area [3]. It is regarded as proven that benzene can induce acute myeloic leukaemia. At present there is increasing discussion whether the chronic intake of benzene could cause other forms of leukaemia, e.g. lymphatic leukaemia.

The reason that benzene manifests carcinogenic effects in contrast to its homologues is thought to be because the aromatic ring is oxidized in human metabolism. Epoxybenzene, which is formed as an intermediate, is considered to be the ultimate carcinogenic agent, as it can form covalent bonds with the genetic material.

In contrast, the aliphatic side chain of the alkylbenzenes is oxidized to form the relatively innocuous aromatic carboxylic acids. They are excreted in the urine in the free form or bound to glycine. The aromatic ring of the alkylbenzenes is only partially oxidized to give alkylphenols [4]. Further details about the toxicity and metabolism of the BTX aromatics can be found in recent monographs [3–9].

There are two principle ways of estimating individual exposure to the BTX aromatics when monitoring exposed persons in occupational medicine. On the one hand, the unchanged aromatic hydrocarbons are determined in blood. This parameter is remarkable for its strict substance specificity as well as its sensitivity. The BTX blood level can indicate exposure even in the environmental range. However, the fact that the solvent level in blood reflects only the current exposure to the harmful substance has proved a drawback.

The concentration of the metabolites in the urine represents a better measure of the exposure situation in the course of time. However, both the phenol and the aromatic carboxylic acids excreted in the urine are diagnostically less specific and less sensitive than the solvent level in the blood. Therefore, determination of the excreted metabolites is suitable only for the concentration range relevant to occupational medicine.

The current possibilities for biological monitoring of persons exposed to BTX aromatics are summarized in a recently published monograph [10].

Preventive examinations for the surveillance of persons exposed to benzene are carried out according to principle G9 laid down by the professional associations, BG principle G 29 is applicable for exposure to the homologues of benzene. The most important data with regard to industrial safety and occupational medicine for the solvents benzene, ethylbenzene, toluene and the xylene isomers are summarized in Table 1. The currently valid BAT and EKA are of particular importance for occupational medicine.

The „normal" level of aromatic hydrocarbons in whole blood is of significance to environmental medicine. In this context Table 2 shows the concentrations of benzene in whole blood given in recent literature, taking smoking habits into account.

Table 3 shows the mean concentrations and ranges of the solvents toluene, ethylbenzene, and the xylenes which are found in the general public using current analytical instrumentation.

Author: *J. Angerer, J. Gündel*
Examiners: *U. Knecht, M. Korn*

Table 1: Important data on exposure to benzene, ethylbenzene, toluene and xylenes for industrial safety and occupational medicine

Parameter	MAK/TRK values 1990		Risk of resorption through the skin	Carcinogens group	Pregnancy group	BAT/EKA values 1990	Sampling time
	mL/m³	mg/m³					
Benzene	5 (TRK)	16	H	III A1	–	EKA: 54 µg/L Total blood TRK-Wert: 5 mL/m³	End of exposure period
Ethylbenzene	100 (MAK)	140	H				
Toluene	100 (MAK)	380			B	BAT: 1.7 mg/L Total blood	End of exposure period
Xylenes	100 (MAK)	440			D	BAT: 1.5 mg/L Total blood	End of exposure period

Table 2: Benzene concentrations in the blood of smokers and non-smokers as a result of environmental exposure. All concentrations are expressed in µg/L.

Smokers Mean ± s Median Range	N	Non-smokers Mean ± s Median Range	N	Smokers + Non-smokers Mean ± s Median Range	N	References
0.584 ± 0.300 0.578 0.109 – 1.136	11			0.800 – n.n – 5.900	250	[11]
		0.127 ± 0.054 0.127 0.049 – 0.191	8			[12]
0.485 ± 0.345 0.390 0.019 – 1.657	33	0.130 ± 0.096 0.125 0.010 – 0.455	25	0.332 ± 0.320 0.235 0.010 – 1.657	58	[13]
0.547 ± 0.195 0.493 0.287 – 0.947	14	0.218 ± 0.096 0.190 0.112 – 0.455	13			[14]
– – 0.130 – 0.430	2	0.176 ± 0.062 0.165 0.080 – 0.300	8			[15]
0.335 ± 0.209 0.280 <0.060 - 0.920	26	0.247 ± 0.154 0.220 <0.060 – 0.780	67			[16]

Table 3: Concentrations of toluene, ethylbenzene and xylenes in human blood as a result of environmental exposure.

Substances [μg/L]		Smokers	N	Non-smokers	N	Smokers + non-smokers	N	Smokers + non-smokers	N
Toluene	Median/Mean	2.0	14	1.1	13	0.57	37	1.5	250
	Range	1.3 – 3.8		0.5 – 4.6		0.02 – 5.0		0.2 – 38	
Ethylbenzene	Median/Mean	0.53	14	0.43	13	1.0	37	1.0	250
	Range	0.98 – 2.7		0.18 – 2.3				< DL*	
Xylenes	Median/Mean	1.85	14	1.4	13			5.2	250
	Range	1.15 – 41		0.68 – 7.1				0.5 – 160	
References		[14]		[14]		[13]		[11]	

* Detection Limit

Benzene and alkylbenzenes (BTX aromatics)

Application   Determination in blood

Analytical principle Capillary gas chromatography Headspace technique

Completed in   May 1993

Contents

Essential Biomonitoring Methods. DFG, Deutsche Forschungsgemeinschaft
Copyright © 2006 WILEY-VCH Verlag GmbH & Co. KGaA, Weinheim
ISBN: 3-527-31478-4

1 General principles

The highly volatile aromatic hydrocarbons contained in blood are determined by means of capillary gas chromatography using the headspace technique. For this purpose the blood samples are warmed to 40 °C in airtight crimp top vials. After distribution of the aromatic hydrocarbons between the liquid and vapour phases has reached equilibrium, an aliquot of the headspace is withdrawn and analysed by gas chromatography with a flame ionisation detector (FID).

Calibration curves are obtained by analysing blood samples to which known quantities of aromatic hydrocarbons have been added. The peak areas are plotted as a function of the concentrations used.

2 Equipment, chemicals and solutions

2.1 Equipment

Gas chromatograph with split-splitless injector, flame ionisation detector, analogous recorder as well as integrator and plotter, if required

Device for automatically injecting headspace samples (headspace autosampler)

Alternatively: Instead of the headspace autosampler the samples can be injected manually. In this case a dry thermostat is required (adjustable to 40 °C with drill holes to accomodate the headspace vials).

1000 µL syringe for gas chromatography (e.g. from Hamilton)

Automatic pipettes (e.g. Eppendorf-Multipette)

Microlitre pipettes, adjustable for dosages between 10 and 100 as well as 100 and 1000 µL (e.g. from Eppendorf)

Magnetic stirrer

10, 20, 50, 100 and 500 mL volumetric flasks

Conical flask with ground glass opening

20 mL crimp top vials with teflon-coated butyl rubber stoppers and aluminium crimp caps as well as crimping tongs for sealing and opening them.
The crimp top vials and especially the stoppers must be heated in the drying cupboard at 110 °C for more than 24 hours. It is advisable to heat the stoppers for as long as a week.

Disposable syringes containing potassium EDTA as an anticoagulant (e.g. potassium EDTA Monovetten® from Sarstedt)

2.2 Chemicals

Chemicals of the highest available purity are used:

Benzene, p.a. (e.g. from Merck)

Toluene, p.a. (e.g. from Merck)

Ethylbenzene, p.a. (e.g. from Merck)

m-Xylene, p.a. (e.g. from Merck)

o-Xylene, p.a. (e.g. from Merck)

Ultrapure water (ASTM type 1) or double distilled water

Sodium chloride

2-Ethoxyethanol p.a. (e.g. from Merck)

Defibrinated sheep's blood (e.g. from GLD, Gesellschaft für Labordiagnostica, Essen)

K_2-EDTA

Purified nitrogen (99.999 %)

Hydrogen/compressed air

2.3 Solutions

0.9 % sodium chloride solution:
4.5 g sodium chloride are dissolved in ultrapure water in a 500 mL volumetric flask. The flask is filled up to the mark with ultrapure water.

2.4 Calibration standards

a) Solution of the aromatic hydrocarbons in 2-ethoxyethanol

Starting solution in 2-ethoxyethanol:
After placing about 2 mL 2-ethoxyethanol in a 10 mL volumetric flask, 0.5 mL benzene and 1 mL each of toluene, ethylbenzene, o- and m-xylene are pipetted into the flask one after another. The mass of the aromatic hydrocarbons added is calculated using their specific weights and checked gravimetrically.

The volumetric flask is filled up to its nominal volume to give a solution containing the individual aromatic hydrocarbons with concentrations between 43.51 and 88.10 g/L. Table 4 shows the volumes of the individual aromatic hydrocarbons added, their corresponding masses as well as their concentrations in this starting solution.

Stock solution in 2-ethoxyethanol:
1 mL of the starting solution is pipetted into a 20 mL volumetric flask and the flask is filled up to the mark with 2-ethoxyethanol. The concentrations of the individual aromatic hydrocarbons are given in Table 4, column 5.

b) Solution of the aromatic hydrocarbons in 0.9% sodium chloride solution

800 μL of the stock solution are pipetted into a 50 mL volumetric flask containing about 15 mL 2-ethoxyethanol. The flask is subsequently filled up to the mark with 0.9% sodium chloride solution. The concentrations of the individual aromatic hydrocarbons are given in Table 4, column 6.

c) Calibration solutions of the aromatic hydrocarbons in blood

Between 0.1 and 1.5 mL of the aromatic hydrocarbons in 0.9% sodium chloride solution are pipetted into a 100 mL volumetric flask and it is filled to the mark with blood (cf. Table 5). After thorough mixing of the calibration solutions aliquots (each 2 mL) are filled into the crimp top vials. The vials are immediately covered with teflon-coated rubber septa and sealed with aluminium crimp caps. These calibration standards can be stored in this form in the deep freezer. No changes in the concentrations could be ascertained after a storage period of one year.

3 Specimen collection and sample preparation

For occupational health surveillance the blood specimen for determination of the aromatic hydrocarbons is generally taken at the end of the exposure period. If the substance in question has been assigned a BAT value or an EKA value the sampling time can be found in the list of BAT values [6]. In order to prevent contamination of the blood sample by solvents, the puncturesite of the test person is cleansed with 3% hydrogen peroxide solution or with soap and water rather than the usual disinfectant. The blood sample is withdrawn using a disposable syringe containing an anticoagulant (e.g. EDTA Monovette® from Sarstedt).
Approximately 5 mg of the anticoagulant potassium EDTA in solid form are placed in the crimp top vials. 2 mL of the blood sample are injected into a prepared crimp top vial immediately after collection. The liquid is vigorously swirled around to dissolve the anticoagulant in the vessel. Thus, coagulation of the blood sample is prevented. The sample can be transported in this state, no further precautions are necessary. Samples can be stored for several days in the refrigerator until the analysis is carried out. If longer storage is necessary deep freezing has proved the best method of keeping the samples. Before the analysis the samples are brought to room temperature.

The crimp top vials are placed in the headspace autosampler or in the drill holes of the thermostat. The samples are incubated at a temperature of 40 °C for at least 5 hours. Then equilibrium has certainly been reached.

4 Operational parameters for gas chromatography

Column:	Material:	Fused silica
	Length:	60 m
	Inner diameter:	0.33 mm
	Stationary phase:	100 % Dimethylpolysiloxane, chemically bonded and cross-linked, (DB1)
		Film thickness 1.0 μm
Detector:	Flame ionisation detector	
Temperatures:	Column:	7 min. at 35 °C,
	Increase:	10 °C per min. until 90 °C; then isotherm for 4 min., then increase 30 °C per min. until 150 °C
	Injection block:	230 °C
	Detector:	300 °C
Carrier gas:	Purified nitrogen with a column pressure of 1 bar (1000 hPa), flow rate 1 mL/min	
Split:	9 mL/min	
Make up gas:	Purified nitrogen, 30 mL/min	
Sample volume:	1 mL	

The following retention times of the individual aromatic hydrocarbons can serve as a guide:

benzene	7.85 min
toluene	11.27 min
ethylbenzene	14.64 min
m-xylene	14.97 min
o-xylene	15.84 min

Figure 1 shows the gas chromatogram of a calibration standard in blood.

5 Analytical determination

After incubating the crimp top vials at 40 °C for 5 hours, 1000 µL of the vapour in the headspace are injected into the gas chromatograph using a headspace autosampler. Alternatively, this can be carried out manually by injection with a pre-warmed airtight syringe.

A maximum of two headspace samples can be taken from each crimp top vial.

6 Calibration and calculation of the analytical result

A sample of each of the calibration standard solutions prepared as described in Section 2.3 is analysed as instructed. A blood sample which has not been spiked with aromatic hydrocarbons serves to determine the reagent blank value. These samples are kept deep frozen until analysis in sealed airtight crimp top vials. The resulting peak areas for the individual aromatic hydrocarbons are plotted as a function of the corresponding concentrations used. If reagent blank values are ascertained they must be previously subtracted. As the peaks of the chromatogram form very narrow bands (cf. Fig. 1), evaluation of the peak areas should be carried out using an integrator. If evaluation is manual it is advisable to evaluate the peak heights rather than the peak areas.

A complete new calibration curve need not be made for each analytical series. It is sufficient to determine an aqueous control standard with every analytical series. A correction factor is calculated from the relationship between the value for this standard in the new analytical series and the value for the equivalent standard obtained from the complete calibration curve. Each assay result read off from the calibration curve is adjusted by this factor. A new calibration curve should be plotted if systematic deviation of the quality control results is discernible.

The peak areas or peak heights are obtained from the signals at the characteristic retention times of the individual aromatic hydrocarbons. Any reagent blank values which may occur are subtracted. With these values the equivalent concentration in µg per litre blood (cf. Fig. 2) are read off from the calibration curve.

In the concentration range given for the calibration standards there is a linear relationship between the gas chromatographic signal or its area and the concentration of each individual aromatic hydrocarbon.

7 Standardization and quality control

Quality control of the analytical results is carried out according to TRgA 410 [17] and the Special Preliminary Remarks. At present no control material containing a specified quantity of aromatic hydrocarbons is commercially available. Control blood is prepared in the laboratory to carry out an internal control. A half year's supply of control material is prepared, aliquots are filled into ampullas and kept deep frozen. The concentration of the aromatic hydrocarbons in this control blood should lie in the middle of the most frequently occurring concentration range.

8 Reliability of the method

8.1 Precision

To determine the within-series imprecision, animal blood was spiked twice with different known quantities of aromatic hydrocarbons to give solutions containing the aromatic hydrocarbons at concentrations between 56.7 and 113.8 μg per litre blood, or between 694 and 718 μg per litre blood, respectively. When the blood samples were determined eight or five times, respectively using an autosampler the relative standard deviations were between 1.5 and 9.9 %, the corresponding prognostic ranges were between 4.2 and 23.4 % (see Table 6).

In addition, the between-day precision was determined. The above-mentioned blood solutions containing concentrations between 694 and 718 μg/L were analysed on eight different days. Using an autosampler, relative standard deviations between 5.8 and 7.5 % were found which are equivalent to prognostic ranges of between 13.6 and 17.6 % (see Table 7).

Blood samples with a mean toluene concentration of 3134 μg per litre blood as well as a mean ethylbenzene concentration of 2639 μg per litre blood were analysed eleven times to test the precision in the series when samples were injected manually. The resulting relative standard deviations were 20.9 and 21.2 % which are equivalent to prognostic ranges of 47.3 and 48.0 %.

Table 6: Precision in the series for the gas chromatographic determination of aromatic hydro-carbons in blood (low concentrations $n = 8$, higher concentrations $n = 5$).

Substance	Expected value µg/L	Standard deviation (rel.) %	Prognostic range %
benzene	56.7	9.1	21.5
	718	2.1	5.7
toluene	57.1	8.0	18.9
	698	2.0	5.4
ethylbenzene	57.6	9.9	23.4
	694	1.5	4.2
m-xylene	113.8 (with p-xylene)	4.7	11.1
	694	2.0	5.6
o-xylene	57.7	6.6	15.6
	706	1.7	4.8

Table 7: Between-day precision for the gas chromatographic determination of aromatic hydrocarbons in blood ($n = 8$ days).

Substance	Expected value µg/L	Standard deviation (rel.) %	Prognostic range %
benzene	718	5.8	13.6
toluene	698	7.5	17.6
ethylbenzene	694	6.3	14.8
m-xylene	694	5.9	13.9
o-xylene	706	6.0	14.1

8.2 Accuracy

Recovery experiments were carried out by the examiners of the method to check its accuracy. For this purpose samples were spiked with specific quantities of BTX aromatics. The results are compared with those obtained from the standard solutions also prepared in blood. Table 8 shows the recovery rates.

Table 8: Recovery rates of the aromatic hydrocarbons in blood

Substance	Recovery rate
benzene	91.0 – 96.5 %
toluene	93.8 – 97.3 %
ethylbenzene	90.7 – 94.6 %
m-xylene	94.2 – 97.9 %
o-xylene	95.7 – 101.4 %

8.3 Detection limit

The detection limits were calculated as three times the signal/background ratio:

Substance	Detection limit [μg/L]
benzene	3
toluene	5
ethylbenzene	8
m-xylene	8
o-xylene	8

8.4 Sources of error

a) Pre-analytical phase

The time at which the blood sample is taken has a considerable effect on the analytical result. The half-life for the elimination of aromatic hydrocarbons from the blood is very short in some cases, so that the solvent level decreases rapidly after exposure. Therefore, it is essential to adhere to the sampling time stipulated in the list of BAT values [6]. This is usually at the end of the exposure period.

Coagulation of the blood sample impedes reproducible distribution of the aromatic hydrocarbons between the biological matrix and the vapour phase at equilibrium. Thus, the blood sample should be vigorously swirled round the crimp top vial to ensure that it is thoroughly mixed with the anticoagulant. Disposable syringes containing solid EDTA (e.g. EDTA Monovettes® from Sarstedt) are recommended for sample collection.

In order to prevent loss of the highly volatile aromatic hydrocarbons during transport and storage, it is important to fill the samples into the crimp top vials and seal them immediately after sample collection. It should be impossible to turn the seals by hand. To ensure this the crimping tongs must be appropriately adjusted.

The use of Vacutainers® for the analysis of highly volatile organic solvents is not recommended. Considerable loss of the aromatic hydrocarbons could be observed due to adsorption on the rubber membrane of the sealing system. Thus, Vacutainers® are neither suitable as sample collection nor as transport vessels for the determination of aromatic hydrocarbons in blood.

Exogenous contamination can influence the results of this method to determine aromatic hydrocarbons in blood. In order to prevent the pollution of the samples with solvents it is necessary to dispense with the usual disinfectant for cleansing the puncturesite before sampling. A 3 % solution of hydrogen peroxide can be used. Thorough cleansing with soap and water has also proved sufficient. Thus, interference in the chromatogram from overlapping of the solvents contained in swabs can be avoided.

Furthermore, the crimp top vials as well as the teflon-coated butyl rubber stoppers should be thoroughly heated in the drying cupboard at 110 °C for three days before use. Any traces of solvents contained in the butyl rubber stoppers or adhering to the glass walls are thus expelled. This is especially important for the analysis of benzene.

b) Analytical phase

It is of decisive importance for the reliability and accuracy of the analytical results to ensure that the distribution equilibrium between the liquid and the vapour phase has definitely been attained. Experience has shown that equilibrium has been reached after 5 hours. Nevertheless, the user of this method should check how long it takes to reach equilibrium under the operational conditions of the apparatus he uses.

The use of an autosampler has meanwhile become standard for headspace analysis. If manual injection is carried out it is important to ensure that the syringe is airtight and that condensation of the sample components in the syringe is avoided. The latter is achieved by warming the airtight syringe to 40 °C before the vapour phase is withdrawn. To prevent aromatic hydrocarbons being carried over from one sample to another the syringe is carefully flushed with nitrogen after every sample injection. Despite these precautions, however, it has been shown that the use of an autosampler distinctly improves the precision in the analysis of the aromatic hydrocarbons.

Tests were carried out to investigate whether various industrially important solvents cause interference with this method. Only n-butanol may interfere with the determination of benzene when a DP-1 column is used. However, cyclohexane, heptane, methanol and isopropanol do not cause interference in the determination of aromatic hydrocarbons. When a DP- 1701 column is used heptane can interfere with the determination of benzene. In case of doubt a sample should be analysed using a column of different polarity.

9 Discussion of the method

The gas chromatographic headspace analysis using a capillary column is the current standard method of determining the BTX aromatic hydrocarbons in blood. Headspace analysis has several advantages beyond other methods described in the literature in which the solvents are eliminated from the matrix or in which the matrix is separated on a heated precolumn. The analysis is simple to carry out because no sample preparation is necessary. As no analytical background interference occurs, very low detection limits can be achieved.

Five different aromatic hydrocarbons can be determined in one analytical operation using this method. For the substances which have already been assigned EKA or BAT values in blood (benzene 40 µg per litre, toluene 1.7 mg per litre, xylene 1.5 mg per litre) concentrations of these solvents between 1/13 and 1/340 of the EKA or BAT value can be detected in blood.

In principle, it is also possible to determine styrene with this method. However, it is difficult to assess exposure to styrene by means of determination of styrene in blood, as enzymatic and chemical changes can lead to a decrease in the level of styrene in vitro. On the other hand a BEI value has been assigned for styrene in blood (0.55 mg per litre). Using the present method 1/55 of the BEI value can be detected.

The reliability criteria given here were determined using a headspace autosampler. Relative standard deviations lower than 10 % can be achieved. Although manual injection is possible, the precision given here cannot be attained. In the case of manual injection it is given in Section 8.4.

The examiners of this method tested the use of an internal standard (2-hexanone, cyclohexane). However, no improvement in the precision could be achieved.

In order to attain a reproducible distribution equilibrium between the biological matrix and the vapour phase, it is essential to prevent coagulation of the blood sample. Thus, in addition to the anticoagulant in the disposable syringes, EDTA is also added to the headspace vials as a precaution. To ensure that the distribution equilibrium is reached the headspace vials should be incubated at 40 °C for at least 5 hours. The analytical results are normally independent of the content of the headspace vials. Erroneous results occur only when they are clearly overfilled.

The method described here cannot separate the m-xylene and p-xylene isomers when a DP-1 or DP-1701 column is used. If these two substances are to be separated it is advisable to use a DP wax column. The retention times of the various columns must be checked in each case.

Instruments used:
Gas chromatograph 3300 with data system DS 600 from Varian as well as an autosampler Dani HSS 3950.

10 References

[1] Römpps Chemie-Lexikon. Franckh'sche Verlagshandlung, Stuttgart, 8th edition 1979.

[2] *K. Weissermehl* and *H.-J. Arpe* (eds.): Industrielle Organische Chemie. VCH, Weinheim, 2nd revised and extended edition 1978.

[3] *Deutsche Forschungsgemeinschaft*: Maximum Concentrations at the Workplace and Biological Tolerance Values for Working Materials 1991. Report No. XXVII of the Commission for the Investigation of Health Hazards and Chemical Compounds in the Work Area. VCH, Weinheim 1991.

[4] *J. Angerer* and *G. Lehnert*: Occupational chronic exposure to organic solvents VIII. Phenolic compounds – metabolites of alkylbenzenes in man. Simultaneous exposure to ethylbenzene and xylenes. Int. Arch. Occup. Environ. Health *43*, 145–150 (1979).

[5] *J. Angerer*: Occupational chronic exposure to organic solvents VII. Metabolism of toluene in man. Int. Arch. Occup. Environ. Health *43*, 63–67 (1979).

[6] *D. Henschler* and *G. Lehnert* (eds.): Biologische Arbeitsstoff-Toleranz-Werte (BAT-Werte). Arbeitsmedizinisch-toxikologische Begründungen. Deutsche Forschungsgemeinschaft, VCH, Weinheim 1991.

[7] *Gesellschaft Deutscher Chemiker. Beratergremium für umweltrelevante Altstoffe*: Umweltrelevante alte Stoffe II. VCH, Weinheim 1988.

[8] NIOSH: Occupational hazard assessment / Criteria for controlling occupational exposure to aromatic hydrocarbons. DHHS Publication No. 84–1501, Cincinnati 1984.

[9] ECETOC, Technical report 29: Concentrations of industrial organic chemicals measured in the environment. European Chemical Industry. Brussels, 1988.

[10] *J. Angerer* and *B. Hörsch*: Determination of aromatic hydrocarbons and their metabolites in human blood and urine. J. Chromatogr. *580*, 229–255 (1992).

[11] *R.S. Antoine, I.R. DeLeon* and *R.M. O'Dell-Smith*: Environmentally significant volatile organic pollutants in human blood. Bull. Environ. Contam. Toxicol. *36*, 364–371 (1986).

[12] *L. Perbellini, G.B. Faccini, G. Pasini, F. Cazzoli, S. Pistooia, R. Rosellini, M. Valscci* and *F. Brugnone*: Environmental and occupational exposure to benzene by analysis of breath and blood. Br. J. Ind. Med. *45*, 345–352 (1988).

[13] *F. Brugnone, L. Perbellini, G.B. Faccini, F. Passini, G. Maranelli, L. Romeo, M. Gobbi* and *A. Zedde*: Breath and blood levels of benzene, toluene, cumene and styrene in non-occupational exposure. Int. Arch. Occup. Environ. Health *61*, 303–311 (1989).

[14] *H. Hajimiragha, U. Ewers, A. Brockhaus* and *A. Boettger*: Levels of benzene and other volatile aromatic compounds in the blood of non-smokers and smokers. Int. Arch. Occup. Environ. Health *61*, 513–518 (1989).

[15] *J. Angerer, G. Scherer, K.H. Schaller* and *J. Müller*: The determination of benzene in human blood as an indicator of environmental exposure to volatile aromatic compounds. Fresenius J. Anal. Chem. *339*, 740–742 (1991).

[16] *J. Angerer, B. Heinzow, D.O. Reinmann, W. Knorz* and *G. Lehnert*: Internal exposure to organic substances in a municipal waste incinerator. Int. Arch. Occup. Environ. Health *64*, 265–273 (1992).

[17] TRgA 410: Statistische Qualitätssicherung. In: Technische Regeln und Richtlinien des BMA zur Verordnung über gefährliche Stoffe. Bek. des BMA vom 3.4.1979, Bundesarbeitsblatt 6/1979, pp. 88 ff.

Author: *J. Angerer, J. Gündel*
Examiners: *U. Knecht, M. Korn*

Table 4: Preparation and concentration of the aromatic hydrocarbons in 2-ethoxyethanol and in 0.9% sodium chloride solution.

Substance	Volume of the individual BTX aromatics	Equivalent to a mass in 10 mL 2-ethoxyethanol of	Starting solution in 2-ethoxyethanol	Stock solution in 2-ethoxyethanol	Solution in 0.9% sodium chloride
	μL	g	g/L	g/L	mg/L
Benzene	500	0.4352	43.51	2.18	34.82
Toluene	1000	0.8695	86.95	4.35	69.56
Ethylbenzene	1000	0.8702	87.02	4.35	69.62
o-Xylene	1000	0.8810	88.10	4.41	70.48
m-Xylene	1000	0.8648	86.48	4.32	69.18

Table 5: Pipetting scheme for the preparation of the calibration solutions in blood.

Calibration solution No.	Volume of the solution in 0.9% sodium chloride mL	Final volume of the calibration solution mL	Concentration of the calibration solution µg/L					
			Benzene	Toluene	Ethylbenzene	o-Xylene	m-Xylene
1	0.10	100	34.82	69.56	69.62	70.48	69.18
2	0.30	100	104.47	208.70	208.88	211.46	207.56
3	0.80	100	278.46	556.54	557.02	563.90	553.50
4	1.5	100	522.35	1043.50	1044.40	1057.31	1037.80

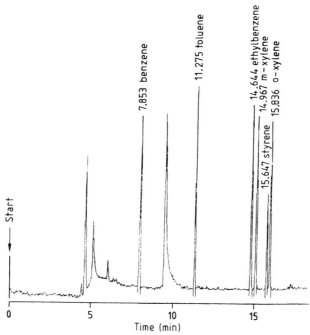

Fig. 1: Gas chromatogram of a calibration standard in blood.
Concentrations: benzene 114.07 µg/L
toluene 217.30 µg/L
ethylbenzene 203.12 µg/L
m-xylene 200.73 µg/L
o-xylene 201.18 µg/L

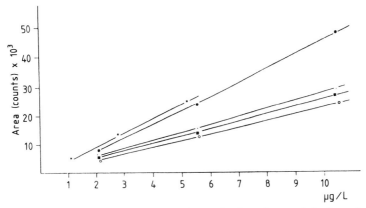

Fig. 2: Examples of calibration curves for the determination of aromatic hydrocarbons in blood.
  * benzene
  ● toluene
  · ethylbenzene
  ■ m-xylene
  ○ o-xylene

Beryllium
Standard Addition Procedure

Application   Determination in urine

Analytical principle Electrothermal atomic absorption spectrometry

Completed in   December 1995

Summary

Using the procedure described here beryllium can be determined in the urine of people who have been occupationally exposed to this element.

Beryllium in urine is determined by means of atomic absorption spectrometry using the grapite furnace technique with a pyrolytically coated graphite tube and L'Vov platform. No previous sample preparation is necessary. In order to reach the optimum mineralization temperature a matrix modifier, a mixture of palladium (II) nitrate and magnesium nitrate, is used.

The interference effects of the sample matrix are minimized by using the Zeeman-effect background compensation and the standard addition procedure.

Within-series imprecision:	Standard deviation (rel.)	$s_W = 3.8\ \%$ (peak height)
		$s_W = 4.5\ \%$ (peak area)
	Prognostic range	$u = 8.0\ \%$ (peak height)
		$u = 10.1\ \%$ (peak area)

At a concentration of 1.0 µg beryllium per litre urine and where $n = 10$ determinations

| Inaccuracy: | Recovery rate | $r = 92{-}98\ \%$ |
| Detection limit: | 0.1 µg beryllium per litre urine | |

Beryllium

Beryllium (atomic mass 9.0122 g/mol, atomic number 4, m. p. 1287 °C) is a silvery white, very hard and brittle lightweight metal. Its chemical behaviour is very similar to that of

Essential Biomonitoring Methods. DFG, Deutsche Forschungsgemeinschaft
Copyright © 2006 WILEY-VCH Verlag GmbH & Co. KGaA, Weinheim
ISBN: 3-527-31478-4

aluminium. With an average content of about 6 ppm in the Earth's crust beryllium is one of the less common lements [1, 2]. Beryllium occurs naturally in the form of compounds with oxygen, as a mineral (beryl) or as an ore (common beryl). The most important deposits are mined in Argentina, Brazil, India, South Africa and the former USSR. As in the production of aluminium, beryllium is extracted by electrolysis of beryllium chloride in the smelting flux or by reduction of beryllium fluoride with manganese [3].

Beryllium and its compounds (especially beryllium oxide) are used for a variety of purposes in many modern branches of industry. Although beryllium alloys are four times lighter, they are as hard as steel, and more rigid than aluminium. Beryllium is employed for the manufacture of extremely hard and resistant metal alloys for motors, engines rocket engines and measuring and control instruments. In addition, beryllium is used in the production of highly fireproof devices and materials, and ceramic pigments. On account of its uncommon rigidity and strength relative to its weight it is an attractive working material for the aerospace industry. As it is permeable to short-wave radiation, e. g. X-rays, beryllium-containing glass is required for the manufacture of X-ray tubes [2, 4, 5].

Beryllium oxide and beryllium are used as moderating materials in nuclear reactors. Another special use of this element is in the form of finely dispersed beryllium as a component of solid fuel for rocket engines.

The physico-chemical properties of beryllium, is compounds and its industrial uses are reviewed in IARC [4].

Occupational exposure to beryllium occurs during the treatment of beryllium-containing minerals and ores, the extraction of beryllium, the further processing and machining, especially of dry beryllium compounds which create dust. In addition, beryllium exposure can take place during the production of metal alloys and in the manufacture of beryllium-containing glass, ceramics and porcelain [3].

Beryllium and its compounds are generally absorbed into the organism by inhalation of dusts and fumes, but percutaneous absorption is also possible. Less than one percent of orally ingested beryllium is absorbed [6].

Illnesses caused by beryllium or its compounds are shown under BK No. 1110 in the current "Berufskrankheitenliste" List of Occupational Health Hazards in the Federal Republic of Germany. The most important of these are acute and chronic beryllium intoxication. The acute form of intoxication, as observed in metal foundry workers, results in a feverish disorder which can last up to two days. This illness is frequently accompanied by inflammation of the skin and mucous membranes, e.g. conjunctivitis or bronchitis. The chronic form of intoxication, a granulomatous inflammation of the lung similar to sarcoidosis, can have fatal consequences. Even exposure to low concentrations of beryllium can lead to so-called berylliosis. It develops after an incubation period of months or years in which no symptoms are apparent. The disease is characterized by a persistent, dry cough, increasing breathlessness and considerable loss of weight. X-rays show tiny spots in the lungs which lead to fibrosis [3, 4, 7].

Beryllium and its compounds have been assigned to category III A2 (shown to be clearly carcinogenic only in animal studies) in the list of carcinogenic substances by the Commission for the Investigation of Health Hazards of chemical Compounds in the Work Area [8]. On the basis of recent epidemiological studies the International Agency for Research on Cancer has classified beryllium and its compounds as carcinogenic for humans (Group 1) [4].

Incorporated beryllium is principally accumulated in the prealbumin and γ-globulin fractions [9]. Addition of beryllium compounds to lymphocyte cultures from persons exposed to beryllium causes an increase in blastogenic transformation. These findings indicate that chronic beryllium intoxication may be assigned to the disorders which manifest a delayed allergic, i.e. cell-induced, immune reaction. Toxicokinetic investigations have shown that beryllium is accumulated in the bones. The sensitization of the T-lymphocytes observed after exposure to beryllium seems to be largely due to the passage of the lymphocytes through the bone marrow. Although there is no direct evidence of a relationship between beryllium exposure and these cellular immune reactions, an increase in the level of γ-glo-bulins was found in berylliosis patients. Biochemical studies have shown that beryllium has an inhibiting effect on some enzyme systems, especially on the alkaline phosphatase system, and on the intranuclear replication of DNA [5].

Biomonitoring of occupational exposure to beryllium is carried out by measuring renal ex-cretion of the substance [10]. The results of investigations using modern analytical methods show that the excretion of beryllium in the urine of people who are not occupationally ex-posed to the element is distinctly below 1 µg/L. Table 1 shows an overview of the beryllium concentration in the urine of non-exposed people.

Table 1. Beryllium concentrations in the urine of people who were not occupationally exposed to the element, measured by graphite furnace atomic absorption spectrometry

Country	No. of Persons	Concentration [µg/L; $\bar{x} \pm s$]	Reference
USA	120	0.9 ± 0.4	[11]
Italy	56	0.6 ± 0.2	[12]
USA	–	0.13	[13]
Italy	163	0.24 ± 0.16 (range < 0.03–0.8)	[14]
Italy	579	0.4 (range < 0.02–0.82)	[15]

Smoking seems to influence beryllium excretion in urine. Beryllium concentrations in the urine of heavy smokers of 0.31 ± 0.17 µg/L were higher than those of non-smokers (0.20 ± 0.14 µg/L [14].

According to Zorn et al. [2] an ambient beryllium concentration of 2 µg/m^3 in the air leads to a corresponding renal beryllium excretion of 7 µg/L urine. Apostoli et al. [14] investigated the renal beryllium excretion of dental technicians. The beryllium concentrations varied between 0.05 and 1.7 µg/L (mean value 0.34 µg/L). Studies on the beryllium excretion of foundry workers also showed significantly higher beryllium excretion compared with non-exposed persons. However, these studies could not show any relationship between the beryllium excretion in urine and the duration of the exposure. Cammarano [16] investigated beryllium elimination in the urine of workers who were employed in the cleaning of oil-fired boilers. They reported a mean beryllium excretion of 1.1 µg/L before the working shift and 1.8 µg/L after the end of the shift.

These investigations indicate that the beryllium excretion in urine can be used as a bioindicator to detect exposure to beryllium [10, 17].

Author: *P. Schramel*
Examiners: *J. Angerer, K. H. Schaller*

Beryllium
Standard Addition Procedure

Application   Determination in urine

Analytical principle Electrothermal atomic absorption spectrometry

Completed in   December 1995

Contents

Essential Biomonitoring Methods. DFG, Deutsche Forschungsgemeinschaft
Copyright © 2006 WILEY-VCH Verlag GmbH & Co. KGaA, Weinheim
ISBN: 3-527-31478-4

1 General principles

Beryllium is determined in urine by means of atomic absorption spectrometry using the graphite furnace technique with a pyrolytically coated graphite tube and L'Vov platform. No previous sample preparation is necessary. In order to reach the optimum mineralization temperature a mixture of palladium(II) nitrate and magnesium nitrate is used as a matrix modifier.

The interfering effects of the sample matrix are minimized by the use of the Zeeman-effect background compensation and the standard addition procedure.

2 Equipment, chemicals and solutions

2.1 Equipment

Atomic absorption spectrometer with Zeeman background correction

Graphite furnace, preferably with autosampler

Data station with high resolution graphics

Electrodeless discharge lamp (EDL) for beryllium with power supply or hollow cathode lamp beryllium

Graphite tube, pyrolytically coated, with L'Vov platform

Automatic microlitre pipettes, adjustable between 10 and 100 µL and between 200 and 1000 µL (e.g. from Eppendorf)

100, and 1000 mL volumetric flasks

5 mL pipette

50 mL pipette

20 mL graduated pipette

Shaker (e.g. Vortex from Cenco, the Netherlands)

Autosampler vessels or disposable reaction vessels for automatic sample injection

2.2 Chemicals

Beryllium assay solution 1 g/L (in the form of beryllium acetate $Be_4O(C_2H_3O_2)_6$ in 3 % HNO_3, e.g. from Spex)

Palladium(II) nitrate solution, 10 g/L ($Pd(NO_3)_2$ in 15 % HNO_3 e.g. from Merck)

Magnesium nitrate solution, 10 g/L ($Mg(NO_3)_2$ in 15 % HNO_3, e.g. from Merck)

Utrapure water (ASTM type 1) or double-distilled water

Argon for spectrometry 99,998 %

65 % HNO_3 p.a. (e.g. from Merck)

Glacial acetic acid

2.3 Solutions

1 M HNO_3 (for cleaning the glassware):
About 400 mL ultrapure water are placed in a 1000 mL volumetric flask. 69 mL of 65 % HNO_3 are pipetted into the flask. The volumetric flask is then filled to the mark with ultrapure water.

$Pd(NO_3)_2$ solution (500 mg/L):
5 mL of the $Pd(NO_3)_2$ solution (10 g/L) are pipetted into a 100 mL volumetric flask containing about 50 mL ultrapure water. The flask is subsequently filled to the mark with ultrapure water.

$Mg(NO_3)_2$ solution (500 mg/L):
5 mL of the $Mg(NO_3)_2$ solution (10 g/L) are pipetted into a 100 mL volumetric flask containing about 50 mL ultrapure water. The flask is then filled to the mark with ultrapure water.

Modifier solution:
50 mL of the $Pd(NO_3)_2$ solution (500 mg/L) are pipetted into a 100 mL volumetric flask. Then the flask is filled to the mark with $Mg(NO_3)_2$ solution (500 mg/L).

These solutions can be stored in the refrigerator at 4 °C for several months.

2.4 Calibration standards

Stock solution (1 mg/L):
0.1 mL of the beryllium solution is pipetted into a 100 mL volumetric flask into which about 50 mL ultrapure water have previously been filled. The flask is subsequently filled to the mark with ultrapure water.
This solution can be stored in the refrigerator at 4 °C for several months.
Calibration standards containing 5 to 15 μg beryllium per litre are prepared from this starting solution by dilution with ultrapure water. These calibration standards must be freshly prepared daily according to the following pipetting scheme (Table 2):

Table 2. Pipetting scheme for preparation of the calibration standards

Volume of the stock solution [mL]	Final volume of the calibration standard [mL]	Concentration of the calibration standard [µg/L]	Designation of the calibration standard
0.5	100	5	I
1.0	100	10	II
1.5	100	15	III

3 Specimen collection and sample preparation

As in all trace element analyses, the reagents must be of the highest possible purity and vessels used must be thoroughly cleaned. This also applies to specimen collection.

In order to prevent possible exogenous contamination each of the plastic vessels used for sample collection must be cleansed by leaving them filled with 1 M nitric acid for at least 2 hours and subsequently rinsing them thoroughly with ultrapure water before drying them. For determination in the range of the detection limit the cleansing process is further improved by using warm nitric acid.

Urine is collected in plastic bottles at the end of a working shift. The urine samples are acidified with acetic acid (1 mL glacial acetic acid/100 mL urine). If they cannot be processed immediately they are stored in the refrigerator for up to 5 days. For longer storage the urine specimens are kept deep-frozen (- 18 °C) until analysis. For analysis the urine samples are thawed in a water bath at 40 °C and subsequently allowed to reach room temperature. Before an aliquot is taken for analysis the samples are thoroughly shaken to ensure they are homogeneous.

Then 0.5 mL of the urine are pipetted into four autosampler vessels or into disposable reaction vessels. 0.4 mL of the matrix modifier solution and 0.1 mL of the calibration standard solutions are added in the order shown in Table 3. The samples are subsequently mixed on the shaker for 20 seconds.

A reagent blank containing ultrapure water instead of urine is included in each sample series.

Table 3. Spiking of the sample solutions with the calibration standards

Urine	Modifier	H$_2$O	Std. I	Std. II	Std. III	Added beryllium concentration, expressed in terms of the urine volume used
[mL]	[mL]	[mL]	[mL]	[mL]	[mL]	[µg/L]
0.5	0.4	0.1	–	–	–	–
0.5	0.4	–	0.1	–	–	1.0
0.5	0.4	–	–	0.1	–	2.0
0.5	0.4	–	–	–	0.1	3.0
–	0.4	0.6	–	–	–	–*

Std. I:	5 µg/L		Std. II:	10 µg/L		Std. III: 15 µg/L

* Reagent blank value

4 Operational parameters for atomic absorption spectrometry

Atomic absorption spectrometer:

Wavelength: 234.9 nm

Background compensation: Zeeman effect

Spectral slit width: 0.7 nm

Lamp current: According to the manufacturer's instructions

Analytical Determination: Determination of the peak heights, or alternatively peak areas

The temperature-time program (Table 4) serves only as a guide. The operational parameters must be optimally adjusted for each individual instrument.

Inert gas: Argon

Injected volume: 20 µL

Furnace type: Pyrolytically coated graphite furnace with L'Vov platform

Table 4. Temperature-time program

Analytical step	Step duration		Temperature
	Ramp time [s]	Hold time [s]	[°C]
Drying	20	10	120
Charring I	10	10	300
Charring II	10	5	1100
Charring III	1	5	1100 Gas Stop
Atomization	0	3	2650
Heating	1	3	2750
Cooling	1	5	30

5 Analytical determination

20 µL aliquots of each assay solution prepared as described in Section 3 are analysed by atomic absorption spectrometry, selecting the appropriate instrumental parameters for the equipment used (cf. Section 4). The peak heights in the atomization step are recorded in each case. Alternatively the peak areas can also be evaluated. Thus 4 values are obtained to calculate the analytical result (cf. Figure 1).

6 Calibration and calculation of the analytical result

The beryllium content of the urine sample is plotted graphically with the help of the standard addition procedure. For this purpose the reagent blank values are subtracted from the peak heights or peak areas of the unspiked and the three spiked samples and the resulting values are plotted as a function of the beryllium concentration. The result is a straight line and its point of interception with the concentration axis gives the beryllium content of the sample (cf. Figure 2).

The linear operational range extends from 0.1 to 5 µg/L. If the signal of the urine sample or the spiked urine solutions does not lie within the linear range the samples must be appropriately diluted with ultrapure water and analysed anew.

7 Standardization and quality control

Quality control of the analytical results is carried out as stipulated in TRgA 410 (Regulation 410 of the German Code on Hazardous Working Materials) [18] and in the special prelimi-nary remarks in this series. Material for internal quality control is commercially available from Nycomed AS, Oslo, marketed by Immuno, Heidelberg, Germany under the name of "Seronorm™ Trace Elements Urine".

If necessary, control material can be prepared in the laboratory. For this purpose urine is spiked with a defined quantity of beryllium. Aliquots of this material can be stored in the deep freezer for up to a year and used for quality control. The mean expected value and the tolerance range of this quality material is obtained in a pre-analytical period (one determination of the control material on 20 different days) [19]).

8 Reliability of the method

8.1 Precision

In order to determine the precision in the series samples were prepared by spiking pooled urine from non-exposed persons with a defined amount of beryllium to give a concentration of 1 µg beryllium per litre urine. The urine samples were subsequently analysed ten times.

The relative standard deviation was calculated as 3.8 % (evaluation of the peak heights) or 4.5 % (evaluation of the peak areas), and the equivalent prognostic ranges were 8 % or 10.1 % respectively).

8.2 Accuracy

Recovery experiments were carried out to check the accuracy of the method. Moreover, the control material described in Section 7 was analysed by means of the method described here.

Recovery rate for the spiked urine samples:
The pooled urine prepared as in Section 8.1 was processed and analysed ten times. The mean value of the measured concentrations was 0.95 µg beryllium per litre urine (range 0.92–0.98 µg/mL). The mean recovery rate was calculated as 95 % (92–98 %).

Analysis of the control material:
Control urine "Seronorm Trace Elements Urine" from Nycomed Pharma AS was processed and analysed ten times. According to the manufacturer this urine was spiked with 5 µg be-ryllium per litre. The certified beryllium content was given as 4.5 µg per litre urine. Using this method a mean beryllium concentration of 5.5 µg per litre urine was determined.

8.3 Detection limit

The detection limit, calculated as three times the standard deviation of the reagent blank value, was 0.1 µg beryllium per litre urine.

8.4 Sources of error

As in all trace analyses, purity of the vessels and chemicals is of primary importance.

Without matrix modification chloride and fluoride from the biological matrix can interfere with the electrothermal atomic absorption spectrometric determination of beryllium in urine. Lower results are obtained. The use of a palladium/magnesium modifier eliminates this interference, as it permits an increase in the mineralization temperature to 1100 °C.

Further possible interference effects from the matrix are eliminated by the use of the standard addition procedure.

9 Discussion of the method

Within the framework of biological monitoring and given the current state of the art, determination of beryllium in biological materials is mainly carried out in practice by atomic absorption spectrometry (AAS) [13] and inductively coupled plasma atomic emission spectrometry (ICP-OES) [20]. Furthermore, gas chromatographic methods have also been used to analyze beryllium in biological materials. In this case beryllium is transformed into a volatile compound by formation of a complex using trifluoroacetylacetone as a ligand [5]. However, this method is considerably more complicated than the AAS or the ICP-OES methods.

A further sensitive and reliable atomic absorption spectroscopic procedure has been included in this series (flameless atomic absorption spectrometry) [21]. However, this procedure requires complete mineralization of the sample material and subsequent formation of a beryllium complex with acetylacetone. Thus the analytical method is relatively time-consuming and expensive to carry out.

Using the standard addition procedure described here in conjunction with a matrix modifier the beryllium concentration in urine can be simply and sensitively determined. No time-consuming sample preparation is necessary and the analysis of beryllium in urine can be carried out directly in the graphite furnace. The risk of contamination is therefore greatly reduced. The use of an autosampler not only improves the practicability of this method, but also enhances its precision.

In addition, the matrix modifier ensures a considerable increase in the beryllium signal and distinctly improves signal background (cf. Figures 3 and 4). Consequently the present method permits determination of beryllium down to a value of 0.1 µg per litre urine.

Despite the relatively good detection limit, determination of beryllium excretion in the normal population is not possible with this method. However, the method permits the surveillance of people exposed to beryllium at the workplace for the purposes of occupational medicine.

Instruments used:
Atomic absorption spectrometer Z 3030 with graphite furnace cuvette HGA 600 and auto-sampling system AS 40 from Perkin-Elmer

10 References

[1] *A.L. Reeves:* Beryllium. In: *L. Friberg, G.F. Nordberg* and *V.B. Vouk* (Eds.): Handbook on the Toxicology of Metals. Elsevier, Amsterdam 1986, pp. 95–100.
[2] *H. Zorn, T. Stiefel, J. Beuers* and *R. Schlegelmilch:* Beryllium: In: *H.G. Seiler* and *H. Sigel* (eds.): Handbook on the Toxicity of Inorganic Compounds. Marcel Dekker, Inc., New York 1988, pp. 105–114.
[3] *G. Zerlett:* Erkrankungen durch Beryllium oder seine Verbindungen. ErgoMed *6*, 6–8 (1982).
[4] *IARC:* International Agency for Research on Cancer. Monographs on the evaluation of the carcinogenic risk of chemicals to humans. Vol. 58: Beryllium, Cadmium, Mercury, and exposures in the glass manufacturing industry. Lyon (1993).
[5] *WHO-IPCS:* International Programme on Chemical Safety. Environmental Health Criteria 106, Beryllium. World Health Organization, Geneva (1990).
[6] *H. Zorn* and *H. Diem:* Die Bedeutung des Beryllium und seiner Verbindungen für den Arbeitsmediziner. Zbl. Arbeitsmed. *1*, 3–8 (1974).
[7] *G. Petzow* and *H. Zorn:* Zur Toxikologie beryliumhaltiger Stoffe. Chemiker Zeitung *5*, 236–241 (1974).
[8] *Deutsche Forschungsgemeinschaft:* List of MAK and BAT Values 1995. Maximum Concentrations and Biological Tolerance Values at the Workplace. Report No. 31 of the Commission for the Investigation of Health Hazards of Chemical Compounds in the Work Area. VCH Verlagsgesellschaft, Weinheim 1995.
[9] *T. Stiefel, K. Schulze, H. Zorn* and *G. Tölg:* Toxicokinetic and toxicodynamic studies of beryllium. Arch. Toxicol. *45*, 81–92 (1980).
[10] *P. Apostoli, S. Porru, C. Minoia* and *L. Alessio:* Beryllium. In: *L. Alessio, A. Berlin, M. Boni* and *R. Roi* (eds.): Biological indicators for the assessment of human exposure to industrial chemicals. Commission of the European Communities. 1989, pp. 7–21.
[11] *D.S. Grewal* and *F.X. Kearns:* A simple and rapid determination of small amounts of beryllium in urine by flameless atomic absorption. At. Absorpt. Newsl. *16*, 131–132 (1977).
[12] *C. Minoia, L. Pozzoli, A. Cavalleri* and *E. Capodaglio:* Definizione dei valori di riferimento di 30 elementi in tracce nei liquidi biologici. In: Atti 48° Congresso Nazionale della Società Italiana di Medicina del Lavoro e Igiene Industriale, pp. 179–183, Moduzzi Ed. (1985).

[13] *D.C. Paschal* and *G.G. Bailey:* Determination of beryllium in urine with electrothermal atomic absorption using the L'Vov platform and matrix modification. At. Spectros. *7,* 1–3 (1986).

[14] *P. Apostoli, S. Porru* and *L. Allesio:* Behaviour of urinary beryllium in general population and in subjects with low-level occupational exposure. Med. Lav. *80/5,* 390–396 (1989).

[15] *C. Minoia, E. Sabbioni, P. Apostoli, R. Pietra, L. Pozzoli, M. Gallorini, G. Nicolaou, L. Alessio* and *E. Capodaglio:* Trace element reference values in tissues from inhabitants of the European Community. I. A study of 46 elements in urine, blood and serum of Italian subjects. Sci. Total Environ. *95,* 89–105 (1990).

[16] *G. Cammarano, G. Catenacci* and *C. Minoia:* Esposizione a metallin in addetti alla pulitura di caldaie ad olio combustibile in una centrale termoelettrica. In: Atti 48° Congresso Nazionale della Società Italiana di Medicina Del Lavoro e Igiene Industriale, pp. 179–183, Monduzzi Ed. (1985).

[17] *R.P. Lauwerys* and *P. Hoet* (eds.): Industrial Chemical Exposure – Guidelines for Biological Monitoring. Lewis Publication, Boca Raton, pp. 29–31, (1993).

[18] *Bundesministerium für Arbeit und Sozialordnung:* TRgA 410. Statistische Qualitätssicherung. In: Technische Regeln und Richtlinien des BMA zur Verordnung über gefährliche Stoffe. Bek. des BMA vom 9. 4. 1979, Bundesarbeitsblatt 5/1979, p. 88 ff.

[19] *J. Angerer* and *K.H. Schaller:* Erfahrungen mit der statistischen Qualitätskontrolle im arbeitsmedizinisch-toxikologischen Laboratorium. Arbeitsmed. Sozialmed. Präventivmed. *1,* 33–35 (1977).

[20] *F.B. Co* and *D.K. Arai:* Biological monitoring of toxic metals in urine by simultaneous inductively coupled plasma-atomic emission spectrometry. Am. Ind. Hyg. Assoc. J. *50(5),* 245–251 (1989).

[21] *H. Zorn:* Beryllium-Verbundverfahren. In: *J. Angerer* and *K.H. Schaller* (eds.): Analytische Methoden zur Prüfung gesundheitsschädlicher Arbeitsstoffe. Band 2: Analysen in biologischem Material, 5th issue. Deutsche Forschungsgemeinschaft, VCH Verlagsgesellschaft, Weinheim 1981.

Autor: *P. Schramel*
Prüfer: *J. Angerer, K.H. Schaller*

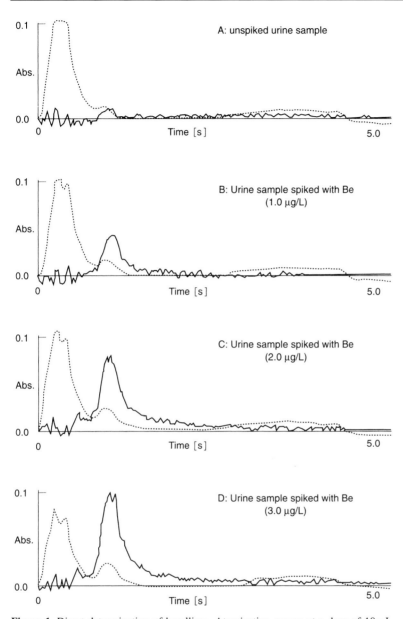

Figure 1. Direct determination of beryllium. Atomization curves at a dose of 10 µL

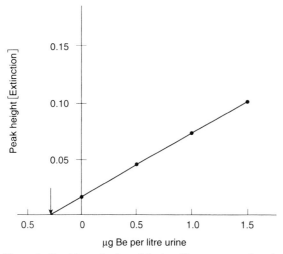

Figure 2. Graphic evaluation of the beryllium content of a urine sample according to the standard addition procedure

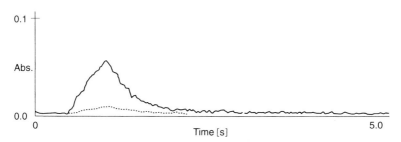

Figure 3. Atomization curve without matrix modification (1.0 µg Be per litre urine)

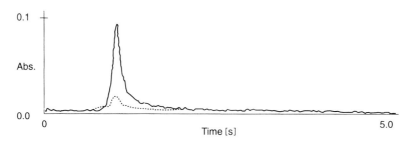

Figure 4. Atomization curve with matrix modification (0.5 µg Be per litre urine)

Butoxyacetic acid

Application   Determination in urine

Analytical principle   Capillary gas chromatography

Completed in   July 1993

Summary

This capillary gas chromatographic method for determination of butoxyacetic acid in urine is a strictly specific test for assessing exposure to butoxyethanol. On account of its sensitivity, the procedure is suitable for the detection of concentrations relevant to occupational medicine as well as those which can occur due to exposure in the so-called „indoor area".

In order to determine the butoxyacetic acid the urine is acidified with hydrochloric acid. The reaction mixture is sucked through a strongly acidic cation exchanger and then through a macroreticular resin, on which the butoxyacetic acid is enriched. After elution butoxyacetic acid is converted to its methyl ester using diazomethane. This derivative is separated by capillary gas chromatography from the other components of urine and detected in a flame ionisation detector. Pentoxyacetic acid serves as an internal standard.

Aqueous standards which are processed like the urine samples and analysed by capillary gas chromatography are used for calibration. The peak areas of the methyl ester of butoxyacetic acid are calculated in relation to the peak areas of the internal standard. Calibration curves are obtained by plotting the resulting quotients as a function of the concentrations of butoxyacetic acid used.

Within-series imprecision: Standard deviation (rel.) $s_w = 1.8\%$
                             Prognostic range $u = 3.9\%$
                             At a concentration of 6.76 mg butoxyacetic acid per litre urine and where $n = 10$ determinations

Between-day imprecision: Standard deviation (rel.) $s = 12.5\%$
                             Prognostic range $u = 28.9\%$
                             At a concentration of 5.26 mg butoxyacetic acid per litre urine where $n = 8$ days

Essential Biomonitoring Methods. DFG, Deutsche Forschungsgemeinschaft
Copyright © 2006 WILEY-VCH Verlag GmbH & Co. KGaA, Weinheim
ISBN: 3-527-31478-4

Inaccuracy:	Recovery rate $r = 90.4-110.9\%$
	At concentrations between 0.22 and 54.72 mg butoxyacetic acid per litre urine and where $n = 10$ determinations
Detection limit:	0.02 mg butoxyacetic per litre urine

Butoxyethanol

Butoxyethanol (MW 118.18 g; b.p. 170.6 °C/743 mm Hg) belongs to the glycol ethers. It is a colourless, oily liquid which can form explosive mixtures with air in its gaseous state [1].

Butoxyethanol is an excellent solvent for substances such as plastics, varnishes and paints. It has the advantage that it is equally miscible with other organic solvents as well as with water. Thus, it is frequently used as a solubilizer. Furthermore, butoxyethanol is less volatile than many other organic solvents and this property is important for many purposes.

As a result of the wide applicability of butoxyethanol, it is frequently used as an organic solvent.

Several review articles give information on the toxicity of butoxyethanol [2, 3].

Butoxyethanol administered to rats caused haemolytic damage, such as swelling and increased osmotic fragility of the erythrocytes. Higher doses lead to haemolysis. The main metabolite – butoxyacetic acid – is primarily responsible for this effect [4–9].

Butoxyethanol can be taken into the human body by means of inhalation [10], after oral ingestion [11, 12] and by absorption of the solvent through the intact skin [13]. Recent investigation results show that butoxyethanol is readily absorbed directly from the vapour phase through the intact skin [14].

In the human organism butoxyethanol is metabolized to butoxyacetic acid by alcohol dehydrogenase which oxidizes the primary hydroxyl groups. This metabolite can be detected in the urine of occupationally exposed persons [15, 16] as well as in those suffering from acute poisoning as a result of a massive intake of butoxyethanol [11]. In contrast, butoxyethanol is not a physiological component of urine.

In cases of acute butoxyethanol poisoning due to attempted suicide, haemoglobinuria, erythropenia as well as diminished haematocrit and haemoglobin values were observed [11, 12].

Under working conditions irritation of the eyes and the nasal epithelium [14, 17] as well as headache were reported [1]. Direct contact with the eyes causes painful irritation and in some cases clouding of the cornea occurs. However, the latter generally clears up after several days [18].

Spermatotoxic and teratogenic effects, such as those caused by the lower homologues of the alkoxyethanols – methoxyethanol and ethoxyethanol – have not yet been reported for butoxyethanol.

Taking into account that the haemolytic effect in humans is weaker than in animal experiments at the same concentrations, the working group for the assignment of MAK values in the Commission for the Investigation of Health Hazards of Chemical Compounds of the Deutsche Forschungsgemeinschaft have defined the MAK value for butoxyethanol as 20 mL/m^3 (100 mg/m^3) [19]. The letter „H" shown for this substance in the MAK value list indicates the risk of cutaneous absorption.

Moreover, butoxyethanol has been assigned to „group C" under the heading „MAK value and pregnancy" [20]. This signifies that there is no reason to fear damage to the developing embryo or foetus when the MAK and BAT values are adhered to.

The results of the urinary analysis of 12 people employed in the production of varnishes showed butoxyacetic acid concentrations between 0.8 and 60.6 mg/L. The mean value was 16.4 mg/L, while the results of the determination of butoxyethanol in air were between 0.1 and 1.0 mg/L, with a mean value of 0.6 mg/L [21]. The discrepancy between the result of ambient monitoring and biological monitoring can be explained by the fact that butoxyethanol is very readily absorbed through the intact skin.

Authors: *J. Angerer, J. Gündel*
Examiners: *G. Kufner, K.H. Schaller*

Butoxyacetic acid

Application   Determination in urine

Analytical principle   Capillary gas chromatography

Completed in   July 1993

Contents

Essential Biomonitoring Methods. DFG, Deutsche Forschungsgemeinschaft
Copyright © 2006 WILEY-VCH Verlag GmbH & Co. KGaA, Weinheim
ISBN: 3-527-31478-4

1 General principles

In order to determine the butoxyacetic acid the urine is acidified with hydrochloric acid. The reaction mixture is first sucked through a strongly acidic cation exchanger and then through a macroreticular resin, on which the butoxyacetic acid is enriched. After elution butoxyacetic acid is converted to its methyl ester derivative using diazomethane. This derivative is separated by capillary gas chromatography from the other components of urine and detected in a flame ionisation detector. Pentoxyacetic acid serves as an internal standard.

Aqueous standards which are processed like the urine samples and analysed by capillary gas chromatography are used for calibration. The peak areas of the methyl ester of butoxyacetic acid are calculated in relation to the peak areas of the internal standard. Calibration curves are obtained by plotting the resulting quotients as a function of the concentrations of butoxyacetic acid used.

2 Equipment, chemicals and solutions

2.1 Equipment

Gas chromatograph with capillary equipment, flame ionisation detector, compensation recorder or integrator

Gas chromatographic column: Length 60 m; inner diameter 0.33 mm; stationary phase silicone oil (dimethylpolysiloxane DB-1), film thickness 0.25 μm (e.g. from J & W Scientific)

Diazomethane apparatus (e.g. from Aldrich)

5 μL syringe for gas chromatography, preferably an autosampler

Soxhlet apparatus

5, 100 and 1000 mL volumetric flasks

5 and 20 mL crimp top vials with PTFE-coated crimp caps as well as crimping tongs

50 mL conical flask with stopper

Work station for vacuum extraction

3 mL cation exchange columns, benzenesulfonic acid (e.g. from Baker)

75 mL cartridges with frits (e.g. from Baker)

3 mL cartridges with frits which can be packed in the laboratory (e.g. from Baker)

Dewar vessel

2.2 Chemicals

As butoxyacetic acid and pentoxyacetic acid are not commercially available, they must be synthesized from sodium, sodium monochloroacetate and n-butyl alcohol and n-propyl alcohol respectively [22].

1-methyl-3-nitro-1-nitrosoguanidine 97% (e.g. from EGA)

Diethyl ether p.a., redistilled over solid KOH

36% hydrochloric acid (Suprapur, e.g. from Merck)

30% sodium hydroxide solution (Suprapur, e.g. from Merck)

Solid sodium hydroxide (Suprapur, e.g. from Merck)

Acetone p.a. (e.g. from Merck)

Methanol p.a. (e.g. from Merck)

Ultrapure water (ASTM type 1) or double distilled water

Anhydrous sodium sulfate p.a. (e.g. from Merck)

XAD-4, purest (e.g. from Serva), particle size 0.2–0.4 mm

Purified nitrogen (99.999%)

Hydrogen (99.90%)

Synthetic air (80% purified nitrogen, 20% oxygen)

Methanol

Dry ice

2.3 Solutions

0.1 M hydrochloric acid:
About 500 mL ultrapure water are placed in a 1000 mL volumetric flask. Then 8.3 mL concentrated hydrochloric acid (36%) are added. After swirling the contents, the flask is filled up to the mark with ultrapure water.

0.1 M sodium hydroxide:
4.0 g sodium hydroxide are weighed, transferred to a 1000 mL volumetric flask and dissolved in ultrapure water. The flask is then filled to the mark with ultrapure water.

Diazomethane solution:
Diazomethane is prepared in a diazomethane apparatus by adding 1 mL 30% sodium hydroxide to 750 mg 1-methyl-3-nitro-1-nitrosoguanidine dissolved in 0.5 mL ultrapure water. The generated diazomethane is collected in 10 mL diethyl ether. During the preparation the diazomethane apparatus is cooled down to between -15 and -20 °C in a Dewar vessel with a methanol/dry ice solution.

Solution of the internal standard:
25 mg pentoxyacetic acid are weighed in a 100 mL volumetric flask. Ultrapure water
is added and swirled around. The flask is filled to the mark with ultrapure water (content: 250 mg/L).

2.4 Calibration standards

Starting solution:
100 mg butoxyacetic acid are weighed in a 100 mL volumetric flask. While shaking occasionally, it is filled to the mark with ultrapure water. (content: 1 g/L). This solution
can be stored for at least six months in the deep freezer.

Stock solution:
10 mL of the starting solution are pipetted into a 100 mL volumetric flask. The flask
is filled to the mark with ultrapure water, while swirling the solution around several
times (content: 100 mg/L).

Calibration standards containing between 0.1 and 50 mg of butoxyacetic acid per litre
(cf. Table 1) are prepared from this stock solution by diluting with ultrapure water.
These solutions must be freshly prepared for every analytical series.

Table 1: Pipetting scheme for the preparation of the calibration standards.

Volume of the stock solution	Final volume of the calibration standard	Concentration of calibration standard
mL	mL	mg/L
0.1	100	0.1
1	100	1
5	100	5
10	100	10
50	100	50

2.5 Preparation of the XAD extraction columns

50 g of the XAD-4 resin are washed with 250 mL acetone in the Soxhlet extractor for
eight hours each time and subsequently dried. This procedure is repeated twice. The
purified resin is filled into brown glass bottles for storage. It can be kept there indefinitely.

The XAD extraction columns are prepared by packing about 1 mL XAD-4 into each of
the empty 3 mL cartridges with frits. They are conditioned by vacuum suction of 10 mL

acetone and three times 10 mL of 0.1 M hydrochloric acid through the columns. After use the XAD extraction columns can be reconditioned in the same way.

2.6 Preparation of the cation exchange columns

The cation exchange columns are conditioned by vacuum suction of 5 mL methanol and 10 mL ultrapure water through the columns. After use the cation exchange columns can be regenerated by sucking 10 mL 0.1 M NaOH, 5 mL methanol and 10 mL ultrapure water through them.

The XAD-4 and the cation exchange columns can be used at least 5 times.

3 Specimen collection and sample preparation

Specimens are collected at the end of a shift after at least three working days. The urine is collected in sealable plastic bottles and stored in the deep freezer until they are processed. Under these conditions the urine can be stored for at least six months.

The urine is thawed in the water bath at about 60 °C. After cooling to room temperature and intensive shaking, 50 mL urine are withdrawn, 0.5 mL of the internal standard solution is added and this mixture is acidified with 30 % hydrochloric acid to a pH of 1–2.

The prepared sample is then sucked through the series of chromatographic columns. The middle column contains the cation exchanger, the lower contains the macroreticular resin XAD-4. The upper column contains a frit which retains the solid products. In addition, the column serves as a solvent reservoir (Fig. 1). After the sample volume has been sucked through, the upper column and the cation exchange column are removed. After use the cation exchange cartridge can be regenerated as described in Section 2.6. The upper column can be reused after careful cleaning.

A work station for vacuum extraction which permits injection of the samples onto the chromatography columns and elution under vacuum conditions is practical for sample preparation by means of ion exchange chromatography and liquid/solid extraction.

The macroreticular resin which absorbs the butoxyacetic acid is washed with 10 mL ultrapure water and then sucked dry in a stream of nitrogen (approx. 30 min). Diethyl ether is used for the elution. The eluate is collected in a 5 mL volumetric flask which is then filled to the mark with diethyl ether. The eluate is dehydrated with anhydrous sodium sulfate. An aliquot of 4 mL is pipetted into a 5 mL crimp top vial and evaporated to dryness by passing a stream of nitrogen over it. The residue is dissolved in 1 mL ethereal diazomethane solution and allowed to react for 30 minutes at room temperature.

4 Operational parameters for gas chromatography

Capillary column:	Material:	Fused silica
	Length:	60 m
	Inner diameter:	0.33 mm
	Stationary phase:	Dimethylpolysiloxane DB-1 Film thickness 0.25 µm
Detector:	Flame ionisation detector	
Temperatures:	Column:	7 min at 40 °C, then increase 5 °C per min until 100 °C; then increase 30 °C per min until 220 °C; 5 min at final temperature
	Injection block:	250 °C
	Detector:	300 °C
Carrier gas:	Purified nitrogen with a column pressure of 1380 hPa	
Split:	30 mL/min	
Sample volume:	1 µL	

The retention times (see Fig. 2) observed under these conditions are:

18.3 min butoxyacetic acid
21.1 min pentoxyacetic acid (internal standard)

5 Analytical determination

For the gas chromatographic analysis 1 µl of each of the prepared samples are injected into the gas chromatograph. If the results lie outside the range of the calibration curve the samples are diluted with ultrapure water and processed anew.

6 Calibration

The aqueous calibration standards (cf. Section 2.4) are processed like the urine samples (Section 3) and analysed by gas chromatography as described in Sections 4 and 5. Calibration curves are obtained by plotting the quotients of the peak areas of butoxyacetic acid and that of the internal standard as a function of the concentrations used. The calibration curves are plotted anew after each analytical series. The calibration curve is linear between 0.1 and 50 mg/L.

7 Calculation of the analytical result

The resulting peak area of butoxyacetic acid is divided by the peak area of the internal standard. The quotient thus obtained is used to read off the relevant concentration of butoxyacetic acid in mg/L urine from the appropriate calibration curve.

8 Standardization and quality control

Quality control of the analytical results is carried out as stipulated in TRg A 410 of the German Arbeitsstoffverordnung (Regulation 410 of the German Code on Hazardous Working Materials) [27] and in the Special Preliminary Remarks. At present no standard material with a specified butoxyacetic acid content is commercially available.

9 Reliability of the method

9.1 Precision

In order to assess the precision in the series the urine of a person who was exposed to butoxyethanol at the workplace was processed and analysed as described in Section 3. A relative standard deviation of 1.8% and a prognostic range of 3.9 were found for n = 10 determinations and a mean butoxyacetic acid concentration of 6.76 mg/L. In addition, mixed urine from persons who were not exposed to butoxyethanol was spiked with different amounts of butoxyacetic acid to determine the precision in the series. Urine samples were obtained containing butoxyacetic acid in concentrations between 0.22 mg/L and 1.36 mg/L (cf. Table 2). Each of the variously spiked urine samples was processed and analysed ten times. The precision in the series lay between 4.2 and 12.6% (cf. Table 2).

To check the between-day precision the urine of a person who was exposed to butoxyethanol at the workplace was processed and analysed on eight different days as described in Section 3. A relative standard deviation of 12.5% and a prognostic range of 28.9 % were found for a mean n-butoxyacetic acid concentration of 5.26 mg per litre urine.

9.2 Accuracy

Recovery experiments were carried out to check the accuracy of the method. A mixture of urine from people who had not been exposed to butoxyethanol was spiked with different amounts of butoxyacetic acid to give samples containing between 0.22 mg/L and 54.72 mg/L of butoxyacetic acid (cf. Table 2). Each of the urine samples was processed and analysed ten times. The mean recovery rates lay between 90.4 and 104.4 % (cf. Table 2).

Table 2: Precision in the series and mean recovery rate for the determination of butoxyacetic acid in urine.

Concentration	No. of determinations	Precision in the series	Mean recovery rate
mg/L	n	%	%
54.72	3	-	104.4
6.76	10	1.8	-
1.36	10	4.2	101.6
0.55	10	12.6	110.9
0.22	10	8.0	90.4

In addition, the losses of butoxyacetic acid and pentoxyacetic acid which occur during sample processing were investigated. For this purpose four spiked urine samples were each processed ten times. The evaluation was carried out using standards which were not processed. Mean losses of butoxyacetic acid of between 10 % and 22.7 % occurred; while no loss of pentoxyacetic acid was discernible.

The author and one of the examiners, who further developed this method for his own purposes, independently analysed seven samples from occupationally exposed persons as part of a series of intercomparison programmes. Their results showed very little deviation from each another.

9.3 Detection limit

Under the conditions described for the sample preparation and the capillary gas chromatographic determination the detection limit was 0.02 mg/L. As no reagent blank values occurred, the detection limit was estimated as three times the background signal.

9.4 Sources of error

Interference caused by the physiological components of urine was investigated by determination of butoxyacetic acid in the urine of a group of 12 people who were not exposed to butoxyethanol at the workplace. No gas chromatographic peak could be found beside or near the characteristic retention time for butoxyacetic acid in any of the investigated urine samples.

10 Discussion of the method

Several methods for the determination of alkoxyacetic acids have been published [10, 23–25]:
Thus, *Jönsson* et al. [23] extracted butoxyacetic acid with ethyl acetate and formed its derivative with N,O-(trimethylsilyl-)trifluoroacetoamide. The drawback of this procedure is that it contains neither an enrichment nor an additional clean-up step. Furthermore, silylation tends to be very susceptible to interference.

Johanson et al. [10] modified a method for the determination of methoxyacetic acid and ethoxyacetic acid from *Smallwood* et al. [24]. In this procedure the urine is extracted three times with methylene chloride. After evaporation of the solvent, the alkoxyacetic acids are converted to their derivatives with pentafluorobenzyl bromide and determined with an FID. The detection limit is 11.4 mg/L for methoxyacetic acid and 5.0 mg/L for ethoxyacetic acid. When the exposure data published for butoxyethanol [25, 26] are considered, the average pulmonary intake as well as the average amount excreted as butoxyacetic acid in the urine, [23], it is assumed that the detection limits achieved by *Johanson* [10] are insufficient to cover the whole concentration range relevant to occupational medicine.
Several investigations have shown that additional clean-up steps are necessary when butoxyacetic acid is converted to its derivative with pentafluorobenzyl bromide before determination with GC/ECD. Otherwise determination by ECD is scarcely possible because of the numerous interfering signals.

Groeseneken et al. [25] first lyophilized the urine samples, then dissolved the residue in methylene chloride and converted it to its derivative with a diazomethane solution. The recovery rate for methoxyacetic acid was 31.4% and 62.5% for ethoxyacetic acid. The high losses during processing lead to great inaccuracy in the method. The detection limits were 0.15 mg/L for methoxyacetic acid and 0.07 mg/L for ethoxyacetic acid. However, these seem rather unrealistic, as no enrichment is carried out during the processing.

The clean-up described in the present method is an improvement over the original instructions [10, 23–25] in that a large proportion of the interfering accompanying substances, such as proteins etc. are retained in the strongly acidic cation exchanger.

The subsequent adsorption of butoxyacetic acid on the macroreticular resin serves to separate further accompanying substances. Thus, this clean-up drastically reduces analytical background interference. The detection limit is lowered, while the specificity is enhanced.

Moreover, the clean-up permits such enrichment of the samples that the concentration range due to environmental exposure can be detected as well as that relevant to occupational medicine. As glycol ether play a significant role in so-called indoor exposure, this aspect is particularly important.

The authors also tested a 60 m capillary column DB-Wax (Carbowax 20 M PEG, chemically cross-linked) as well as the 60 m DB-1 capillary column. It also proved suitable for the separation. However, the DB-1 phase was preferred because its sensitivity to water and oxygen is lower.

Instruments used:
Gas chromatograph 3300 with FID and data system DS 604, Varian

11 References

[1]   *W. Baumann* and *B. Herberg-Liedtke*. In: Druckereichemikalien, Daten und Fakten zum Umweltschutz. Springer Verlag, Berlin 1991, pp. 499–500.
[2]   ECETOC Technical Report No. 17: The toxicology of glycol ethers and its relevance to man: an updating of ECETOC Technical Report No. 4. Brussels 1985.
[3]   HSE (Health and Safety Executive): Glycol ethers. Toxicity review 10 (1985).
[4]   *C.P. Carpenter, U.C. Pozzani, C.S. Weil, J.H. Nair, G.A. Keck* and *H.F. Smyth, Jr.*: The toxicity of butylcellosolve solvent. Arch. Ind. Health *14*, 114–131 (1956).
[5]   *D.E. Dodd, W.M. Smellings, R.R. Maronpot* and *B. Ballantyne*: Ethylene glycol monobutyl ether: acute 9-day and 90-day vapour inhalation studies in Fischer 344 rats. Toxicol. Appl. Pharmacol. *68*, 405–414 (1983).
[6]   *D. Grant, S. Sulsh, H.B. Jones, S.D. Gangolli* and *W.H. Butler*: Acute toxicity and recovery in the hemopoietic system of rats after treatment with ethylene glycol monomethyl and monobutyl ethers. Toxicol. Appl. Pharmacol. *77*, 187–200 (1985).
[7]   *W.J. Krasavage*: Subchronic oral toxicity of ethylene glycol monobutyl ether in male rats. Fundam. Appl. Toxicol *6*, 349–355 (1986).
[8]   *B.I. Ghanayem*: Metabolic and cellular basis of 2-butoxyethanol-induced hemolytic anemia in rats and assessment of human risk in vitro. Biochem. Pharmacol. *38*, 1679–1684 (1989).
[9]   *B.I. Ghanayem, J.M. Sanders, A.-M. Clark, J. Bailer* and *H.B. Matthews*: Effects of dose, age, inhibition of metabolism and elimination on the toxicokinetics of 2-butoxyethanol and its metabolites. J. Pharmacol. Exp. Ther. *253*, 136–143 (1990).
[10] *G. Johanson, H. Kronberg, P.H. Näslund* and *M. Byfält Nordqvist*: Toxicokinetics of inhaled 2-butoxyethanol (ethylene glycol monobutyl ether) in man. Scand. J. Work Environ. Health *12*, 594–602 (1986).

[11] *M.O. Rambourg-Schepens, M. Buffet, R. Bertault, M. Jaussaud, B. Journe, R. Fay* and *D. Lamiable*: Severe ethylene glycol butyl ether poisoning. Kinetics and metabolic pattern. Hum. Toxicol. 7, 187–189 (1988).

[12] *F.P. Gijsenbergh, M. Jenco, H. Veulemans, D. Groeseneken, R. Verberckmoes* and *H.H. Delooz*: Acute butylglycol intoxication: A case report. Hum. Toxicol. 8, 243–245 (1989).

[13] *G. Johanson, A. Boman* and *B. Dynesius*: Percutaneous absorption of 2-butoxyethanol in man. Scand. J. Work Environ. Health *14*, 101–109 (1988).

[14] *G. Johanson, A. Boman* and *B. Dynesius*: Percutaneous absorption of 2-butoxyethanol vapour in human subjects. Br. J. Ind. Med. *48*, 788–792 (1991).

[15] *G. Johanson*: Analysis of ethylene glycol ether metabolites in urine by extractive alkylation and electron-capture gas chromatography. Arch. Toxicol. *63*, 107–111 (1989).

[16] *J. Angerer, E. Lichterbeck, J. Begerow, S. Jekel* and *G. Lehnert*: Occupational chronic exposure to organic solvents. XIII. Glycol ether exposure during the production of varnishes. Int. Arch. Occup. Environ. Health 62, 123–126 (1990).

[17] *E. Browning*: Toxicity and metabolism of industrial solvents. Elsevier Publ. Co, Amsterdam – London – New York 1965, p. 608.

[18] *V.K. Rowe*. In: *F.A. Patty*: Industrial hygiene and toxicology, Vol. II. Interscience Publishers, John Wiley & Sons, New York – London 1962.

[19] *D. Henschler* (ed.): „2-Butoxyethanol". Gesundheitsschädliche Arbeitsstoffe. Toxikologisch-arbeitsmedizinische Begründung von MAK-Werten. Deutsche Forschungsgemeinschaft, VCH Verlagsgesellschaft, Weinheim 1984, 9th issue.

[20] *D. Henschler* (ed.): Gesundheitsschädliche Arbeitsstoffe. Toxikologisch-arbeitsmedizinische Begründung von MAK-Werten. Deutsche Forschungsgemeinschaft, VCH Verlagsgesellschaft, Weinheim 1986, 12th issue.

[21] *B. Söhnlein, S. Letzel, D. Weltle, H.W. Rüdiger* and *J. Angerer*: Occupational chronic exposure to organic solvents. XIV. Examinations concerning the evaluation of a limit value for 2-ethoxyethanol and 2-ethoxyethyl acetate and the genotoxic effects of these glycol ethers. Int. Arch. Occup. Environ. Health *64*, 479–484 (1993).

[22] *G. Rule, W. Hay* and *J. Paul*: Optical activity and the polarity of substituent groups. Part VIII. Growing-chain effects and the ortho-effect in benzoic esters. J. Chem. Soc. 1347–1361 (1928).

[23] *A.K. Jönsson* and *G. Steen*: n-Butoxyacetic acid, a urinary metabolite from inhaled n-butoxyethanol (butylcellosolve). Acta Pharmacol. Toxicol. 42, 354–356 (1978).

[24] *A.W. Smallwood, K.E. DeBord* and *L.K. Lowry*: Analyses of ethylene glycol monoalkyl ethers and their proposed metabolites in blood and urine. Environ. Health Perspect. 57, 249–253 (1984).

[25] *D. Groeseneken, E. van Vlem* and *H. Veulemans*: Gas chromatographic determination of methoxyacetic and ethoxyacetic acid in urine. Br. J. Ind. Med. *43*, 62–65 (1986).

[26] *D.E. Clapp, D.D. Zaebst* and *R.F. Herrick*: Measuring exposure to glycol ethers. Environ. Health Perspect. 57, 91–95 (1984).

[27] TRg A 410: Statistische Qualitätssicherung. In: Technische Regeln und Richtlinien des BMA zur Verordnung über gefährliche Stoffe. Bek. des BMA vom 3.4.1979, Bundesarbeitsblatt 6/1979, pp. 88 ff.

Authors: *J. Angerer, J. Gündel*
Examiners: *G. Kufner, K.H. Schaller*

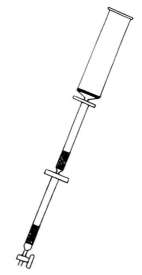

Fig. 1: Sketch of the columns used for solid phase extraction of butoxyacetic acid (above: empty 75 mL cartridge; middle: 3 mL cation exchange column; below: 3 mL cartridge packed with XAD-4).

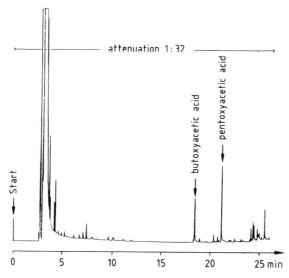

Fig. 2: Gas chromatogram of a urine sample spiked with 1.36 mg butoxyacetic acid and 1.62 mg pentoxyacetic acid per litre urine.

Cobalt

Application	Determination in blood
Analytical principle	Electrothermal atomic absorption spectrometry
Completed in	October 1985

Summary

With this procedure cobalt may be determined reliably in whole blood by means of electrothermal atomic absorption spectrometry. The accuracy of the method is confirmed by comparison with a differential pulse polarography method which is independent at every step. The procedure is suitable for the biological monitoring of persons occupationally exposed to cobalt metal, its alloys or compounds. The detection limit, however, is not low enough to determine concentrations of this essential metal in the physiological range.

The blood samples are diluted eight times and, without further processing, are subjected directly to a charring program with a number of temperature steps in the graphite furnace. The quantitative evaluation is carried out using the standard addition procedure.

Within-series imprecision:	Standard deviation (rel.)	$s_w = 1.1 - 6.8\%$
	Prognostic range	$u = 2.5 - 15.4\%$
	At concentrations ranging from $1.9 - 11.4\ \mu g$ cobalt per litre blood and where $n = 10$ determinations	
Between-day imprecision:	Standard deviation (rel.)	$s = 6.2\%$
	Prognostic range	$u = 14.0\%$
	At a concentration of $11.3\ \mu g$ cobalt per litre blood and where $n = 10$ days	
Imprecision of duplicate analyses:	Standard deviation (rel.)	$s_d = 2.8\%$
	At concentrations ranging from $<1 - 31.1\ \mu g$ cobalt per litre blood and where $n = 12$ duplicate analyses	
Inaccuracy:	Estimated by comparison with an differential pulse polarography method which is independent at every step	
	Recovery rate	$r = 90 - 113\%$
Detection limit:	$1\ \mu g$ Cobalt per litre blood	

Essential Biomonitoring Methods. DFG, Deutsche Forschungsgemeinschaft
Copyright © 2006 WILEY-VCH Verlag GmbH & Co. KGaA, Weinheim
ISBN: 3-527-31478-4

Cobalt

Cobalt (atomic mass 58.93 g/mol, mp 1495 °C, bp 2870 °C) is a steel-grey, magnetic metal of the iron group. It occurs in compounds in oxidation states II and III. The earth's crust contains 0.001 % cobalt. Cobalt occurs naturally mostly in the form of sulfides, arsenides and carbonates. The element is obtained, almost as a by-product, during the smelting of copper ores containing cobalt and nickel.

Cobalt is used in the production of various alloys and special steels which are characterized by temperature stability, corrosion resistance, durability and ferromagnetic properties [1]. A large proportion of the cobalt produced is used in the manufacture of hard metals which consist of metal carbides, usually tungsten carbide, with cobalt as binding agent. These particularly durable hard metals are suitable for applications such as special tools and endoprostheses. Cobalt compounds have a variety of applications e.g. colouring in ceramic products, siccatives in paints and varnishes, as catalysts in organic syntheses, etc.

Although the annual production figures and use of cobalt appear relatively moderate – in the Federal Republic of Germany about 2000 t, in the United States of America 8000 t – the wide variety of applications of the metal are indicative of its wide industrial and consequently environmental distribution. Therefore it may be assumed that many people handle cobalt at work and so their health may be affected by the metal. According to a survey carried out by the Berufsgenossenschaft der chemischen Industrie (Professional Association of the German Chemical Industry) 2300 persons in 70 places of work are exposed to cobalt in the Federal Republic of Germany [2]. Moreover, it is believed that in the United States of America about 1.4 million people are involved in the handling of cobalt, cobalt oxides and cobalt-containing siccatives [3]. The monitoring of the health of persons exposed to cobalt is thus of considerable interest to occupational medicine.

Interest is focussed on the effect of cobalt on the lungs as well as on the question of the mutagenic and carcinogenic effects of this element and its compounds. In addition, it is suspected that cobalt may have been the cause of cardiomyopathy which was observed in individuals after excessive indulgence in beer containing cobalt salts to stabilize the froth. Larger doses of cobalt salts induce an increase in the erythrocyte count as well as the haemoglobin level in blood. These erythropoietic effects have been made use of at times in the treatment of anaemia. The time-dependent and dose-dependent increase in the blood sugar levels which occurred in the course of this treatment is indicative of an effect of the metal on the pancreas. Finally it should be mentioned that repeated administration of cobalt has resulted in hypothyroidism which can be accounted for by the reduced iodine uptake induced by cobalt.

A number of monographs and reviews contain information as to the toxicity of cobalt [3 – 9].

Preventive measures against occupationally induced illness caused by cobalt and its compounds are concerned these days mainly with lung fibrosis, which is observed in particular in connection with the production and processing of hard metals. They are also concerned with the potential mutagenic and carcinogenic effects of cobalt and its compounds, although these effects have yet to be demonstrated. In the Federal Republic of Germany, cobalt and its compounds are included in the MAK Values List (1986) as substances which have proved to be unmistakably carcinogenic in animal experimentation (Section III A 2). The Technische Richtkonzentrationen (Technical Guiding Concentrations) established by the Ausschuß für Gefahrstoffe beim Bundesministerium für Arbeit und Sozialordnung (Committee for Hazardous Working Materials in the Federal Ministry of Labor and Social Affairs) are 0.5 mg/m³ for the production of cobalt powder and catalysts and 0.1 mg/m³ for other installations.

Cobalt is an essential trace element. It is a constituent of vitamin B_{12} which is involved in the formation of the red pigment in blood. A person must take in about 3 μg vitamin B_{12} each day to avoid the deficiency symptoms which lead to so-called pernicious anaemia. This requirement for vitamin B_{12} is covered by the amount of cobalt ingested daily with food and should be between 40 and 50 μg [10]. The majority of the cobalt taken in with food, however, is excreted unchanged in the urine.

Toxic symtoms are induced by cobalt only on ingestion of about 100 times the amount normally eaten in food. The effects on thyroid and blood count described above are registered after doses above 30 mg per day.

According to the most recent atomic absorption spectrometric analyses, the cobalt concentrations in the blood and urine of normal individuals are lower than previously assumed [11 – 14]. 95 % of all urine levels measured in a group of 79 persons who were not involved with cobalt at work were below 0.86 μg/L [15]. Analytical procedural difficulties make it impossible at present to quote a normal cobalt level for blood. The majority of the values cited in the literature are in the range between 0.1 and 146 μg cobalt per litre blood [11] and are considered to be too high.

Occupational exposure to cobalt leads to an increase of the cobalt concentrations in blood and urine which are proportionally related to each other. According to the literature the urine cobalt level seems to be between 7.5 and 10 times that in blood [16 – 18]. Both matrices are therefore in principle suitable for occupational medical screening of persons exposed to cobalt at work. It is important to note that whole blood is used for determining the cobalt level because the proportion of the total cobalt present in the serum is interindividually variable [19].

As our picture of the relationship between the internal cobalt levels and their effects on health is still far from complete, it is at present not possible to state a safe level for cobalt in blood and urine. Were it to be demonstrated that cobalt is carcinogenic for man as well as animals, then such levels could not be determined on principle. Nevertheless, it is possible to estimate from the relationship between the work place levels and those in biological material what level of internal cobalt stress arises under the conditions given by the Technical Guiding Concentrations (Technische Richtkonzentrationen). Such biological correlations have been evaluated by the Senatskommission zur Prüfung gesundheitsschädlicher Arbeitsstoffe (Commission for the Investigation of Health Hazards of Chemical Compounds in the Work Area) [20]. Thus a cobalt concentration of 100 or 500 μg/m³ in the air is equivalent to a blood cobalt level of 5 or 25 μg/L, respectively. Under these exposure conditions urine cobalt concentrations of 60 or 300 μg/L, respectively, are to be expected.

Author: *J. Angerer*
Examiners: *M. Fleischer, K. H. Schaller*

Cobalt

Application   Determination in blood

Analytical principle Electrothermal atomic absorption spectrometry

Completed in   October 1985

Contents

1 General Principles

Cobalt is determined directly in whole blood without any further sample treatment by means of electrothermal atomic absorption spectrometry. Interference from the biological matrix is eliminated by diluting the blood eight times and by using an optimized multistage charring program. The quantitative evaluation is carried out using the standard addition procedure.

Essential Biomonitoring Methods. DFG, Deutsche Forschungsgemeinschaft
Copyright © 2006 WILEY-VCH Verlag GmbH & Co. KGaA, Weinheim
ISBN: 3-527-31478-4

2 Equipment, chemicals and solutions

2.1 Equipment

Atomic absorption spectrometer with background correction at 240.7 nm
Graphite furnace
Chart recorder
Monoelement cobalt hollow cathode lamp
Graphite tube, pyrolytically coated
Roller mixer (e.g. from Denley)
Vortex mixer (e.g. from Cenco)
Disposable polyethylene tubes with stoppers (approx. 12 ml)
50, 100, 500 and 1000 mL Volumetric flasks
Automatic pipettes, adjustable between 20 – 200 and 200 – 1000 μL (e.g. Pipetman from
Gilson, obtainable from Abimed)
Disposable syringes containing anticoagulant (e.g. K-EDTA Monovetten®, Sarstedt,
Nümbrecht, FRG)

2.2 Chemicals

Cobalt standard (e.g. Fixanal from Riedel-de Haën) containing 0.1 g cobalt as cobalt
chloride
65 % Nitric acid (e.g. Suprapur from Merck)
Triton X-100 (e.g. from Merck)
1-Octanol
Ultrapure water (ASTM type 1) or double-distilled water
Argon (99.998 %)

2.3 Solutions

Aqueous Triton X-100 solution (approx. 0.01 %):
About 50 mL ultrapure water is pipetted into a 100 ml volumetric flask. 0.01 mL Triton
X-100, warmed to about 40 °C, is then added. After thorough mixing the flask is filled to
the mark with ultrapure water.

0.01 M Nitric acid:
0.36 mL 65 % nitric acid is pipetted into about 200 mL ultrapure water in a 500 mL
volumetric flask. After thorough mixing, the flask is filled to the mark with ultrapure
water.

1 M Nitric acid (for cleaning glassware and tubes):
72 mL 65 % nitric acid is pipetted into a 1000 mL volumetric flask containing about 400
mL ultrapure water. After thorough mixing, the flask is filled to the mark with ultrapure
water.

2.4 Calibration standards

Stock solution:
The cobalt standard containing 0.1 g Co is diluted to the mark with ultrapure water in a 1000 mL volumetric flask (0.1 g/L).

The calibration standards are prepared by diluting the stock solution with 0.01 M nitric acid. They must be freshly prepared for each analytical series.

Volume of stock solution µL	Final volume of calibration standard mL	Concentration of calibration standard µg/L	Designation of calibration standard
25	50	50	I
62.5	50	125	II
100	50	200	III

3 Specimen collection and sample preparation

The blood specimens are drawn from the arm vein using disposable syringes and transferred to disposable plastic tubes which contain an anticoagulant (e.g. potassium EDTA). Alternatively, commercially available syringes containing anticoagulant may be used. The specimens are mixed thoroughly to prevent clotting.

In this form the specimens may be dispatched and if necessary stored for up to seven days in the refrigerator. For longer storage the specimens should be deep frozen.

Before processing the specimens are allowed to come to room temperature, preferably by taking them out of the refrigerator on the evening previous to analysis. They are then homogenized for at least 30 min on a roller mixer and the aliquots for analysis taken immediately.

Disposable polyethylene tubes are used for sample processing after they have been rinsed first with 1 M nitric acid, then three times with ultrapure water and dried at room temperature. 850 µL of the 0.01 % Triton X-100 solution is pipetted into each of four such prepared tubes and mixed with 40 µL 1-octanol. To each tube 125 µL of the homogenized blood specimen is added using an automatic pipette and allowing the blood to run down the wall of the tube immediately above the surface of the liquid. Then the pipette tip is rinsed by drawing up liquid from the tube. The samples are mixed for 20 s on the vortex mixer. 25 µL 0.01 M nitric acid with or without the appropriate amount of cobalt is added to each tube as shown in the following table:

Sample designation	0.01 % Triton soln.	1-Octa-nol	Blood	0.01 M Nitric acid	Calibration standards 50 μg/L	125 μg/L	200 μg/L	Spiked cobalt concentration expressed in terms of blood volumne used
					I	II	III	
	μL	μL	μL	μL	μL	μL	μL	μg/L
Sample without standard	850	40	125	25	-	-	-	-
+ standard I	850	40	125	-	25	-	-	10
+ standard II	850	40	125	-	-	25	-	25
+ standard III	850	40	125	-	-	-	25	40

After adding the calibration standards the samples are mixed again for 20 s on the vortex mixer.

At least one sample blank is assayed with each set of samples. For this purpose ultrapure water is used instead of blood.

4 Operational parameters for atomic absorption spectrometry

Atomic absorption spectrometer:
Wavelength: 240.7 nm
Background correction: Deuterium lamp
Spectral slit width: 0.2 nm
Lamp current: According to manufacturer's instructions
Analytical determination: Maximum extinction recorded during the atomization step

The temperature program shown in the following table is only intended as a guide. The optimization of the program must be carried out for each individual instrument.

Temperature-time-program:

Analytical step	Step duration Ramp time	Hold time	Temperature
	s	s	°C
Drying	10	25	100
Charring I	15	15	350
Charring II	20	15	500
Charring III	5	15	1000
Atomization	0	10	2650, 30 mL/min
Heating	1	2	2700

Inert gas: Argon
Injected volume: 25 μL

5 Analytical determination

25 μL of each sample solution is injected into the graphite tube and the extinction of the atomization peak is recorded. The value for the unspiked sample should not be more than 0.1 (equivalent to about 30 μg cobalt per litre blood). If the concentration is more than 20 μg/L (with an extinction of about 0.065 for the unspiked sample) a smaller volume of blood (e.g. 50 μL) must be used for the analysis.

6 Calibration and calculation of the analytical result

The cobalt concentration of the blood sample is determined graphically. The extinction of the reagent blank is subtracted from the extinction values of the unspiked sample as well as from the three spiked samples which are then plotted as a function of the added cobalt concentration in blood. The intercept of the resulting straight line with the concentration axis gives the cobalt concentration in μg/L (see Fig. 1). Any dilution of the blood sample before treatment must be allowed for arithmetically.
The calibration curve was shown to be linear for up to about a concentration of 90 μg/L.

7 Standardization and quality control

Quality control of the analytical results is carried out as stipulated by Article TRgA 410 (Regulation 410 of the German Code on Hazardous Working Materials) [21]. Until standards become commercially available, they must be prepared in the laboratory.

8 Reliability of the method

8.1 Precision

For the determination of the within-series imprecision, pooled blood was spiked with three different cobalt concentrations in the range from 2 – 10 μg/L. Each sample was analysed ten times (see Tab. 1). The relative standard deviation varied from 1.1 to 6.8 % and the corresponding prognostic ranges from $u = 2.5$ to 15.4 %.
The between-day imprecision was determined using whole blood to which 10 μg cobalt per litre had been added. The results of analyses on ten different days yielded a standard deviation of $s = 6.2$ % (see Tab. 1).
In addition the precision of the procedure was determined using blood samples from twelve persons occupationally exposed to cobalt. The cobalt levels ranged from < 1 to 31.1 μg/L, on average 10.8 μg/L. The samples were all analysed in duplicate. A stan-

dard deviation may be determined from the differences between duplicate analyses [22]. In our case a value of 0.3 μg/L was obtained. If this is expressed in terms of the average concentration of the samples it is equivalent to a variation of 2.8 %.

8.2 Accuracy

Using blood samples from persons occupationally exposed to cobalt the accuracy of the procedure described here was tested by comparison with a second method. This procedure, using a differential pulse polarography determination after ashing the blood samples, was independent of our procedure at every step. As shown in Fig. 2, the results of the two procedures are highly correlated ($r = 0.986$). The slope of 0.95 indicates that the two methods produce results which may be taken as identical within the respective error limits.

Recovery experiments were also carried out. This procedure is the same as for calibration (standard addition) and serves only as an internal laboratory check. The samples are prepared as described in Section 8.1. Since the cobalt concentration in the pooled blood used was below the detection limit, the recovery could not be determined accurately. There are, however, various indications that the cobalt concentration in the blood of normal individuals is very probably less than 0.1 μg/L. Therefore this value was used in the calculation of the recovery rates, which were found to range from $r = 90$ to 113 %. The individual results are given in Tab. 1.

8.3 Detection limit

With the method described here, blood concentrations above 1 μg/L may be determined with adequate precision.

8.4 Sources of error

The method described here is relatively resistant to interference. If cleaning and rinsing of tubes is carried out as described, the risk of contamination is small compared to that in assays of other metals (e.g. nickel).

9 Discussion of the method

With the method described here [24], cobalt may be determined directly in whole blood by means of electrothermal atomic absorption spectrometry.

The cobalt level may be determined without interference because of the considerable dilution of the biological matrix and the three temperature steps in the charring program. The dilution step results in a drastic reduction in interference from the matrix. By using a multistage temperature-time program it is possible to control charring and eva-poration of the organic components so that losses of cobalt do not occur. This prevents uncontrolled smoke formation which, in our experience, can carry part of the cobalt mechanically out of the sample and so out of the assay system.

The final temperature during the charring program, 1000 °C, is high enough to ash or evaporate organic components which could interfere with the atomic absorption spectro-metric determination without danger of losses of cobalt.

Gas stop conditions during the atomization step result in a slight increase in sensitivity but, at the same time, in a markedly higher background. For this reason an internal gas flow of 30 mL/min is used.

The results of the precision and recovery determinations demonstrate that the method yields analytically valid results for cobalt concentrations which are higher than the nor-mal levels. The between-day imprecision of about 6 % fulfills the requirements of sta-tistical quality control. The good reliability of the method may be seen also in the results of the duplicate analyses of blood samples from occupationally exposed persons. It may thus be concluded that interindividual differences in blood composition appear to have no disadvantageous effects on the reliability of the procedure.

The accuracy of the method was checked and confirmed with an independent diffe-rential pulse polarography method [23]. Since the latter determination is preceded by a total ashing of the blood, these results demonstrate additionally that the direct method presented here determines all the cobalt in the blood samples independent of its chemical form.

Under the conditions described here, cobalt concentrations of about 1 μg/L lead to a measured extinction of about 0.004, which is normally given as the detection limit of electrothermal atomic absorption spectrometric methods.

The results demonstrate that the method introduced here is superior in terms of sensiti-vity and analytical reliability to older methods of blood plasma analysis [25, 26]. In a procedure published by *Delves* et al. [27], whole blood is diluted in the ratio 1:2 with a mixture of hydrochloric acid and ammonium dihydrogen phosphate and then analysed by means of ETAAS. Residual matrix components are diminished by flushing with oxygen gas during the charring step in the graphite tube. Usually this results in a drastic-ally shortened life span for the graphite tube and a rapid reduction in sensitivity of the cobalt assay.

Deproteinization procedures, such as have proved of value for the determination of, e.g., cadmium in blood [28], are not successful in the case of cobalt. With a method des-cribed by *Christensen* et al. [14], in which whole blood is deproteinized with nitric acid and the mixture then warmed, only an average of 30 % of the actual cobalt content is determined [29]. This may be accounted for by the fact that a fraction of the cobalt is bound to plasma proteins and a further part to haemoglobin [19].

To summarize, the cobalt content of blood may be determined reliably with the method described here. This applies for cobalt concentrations which are increased above the physiological range e.g. as a result of occupational exposure. The physiological concen-tration range, which, according to our results, lies under the detection limit obtained here (1 μg/L), cannot be determined with this method.

Instruments used:
Atomic absorption spectrometer 4000 with graphite furnace HGA 500 from Perkin Elmer

10 References

[1] *W. Betteridge:* Cobalt and its alloys. John Wiley & Sons, New York 1982.

[2] *Berufsgenossenschaft der chemischen Industrie:* Ergebnisse der Erfassung von gefährlichen Arbeitsstoffen. Sichere Chemiearbeit 8/1982, p. 60 – 62.

[3] *NIOSH:* Occupational hazard assessment/Criteria for controlling occupational exposure to cobalt. DHHS Publication No. 82 – 107, Cincinnati 1981.

[4] *C. G. Elinder and L. Friberg:* Cobalt. In: *L. Friberg, G. F. Nordberg,* and *B. Vouk* (eds.): Handbook on the toxicology of metals. Elsevier, Amsterdam 1980, p. 399 – 410.

[5] *M. D. Kipling:* Cobalt. In: *H. A. Waldron* (ed.): Metals in the environment. Academic Press 1980, p. 133 – 153.

[6] *G. N. Schrauzer:* Cobalt. In: *E. Merian* (ed.): Metalle in der Umwelt. Verlag Chemie, Weinheim 1984, p. 425 – 433.

[7] *I. C. Smith* and *B. L. Carson*: Trace metals in the environment. Vol. 6, Cobalt. Ann Arbor Science, Ann Arbor 1981.

[8] *H. E. Stokinger:* Cobalt. In: *G. D. Clayton* and *F. E. Clayton* (eds.): Patty's industrial hygiene and toxicology. Vol. II A, John Wiley & Sons, New York 1981, p. 1605 – 1619.

[9] *J. Angerer* and *R. Heinrich:* Cobalt. In: *H. G. Seiler* and *H. Sigel:* Handbook on the toxicity of inorganic compounds. Marcel Dekker, New York, in press.

[10] *R. Schelenz:* Dietary intake of 25 elements by man estimated by neutron activation analysis. J. Radioanal. Chem. *37,* 539 – 548 (1977).

[11] *J. Versieck* and *R. Cornelis:* Normal levels of trace elements in human blood plasma or serum. Anal. Chim. Acta *116,* 217 – 254 (1980).

[12] *V. V. Lidums:* Determination of cobalt in blood and urine by electrothermal atomic absorption spectrometry. At. Absorp. Newslett. *18,* 71 – 72 (1979).

[13] *M. Hartung, K. H. Schaller, M. Kentner, D. Weltle,* and *H. Valentin:* Untersuchungen zur Cobalt-Belastung in verschiedenen Gewerbezweigen. Arbeitsmed. Sozialmed. Präventivmed. *18,* 73 – 75 (1983).

[14] *J. M. Christensen, S. Mikkelsen,* and *A. Skov:* A direct determination of cobalt in blood and urine by Zeeman atomic absorption spectrophotometry. In: Chemical toxicology and clinical chemistry of metals. IUPAC 1983.

[15] *R. Heinrich* and *J. Angerer:* Cobalt und Nickel in biologischem Material – Analysenverfahren für physiologische und expositionsbedingte Konzentrationen. In: *B. Welz* (ed.): Fortschritte in der atomabsorptionsspektrometrischen Spurenanalytik. Vol. 2, VCH Verlagsgesellschaft, Weinheim 1986, p. 159 – 170.

[16] *J. Angerer, R. Heinrich, F. X. Felixberger, D. Szadkowski,* and *G. Lehnert:* Cobalt-Aufnahme und -Ausscheidung bei beruflicher Exposition. In: *H. Konietzko* and *F. Schuckmann:* Bericht über die 24. Jahrestagung der Deutschen Gesellschaft für Arbeitsmedizin e. V., Mainz 2. – 5. Mai 1984, Gentner Verlag, Stuttgart 1984, p. 293 – 296.

[17] *J. Angerer, R. Heinrich, D. Szadkowski,* and *G. Lehnert:* Occupational exposure to cobalt powder and salts – biological monitoring and health effects. In: *T. D. Lekkas* (ed.): Heavy Metals in the Environment. Proceedings Int. Conf. Athens, September 1985, Vol. 2, p. 11 – 13.

[18] *Y. Ichikawa, Y. Kusaka,* and *S. Goto:* Biological monitoring of cobalt exposure, based on cobalt concentrations in blood and urine. Int. Arch. Occup. Environ. Health *55*, 269 – 276 (1985).

[19] *E. Schumacher-Wittkopf:* Characterization of cobalt-binding proteins in occupational cobalt exposure. Toxicol. Environ. Chem. *8*, 1 – 9 (1984).

[20] *J.Angerer:* Cobalt. In: *D. Henschler* and *G. Lehnert* (eds.): Biologische Arbeitsstoff-Tole-ranz-Werte (BAT-Werte). Arbeitsmedizinisch-toxikologische Begründungen. Deutsche For-schungsgemeinschaft, Verlag Chemie, Weinheim, 3rd issue 1986.

[21] TRgA 410: Statistische Qualitätssicherung. Ausgabe April 1979. In: *W. Weinman* and *H.-P. Thomas:* Gefahrstoffverordnung Teil 2: Technische Regeln (TRGS) und ergänzende Bestim-mungen zur Verordnung über gefährliche Stoffe. Carl Heymanns Verlag K. G., Köln, 10th issue, 1984.

[22] see p. 8, Vol. 1 of this series, 1985.

[23] *R. Heinrich* and *J. Angerer:* Determination of cobalt in biological materials by voltammetry and electrothermal atomic absorption spectrometry. Int. J. Environ. Anal. Chem. *16*, 305 – 314 (1984).

[24] *J. Angerer* and *R. Heinrich:* Bestimmung von Cobalt in Blut mit Hilfe der elektrothermalen Atomabsorptionsspektrometrie. Fresenius' Z. Anal. Chem. *318*, 37 – 40 (1984).

[25] *F. J. M. J. Maessen, F. D. Posma,* and *J. Balke:* Direct determination of gold, cobalt and lithium in blood plasma using the mini-Massmann carbon rod atomizer. Anal. Chem. *46*, 1445 – 1449 (1974).

[26] *R. A. A. Muzzarelli* and *R. Rocchetti:* Atomic-absorption determination of manganese, cobalt and copper in whole blood and serum with a graphite atomizer. Talanta *22*, 683 – 685 (1975).

[27] *H. T. Delves, R. Mensikov,* and *L. Hinks:* Direct determination of cobalt in whole-blood by electrothermal atomization and atomic absorption spectroscopy. In: *P. Brätter* and *P. Schramel* (eds.): Trace Element – Analytical Chemistry in Medicine and Biology. Vol. 2, Walter de Gruyter & Co., Berlin – New York 1983, p. 1123 – 1127.

[28] *M. Stoeppler, J. Angerer, M. Fleischer,* and *K. H. Schaller:* Cadmium. Vol. 1 of this series, 1985 p. 79 – 92.

[29] *R. Heinrich* and *J. Angerer:* ETAAS-determination of cobalt in whole blood – comparison of direct and deproteinization procedures. Fresenius' Z. Anal. Chem. *322*, 772 – 774 (1985).

Author: *J. Angerer*
Examiners: *M. Fleischer, K. H. Schaller*

Tab. 1: Imprecision and recovery rates determined using human blood samples spiked with cobalt.

Imprecision	n	Expected value (Reagent blank + added conc.) μg/L	Measured value $\bar{x}$ μg/L	s %	u %	Recovery rate %
Series	10	0.1* + 2.0	1.9	6.8	15.4	90
Series	10	0.1* + 5.0	5.7	4.2	9.5	112
Series	10	0.1* + 10.0	11.4	1.1	2.5	113
Day to day	10	0.1* + 10.0	11.3	6.2	14.0	112

* The cobalt concentration of the unspiked blood sample was set at 0.1 μg/L for the calculation of the recovery rates.

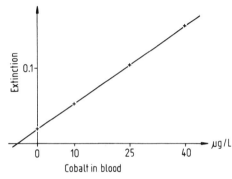

Fig. 1: Example of a linear standard addition graph for the atomic absorption spectrometric determination of cobalt in blood using a pyrolytically coated graphite furnace.

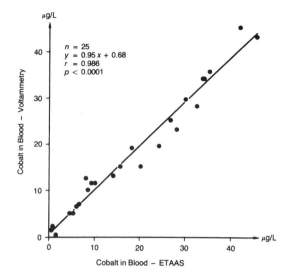

Fig. 2: Correlation curve for the parallel analysis of 25 blood samples from persons exposed to cobalt using the direct method described here and a differential pulse polarography procedure [23] which was independent at every step.

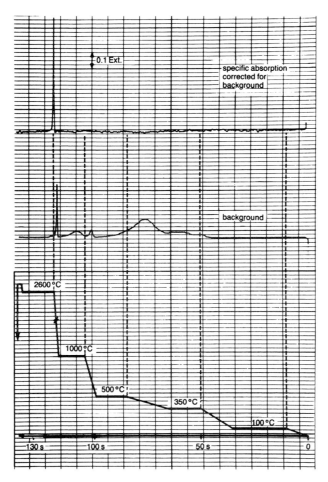

Fig. 3: Time course of the background signal and the corrected specific absorption during the direct determination of cobalt in whole blood.

Cotinine

Application   Determination in urine to ascertain passive exposure to smoking

Analytical principle Capillary gas chromatography/
                         mass spectrometric detection (MS)

Completed in   May 2001

Summary

The method described here serves to quantify cotinine, one of the main metabolites of nicotine excreted in urine. Sensitive determination of cotinine in urine enables differentiation between non-smokers, passive smokers and smokers with a low tobacco consumption.

Deuterated cotinine (cotinine-d_3) is added to a urine sample as an internal standard. Sodium hydroxide is also added, and then extraction is carried out using dichloromethane. The extract is dried under a stream of nitrogen, dissolved in toluene and after capillary gas chromatographic separation, the analyte is quantified by means of mass selective detection in the selective ion monitoring (SIM) mode. Calibration is performed using calibration standards which are prepared in pooled urine and are treated in the same manner as the samples to be analysed.

Cotinine

Within-series imprecision: Standard deviation (rel.) s_w = 6.1% or 1.3%
                           Prognostic range u = 12.9% or 2.7%
                           at a concentration of 10 µg or 100 µg cotinine per litre urine
                           and where n = 17 determinations

Between-day imprecision: Standard deviation (rel.) s_w = 6.4% or 2.8%
                         Prognostic range u = 13.6% or 5.9%
                         at a concentration of 10 µg or 100 µg cotinine per litre urine
                         and where n = 15 determinations

Accuracy: Recovery rate r = 95%

Detection limit: 1 µg cotinine per litre urine

Essential Biomonitoring Methods. DFG, Deutsche Forschungsgemeinschaft
Copyright © 2006 WILEY-VCH Verlag GmbH & Co. KGaA, Weinheim
ISBN: 3-527-31478-4

Cotinine

Cotinine is one of the main metabolites of nicotine in the mammalian organism. Determination of cotinine is preferable to determination of nicotine to evaluate a person's smoking status. In contrast to the mother substance, nicotine, exogenous contamination can be virtually excluded in the case of the cotinine metabolite. The cotinine concentration limit between non-smokers who may be passively exposed to smoking and (occasional) smokers is given as approximately 100 µg/L in urine [1] (see also Table 1). It should be noted that cotinine determination enables monitoring of only a relatively short period (3–4 days) of previous exposure to tobacco smoke. Measurement of cotinine does not detect smoking that took place before this exposure period.

Table 1. Mean value ranges for cotinine concentrations in the urine of non-smokers, passive smokers and smokers [4, 5]

	Non-smokers	Passive smokers	Smokers	Cut off*
Cotinine in urine [µg/L]	1–10	8–25	1300–1700	60–120

* The cut-off range serves to differentiate between smokers and non-smokers

General characteristics and fundamentals on the toxicology of nicotine and cotinine have already been discussed in this series. Therefore the reader should refer to the 14th issue of "Analysen in biologischem Material" or the 7th volume of "Analysis of Hazardous Substances in Biological Materials" [2, 3].

Author: *M. Müller*
Examiners: *R. Heinrich-Ramm, H.-W. Hoppe*

Cotinine

Application   Determination in urine

Analytical principle Capillary gas chromatography/
                         mass spectrometric detection (MS)

Completed in   May 2001

Contents

1 General principles

5 M sodium hydroxide and then deuterated cotinine (cotinine-d_3) as an internal standard are added to 2 mL of urine, and extraction is carried out using dichloromethane.
The extract is dried under a stream of nitrogen, dissolved in toluene and after capillary gas chromatographic separation, the analyte is quantified by means of mass se-

Essential Biomonitoring Methods. DFG, Deutsche Forschungsgemeinschaft
Copyright © 2006 WILEY-VCH Verlag GmbH & Co. KGaA, Weinheim
ISBN: 3-527-31478-4

lective detection in the selective ion monitoring (SIM) mode. Calibration is performed using calibration standards which are prepared in pooled urine and are treated in the same manner as the samples to be analysed.

2 Equipment, chemicals and solutions

2.1 Equipment

Capillary gas chromatograph with split/splitless injector, mass selective detector (MSD) and data processing system

Capillary gas chromatographic column:
Length: 30 m, inner diameter: 0.25 mm; stationary phase: 5% phenylmethylpolysiloxane; film thickness: 0.25 µm (e.g. from Hewlett-Packard)

10 µL Syringe for gas chromatography, but the use of an autosampler is preferable

Glass centrifuge tubes 12 mL (e.g. from Schott) with polyethylene stoppers

Test-tube shaker (Vortex mixer)

Microlitre pipettes, adjustable between 10 and 100 µL, and between 100 and 1000 µL (e.g. from Eppendorf)

Finn pipette 1–5 mL

Laboratory centrifuge

2 mL Autosampler vials

10, 100 and 1000 mL Volumetric flasks

250 mL Glass beaker

Polyethylene sample vessels (2 mL) (e.g. from Eppendorf)

Sterile polyethylene vessels for collection of urine

Polyethylene tubes (10 mL) for portioning

Magnetic stirrer

2.2 Chemicals

Cotinine 98% p.a. (e.g. from Aldrich)

Cotinine-d_3 98 atom % D (e.g. from Sigma)

Dichloromethane p.a. (e.g. from Baker)

Toluene p.a. (e.g. from Fluka)

Sodium sulphate anhydrous p.a. (e.g. from Fluka)

Sodium hydroxide p.a. (e.g. from Fluka)

Concentrated hydrochloric acid p.a. (e.g. from Merck)

Bidistilled water

Helium 4.6 (e.g. from Linde)

Nitrogen 4.0 (e.g. from Linde)

2.3 Solutions

5 M NaOH:
Approximately 150 mL bidistilled water is placed in a 250 mL glass beaker. After
the addition of 40 g sodium hydroxide, the solution is mixed with a magnetic stirrer
until it is clear, and then the solution is filled to a total volume of 200 mL with bidistilled water.

0.1 M Hydrochloric acid:
10 mL of 37% HCl are placed into a 1000 mL volumetric flask and the flask is filled
to its nominal volume with bidistilled water.

Solution of the internal standard

Stock solution:
Approximately 10 mg cotinine-d_3 are weighed exactly into a 100 mL volumetric
flask. The flask is subsequently filled to its nominal volume with 0.1 M hydrochloric
acid (100 mg/L).

Working solution:
100 µL of the stock solution of the internal standard are pipetted into a 10 mL volumetric flask. The flask is subsequently filled to its nominal volume with 0.1 M hydrochloric acid (1 mg/L).

2.4 Calibration standards

The calibration standard solutions are prepared in pooled urine from non-smoking test
persons. For the purpose of preparing pooled urine spontaneous urine samples are collected from the test persons in a suitable vessel, thoroughly mixed and stored at –18 °C
until the standards and the control material are prepared. If necessary, the cotinine content of the individual urine samples used to prepare the pooled urine is checked.
The working solution and the calibration standards must be freshly prepared before
each analytical series.

Stock solution:
Approximately 10 mg cotinine are weighed exactly into a 100 mL volumetric flask. The flask is subsequently filled to its nominal volume with 0.1 M hydrochloric acid (100 mg/L).

Working solution:
100 µL of the cotinine stock solution are pipetted into a 10 mL volumetric flask. The flask is subsequently filled to its nominal volume with 0.1 M hydrochloric acid (1 mg/L).

Calibration standards:
Calibration standards in pooled urine are prepared from the cotinine working solution and the working solution of the internal standard in accordance with the following pipetting scheme (Table 2).

Table 2. Pipetting scheme for the preparation of the calibration standards

Volume of pooled urine [mL]	Volume of the working solution of the internal standard [µL]	Volume of the cotinine working solution [µL]	Volume of 0.1 M hydrochloric acid [µL]	Concentration of the calibration standard [µg/L]
2	100	–	1400	0
2	100	10	1390	5
2	100	50	1350	25
2	100	100	1300	50
2	100	200	1200	100
2	100	500	900	250

3 Specimen collection and sample preparation

Spontaneous urine samples are collected in sealable polyethylene bottles and stored in the deep-freezer at approx. –18 °C until sample processing for cotinine determination is carried out. The urine can be stored for at least six months under these conditions.

3.1 Sample preparation

Before analysis, the samples are thawed and thoroughly mixed. 100 µL of the working solution of the internal standard and 1400 µL 0.1 M hydrochloric acid are added to 2 mL urine in a glass centrifuge tube and shaken on a Vortex mixer. 2 mL of 5 M sodium hydroxide are added using a pipette and the sample is mixed again (Vortex). Then 5 mL dichloromethane are added to the sample, and the analyte is extracted by shaking for 30 seconds on the Vortex mixer. The phases of the sample contained in the tube sealed with a polyethylene stopper are separated by centrifugation at room temperature for 5 minutes at 1400 g. After carefully withdrawing the upper aqueous phase using a Finn pipette, this phase is discarded. The organic phase is dried by adding 3 g of anhydrous sodium sulphate and by shaking the sample briefly on the Vortex mixer. The sealed sample is subsequently centrifuged anew at room temperature for 5 minutes at 1400 g. The supernatant is transferred to a fresh centrifuge tube and the contents are evaporated carefully to dryness in a stream of nitrogen. Finally the residue is dissolved in 1 mL toluene and transferred into a 2 mL autosampler vial.

4 Operational parameters

4.1 Operational parameters for gas chromatography and mass spectrometry

Capillary column:	Material:	Fused silica
	Stationary phase:	DB-5
	Length:	30 m
	Inner diameter:	0.25 mm
	Film thickness:	0.25 μm

Detector:	Mass selective detector (MSD)	
Temperatures:	Column:	Initial temperature 95 °C for 3 minutes isothermal, then increase at a rate of 5 °C/min to 110 °C, then at a rate of 30 °C/min to 280 °C, then 5 min at the final temperature
	Injector:	250 °C
	Transfer line:	250 °C

Carrier gas: Helium 4.6 with a constant flow of 1.2 mL per minute

Split: Splitless, split on after 30 s

Sample volume: 2 μL

Ionisation type: Electron impact ionisation (EI)

Ionisation energy: 70 eV

Dwell time: 50 ms

Electron multiplier: 1400 V + 600 V rel.

All other parameters must be optimised in accordance with the manufacturer's instructions.

5 Analytical determination

In each case 2 μL of the toluene solution are injected into the gas chromatograph for the analytical determination of the urine samples processed as described in Section 3.1.

The temporal profiles of the ion traces shown in Table 3 are recorded in the SIM mode.

Table 3. Retention times and masses

Compound	Retention time [min]	Masses
Cotinine	10.2	176
		98 *
Cotinine-d_3 (IS)	10.2	179
		101 *

The masses marked with * are used for quantitative evaluation.

The retention times shown in Table 3 serve only as a guide. Users of the method must satisfy themselves of the separation power of the capillary column used and the resulting retention behaviour of the substances. Figure 1 shows an example of a chromatogram of the processed urine from a light smoker (5 cigarettes per day).

If the measured values are above the linear range of the calibration graphs (>250 μg/L), the urine samples are diluted with water in the ratio of 1:10, processed and injected anew.

Two quality control samples are analysed with each analytical series.

6 Calibration

The calibration standards are processed in the same manner as the urine samples (Section 3.1) and analysed by means of gas chromatography/mass spectrometry as described in Sections 4 and 5. Linear calibration graphs are obtained by plotting the quotients of the peak areas of cotinine and that of the internal standard as a function of the concentrations used. It is unnecessary to plot a complete calibration curve for every analytical series. It is sufficient to analyse one calibration standard for every analytical series. The ratio of the results obtained for these standards and the result for the equivalent standard in the complete calibration graph is calculated. Using this quotient, each result read off the calibration graph is corrected.

New linear calibration graphs should be plotted if the precision control results indicate systematic deviations.

The calibration graph is linear between the detection limit and 250 μg per litre urine.

7 Calculation of the analytical result

Quotients are calculated by dividing the peak areas of the analyte by that of the internal standard. These quotients are used to read off the pertinent concentration of cotinine in μg per litre from the relevant calibration graph. If the pooled urine used to prepare the calibration standards exhibits a background signal, the resulting calibration graph must be shifted in parallel so that it passes through the zero point of the coordinates. (The concentrations of the background exposure can be read off from the point where the graph intercepts the axis before parallel shifting in each case.)

8 Standardisation and quality control

The guidelines given by the Bundesärztekammer (German Medical Association) [6, 7] and the special preliminary remarks to this series are to be followed to ensure the quality of the analytical results. In order to determine the precision of the method a urine sample containing a constant concentration of cotinine is analysed. As material for quality control is not commercially available, it must be prepared in the laboratory. It is advisable to use the urine or pooled urine of passive smokers for this purpose. The concentration of this control material should lie in the range of the cut-off value of 100 μg/L. A six-month supply of this control material is prepared, divided into aliquots in 10 mL polyethylene tubes which are stored in the deep-freezer. The theoretical value and the tolerance range for this quality control material are determined in a preliminary period (one analysis of the control material on each of 20 different days) [8, 9].

External quality assurance to check the reliability of the method can be achieved by participation in round-robin experiments. The Deutsche Gesellschaft für Arbeits- und Umweltmedizin (German Association for Occupational and Environmental Medicine) offers cotinine as a parameter for toxicological occupational and environmental analyses in their round-robin programme [10].

9 Reliability of the method

9.1 Precision

Pooled urine spiked with 10 μg/L and 100 μg/L cotinine were processed and analysed to check the precision. Seventeen replicate determinations of this urine sample yielded the precision in the series shown in Table 4.

Table 4. Precision for the determination of cotinine

	n	Concentration [µg/L]	Standard deviation (rel.) [%]	Prognostic range [%]
In the series	17	10	6.1	12.9
	17	100	1.3	2.7
From day to day	15	10	6.4	13.6
	15	100	2.8	5.9

In addition, the precision from day to day was determined. The same material was used as for the determination of the precision in the series. This urine was processed and analysed on each of 15 different days. The precision results are also shown in Table 4.

9.2 Accuracy

The loss due to processing was determined to check the accuracy of the method. For this purpose reference standards prepared in water and urine were processed and analysed. Toluene standards were simultaneously prepared with the same cotinine and cotinine-d_3 concentrations as the reference standards in their respective matrix. These toluene standards were injected into the GC and analysed without further treatment. Mean absolute recovery rates of 98.3% (water) and 95.0% (pooled urine) were obtained by comparison of the toluene calibration graphs with the calibration graphs for water and pooled urine. This means that 1.7% cotinine (and cotinine-d_3) are lost during the processing and analysis of aqueous cotinine samples. In the case of urine samples, the losses due to processing are about 5%. As the analyses are carried out using deuterated cotinine as an internal standard, the relative recovery is generally about 100%.

9.3 Detection limits

Under the conditions given here the detection limit, calculated as three times the signal/noise ratio of the analytical background noise in the temporal vicinity of the analyte signal, is approximately 1 µg/L.

9.4 Sources of error

Interference due to matrix effects was not observed for the ion traces m/z 98, 101, 176 and 179 under the conditions given here. The deuterated internal standard cotinine-d_3 must be checked to ensure it does not contain impurities of non-deuterated cotinine.

As already ascertained in the alternative method published as part of this series, a liner for the GC injector which has not been properly deactivated or has been contaminated can greatly reduce the sensitivity of the method and can lead to chromatographic problems. Interference to the analysis due to peaks with retention times similar to those of the analyte and internal standard was not observed during the analysis of more than 550 individual urine samples [11].

10 Discussion of the method

The procedure presented here is a further advance on the method devised by Skarping and co-workers in 1988 [12]. It permits the sensitive, reliable and, due to the use of mass spectrometry, extremely specific analysis of cotinine in the concentration range between 1 and 250 µg/L. The procedure enables the determination of cotinine in the urine of non-smokers, passive smokers and persons with a low consumption of tobacco (see also Figure 1).

The urine samples need only be adjusted to the alkaline range, subjected to liquid/liquid extraction and finally analysed. No derivatisation of the analyte is necessary.

The analytical reliability criteria of the method were checked using control material prepared in the laboratory. For this purpose pooled urine of non-smokers was spiked with two different concentrations of cotinine, processed and analysed several times. The precision achieved is excellent. Among other factors this is due to the use of deuterated cotinine as an internal standard. Moreover, the losses due to processing of only 5% in the case of urine and 1.7% for aqueous samples confirm the accuracy of the analytical results. Therefore matrix effects are not to be expected in the analysis of urine samples containing creatinine concentrations in the range between 0.5 and 2.5 g/L. On account of the very slight losses due to processing and the almost 100% relative recovery direct calibration can be carried out without problems instead of a matrix-based calibration.

The method presented here was developed and optimised in order to enable quantification of cotinine in urine without interference in the concentration range between 1 and 250 µg/L. Neither the author nor the examiners observed any interfering peaks, which occasionally occur when a non-specific NPD detector is used in this concentration range [2, 3]. As the cotinine concentration in the urine of smokers is more than 1000 µg/L as a rule, the urine of smokers must be diluted before analysis. For this reason it is advisable to clarify the smoking status of the test person by anamnesis prior to analysis. The linearity range was thoroughly checked by the examiners of the method. Linearity is given up to approx. 1500 µg/L. If calibration is carried out accordingly, many urine samples of smokers can be quantified without prior dilution. The use of mass spectrometry has the advantage of high specificity. This permits a short analysis time.

In a recent study the cotinine values of 569 employees in the construction industry were investigated using this method. There was a good correlation between the smoking habits obtained by anamnesis and the cotinine levels found in the urine of the test subjects. The method described here has proved successful on account of its practicability for routine investigations and its robustness [11].

Instruments used:
Gas chromatograph 6890 II with mass selective detector 5973, autosampler 6890 and data system from Hewlett-Packard.

11 References

[1] *V. Haufroid* and *D. Lison:* Urinary cotinine as a tobacco-smoke exposure index: A minireview. Int. Arch. Occup. Environ. Health 71, 162–168 (1998)

[2] *G. Scherer* and *I. Meger-Kossien:* Cotinine in urine, plasma or serum. In: J. *Angerer* and *K.H. Schaller (eds.)* Analysis of Hazardous Substances in Biological Materials, Vol. 7. Wiley-VCH, 171–189 (2001)

[3] *G. Scherer* and *I. Meger-Kossien:* Cotinin in Urine, Plasma oder Serum. In: *J. Angerer* and *K.H. Schaller (eds.)* Analysen in Biologischem Material. Loose-leaf collection, 14th Issue. Wiley-VCH (2000)

[4] *N. L. Benowitz:* Cotinine as a biomarker of environmental tobacco smoke exposure. Epidemiol. Rev. 18, 188–204 (1996)

[5] *G. Scherer* and *E. Richter:* Biomonitoring exposure to environmental tobacco smoke (ETS): A critical reappraisal. Hum. Exp. Toxicol. 16, 449–459 (1997)

[6] *Bundesärztekammer:* Qualitätssicherung der quantitativen Bestimmungen im Laboratorium. Neue Richtlinien der Bundesärztekammer. Dt. Ärztebl. 85, A699–A712 (1988)

[7] *Bundesärztekammer:* Ergänzung der „Richtlinien der Bundesärztekammer zur Qualitätssicherung in medizinischen Laboratorien". Dt. Ärztebl. 91, C159–C161 (1994)

[8] *G. Lehnert, J. Angerer* and *K.H. Schaller:* Statusbericht über die externe Qualitätssicherung arbeits- und umweltmedizinisch-toxikologischer Analysen in biologischen Materialien. Arbeitsmed. Sozialmed. Umweltmed. 33(1), 21–26 (1998)

[9] *J. Angerer* and *G. Lehnert:* Anforderungen an arbeitsmedizinisch-toxikologische Analysen – Stand der Technik. Dt. Ärztebl. 37, C1753–C1760 (1997)

[10] *Ringversuch Nr. 28*: Qualitätsmanagement in der Arbeits- und Umweltmedizin, Projektgruppe Qualitätssicherung, Organisation: Institut für Arbeits-, Sozial- und Umweltmedizin der Universität Erlangen-Nürnberg (2001)

[11] *M. Müller, P. Ruhnau, C. Caumanns, R. Böhm* and *E. Hallier:* Cotininbestimmung aus dem Humanurin zur Objektivierung des Confounders Rauchen in potentiell belasteten Kollektiven mit GC/MS. Verh. Dt. Ges. Arbeitsmed. 39, 651–653 (1999)

[12] *G. Skarping, S. Willers* and *M. Dalene:* Determination of cotinine in urine using glass capillary gas chromatography and selective detection, with special reference to the biological monitoring of passive smoking. J. Chromatogr. 454, 293–301 (1988)

Author: *M. Müller*
Examiners: *R. Heinrich-Ramm, H.-W. Hoppe*

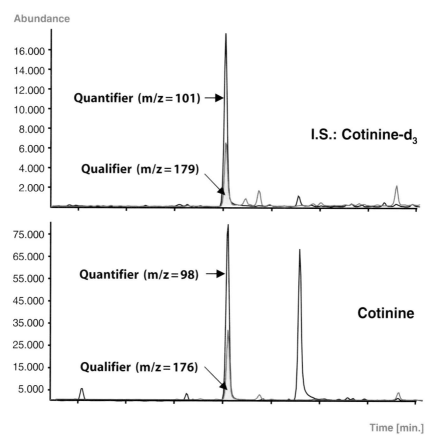

Fig. 1. Example of a chromatogram of a light smoker (5 cigarettes per day) with a cotinine concentration of 227 µg/L urine

N-2-Cyanoethylvaline, N-2-Hydroxyethylvaline, N-Methylvaline
(as evidence of exposure to acrylonitrile, ethylene oxide as well as methylating agents)

Application   Determination in blood

Analytical principle   Gas chromatography/mass spectrometry

Completed in   February 1996

Summary

The adducts of acrylonitrile, ethylene oxide and methylating agents (e. g. methyl halides) on the N-terminal valine of haemoglobin can be sensitively and reliably determined using this gas chromatographic/mass spectrometric method. Due to its sensitivity the method is suitable not only for determining concentrations relevant to occupational medicine but also those which occur in the ecological range.

In order to determine the adducts of the N-terminal valine in haemoglobin, the erythrocytes are separated from whole blood and the cells are destroyed through lysis. The globin is precipitated out of the haemoglobin solution, the internal standard N-2-ethoxyethylvaline-alanine-anilide is added and the alkylated N-terminal valines N-2-cyanoethylvaline (CEV), N-2-hydroxyethylvaline (HEV) or N-methylvaline (MEV) are converted to their derivatives using pentafluorophenyl isothiocyanate and cleaved from the protein by means of a modified Edman degradation. Alternatively, DL-pipecolinic acid (PCA) can be used as an internal standard.

The derivatives of N-2-cyanoethylvaline, 1-(2-cyanoethyl)-5-isopropyl-3-pentafluoro-phenyl-2-thiohydantoin (N-2-cyanoethylvaline-PFPTH), of N-2-hydroxyethylvaline, 1-(2-hydroxyethyl)-5-isopropyl-3-pentafluorophenyl-2-thiohydantoin (N-2-hydroxyethylvaline-PFPTH), and of N-methylvaline, 1-(2-methyl)-5-isopropyl-3-pentafluorophenyl-2-thiohydantoin (N-methylvaline-PFPTH), are extracted, washed several times and separated from the other components of the sample by means of capillary gas chromatography. The quantitative determination is carried out by mass spectrometry using selected ion monitoring based on the masses m/e = 335 and m/e = 377 (N-2-cyanoethylvaline), m/e = 308 and m/e = 350 (N-2-hydroxyethylvaline) and m/e = 310 and m/e = 338 (N-methylvaline). Calibration is carried out using solutions of the dipeptides N-2-cyanoethylvaline-leucine-anilide, N-2-hydroxyethylvaline-leucine-anilide and N-methylvaline-leucine-anilide with added pooled human globin in formamide which are processed in the same manner as the protein samples.

N-2-Cyanoethylvaline

Within-series imprecision:	Standard deviation (rel.) $s_w = 7.7\,\%$ Prognostic range $u = 17.2\,\%$ At a concentration of 10 µg N-2-cyanoethylvaline per litre blood (408 pmol per gram globin) and where $n = 10$ determinations
Between-day imprecision:	Standard deviation (rel.) $s = 13\,\%$ Prognostic range $u = 29\,\%$ At a concentration of 10 µg N-2-cyanoethylvaline per litre blood (408 pmol per gram globin) and where $n = 10$ days
Detection limit:	0.3 µg N-2-cyanoethylvaline per litre blood (12 pmol per gram globin)

N-2-Hydroxyethylvaline

Within-series imprecision:	Standard deviation (rel.) $s_w = 7\,\%$ Prognostic range $u = 15.8\,\%$ At a concentration of 2.2 µg N-2-hydroxyethylvaline per litre blood (95 pmol pergram globin) and where $n = 9$ determinations
Between-day imprecision:	Standard deviation (rel.) $s = 12\,\%$ Prognostic range $u = 29.3\,\%$ At a concentration of 2.3 µg N-2-hydroxyethylvaline per litre blood (99 pmol per gram globin) and where $n = 6$ days
Detection limit:	0.4 µg N-2-hydroxyethylvaline per litre blood (19 pmol per gram globin)

N-Methylvaline

Within-series imprecision:	Standard deviation (rel.)	$s_w = 8.8\,\%$
	Prognostic range	$u = 19.6\,\%$

At a concentration of 10 µg N-methylvaline per litre blood (530 pmol per gram globin) and where $n = 10$ determinations

Between-day imprecision:	Standard deviation (rel.)	$s = 11\,\%$
	Prognostic range	$u = 26.9\,\%$

At a concentration of 2.2 µg N-methylvaline per litre blood (265 pmol per gram globin) and where $n = 6$ days

Detection limit: 0.2 µg N-methylvaline per litre blood (12 pmol per gram globin)

Acrylonitrile

Acrylonitrile (molecular mass 53.06 g, boiling point 77 °C) is a colourless, flammable, pungent liquid which is barely soluble in water, but highly soluble in almost all organic solvents.

Large-scale industrial synthesis of acrylonitrile is mainly carried out using acetylene and hydrogen cyanide. It is of great industrial importance as a raw material for the manufacture of plastics (acrylonitrile-butadiene-stryrene copolymers, stryrene-acrylonitrile copolymers and nitrile rubber) and synthetic fibres (polyacrylic fibres). In addition, acrylonitrile is used as an insecticide [1].

Acrylonitrile is rapidly absorbed after oral or dermal intake, and after inhalation. It is mainly excreted with the urine.

Figure 1 shows a schematic representation of the metabolism of acrylonitrile.

Acrylonitrile is acutely toxic, whereby the symptoms of poisoning such as cyanosis, dizziness, vomiting and headache are similar to those of poisoning with hydrogen cyanide. In liquid form and as a vapour, acrylonitrile causes irritation to the skin and mucous membranes.

In animal studies a tumorigenic effect was observed after oral intake or inhalation of acrylonitrile. The results of in vitro studies on gene toxicity are predominantly positive, whereas negative results have been described for in vivo investigations on mammals.

Epidemiological surveys carried out where acrylonitrile is processed show contradictory findings and it is impossible to make final conclusions as to the carcinogenic effect on humans at the present time [3].

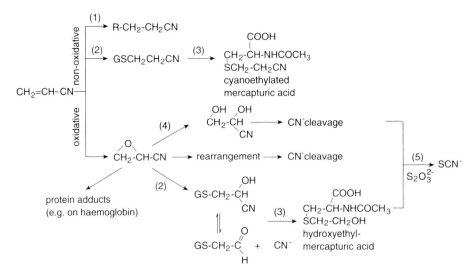

Abbreviations: (1): non-enzymatic conjugation; (2): GSH transferase;
(3): glutathionase; (4): epoxide hydrolase; (5): rhodanase

Figure 1. Metabolism of acrylonitrile (modified from [2])

Acrylonitrile has been assigned to category III A2 in the list of substances shown to be clearly carcinogenic only in animal studies by the Deutsche Forschungsgemeinschafts Commission for the Investigation of Health Hazards of Chemical Compounds in the Work Area. Due to the risk of absorption through the skin acrylonitrile has been marked with an "H" in the List of MAK and BAT values [4].

Ethylene oxide

Ethylene oxide (molecular mass 44.05 g) is a gas with a sweet ether-like smell which is synthesized by direct oxidation of ethylene using silver as a catalyst. About 490,000 tonnes were produced by this process in the Federal Republic of Germany in 1985.

Only a small amount of ethylene oxide is directly used, e. g. as a pesticide as well as an antibacterial and fermentation-inhibiting agent. Its primary importance lies in the re-activity of the epoxy ring, which renders it a key substance in the formation of a multitude of further intermediate and final products. The most significant co-reactants of ethylene oxide together with the most important reaction products and secondary products are shown in Table 1 [1].

Table 1. The most significant co-reactants of ethylene oxide with the most important reaction products and secondary products.

Co-reactants	Reaction products	Secondary products
Water	Ethylene glycol	Glyoxal
	Diethylene glycol	Dioxolan
	Polyethylene glycols	Dioxane
Alkylphenols	Polyethoxylates	
Alkanols		
Fatty acids		
Fatty amines		
Ammonia	2-Aminoethanol	Ethyleneimine
	Diethanolamine	Morpholine
	Triethanolamine	
Alcohols ($R-CH_2OH$)	Glycol monoalkyl ether	Glycol dialkyl ether
$R = H, CH_3, n-C_3H_7$	Diglycol monoalkyl ether	Ester of glycol-monoalkyl ether

Ethylene oxide poisoning is usually a result of inhaling the substance. As it has an olfactory detection threshold of 700 ppm, it cannot be perceived early enough at the concentrations normally present at workplaces. Acute intoxication is characterized by local irritation of the skin and mucous membranes, but also by systemic effects on the central nervous system, the heart and other organs. The main symptoms are headaches, nausea and periodical vomiting which generally persists for a longer time. In addition, dyspnoea, irritation of the eyes and the mucous membranes of the upper respiratory tract, damage to the heart, states of excitation, bewilderment, dizziness as well as unconsciousness could be observed. Gaseous ethylene oxide or aqueous ethylene solution can also be absorbed through the skin and can cause poisoning in this way. After a latent period of up to several hours, these victims suffer from nausea, vomiting and headaches lasting about two hours and later, after 12 to 24 hours, local symptoms such as the pronounced formation of blisters are observed [3, 5].

The central and peripheral nervous systems are especially affected in cases of chronic intoxication [3, 5].

Ethylene oxide is directly mutagenic in *Salmonella typhimurium* TA100 and TA1535. It induces chromosomal aberrations and sister chromatid exchange in mammalian cells in vitro and in vivo. A significant increase in chromosomal aberrations and sister chromatid exchange was observed in peripheral lymphocytes of monkeys exposed to ethylene oxide.

As an electrophilic alkylating agent, ethylene oxide reacts with nucleophilic groups, such as the amino, carboxyl, phenolic hydroxyl and thiol groups. The formation of haemoglobin adducts was observed in rats and humans.

The main adducts are S-(2-hydroxyethyl)-L-cysteine, N-1-(2-hydroxyethyl)histidine as well as N-3-(2-hydroxyethyl)histidine and N-2-hydroxyethylvaline.

After a latent period of more than ten years, an increased mortality rate due to leukaemia and stomach tumours was observed in 89 men who had been occupationally exposed to ethylene oxide. In this case the former ethylene oxide concentration in the respiratory air was estimated to be 10 to 50 ppm. Recently 10 ppm were measured at the same workplace. It should be noted, however, that the test persons were also intermittently exposed to other compounds such as 2-chloroethanol, ethylene dichloride, 2,2'-chloroethyl ether and propylene oxide.

Three cases of leukaemia occurred in a group of 230 persons who were occupationally exposed to ethylene oxide. For people who are not exposed to the substance the expected statistical rate would be 0.2. The average ethylene oxide concentration is given as 20 ± 10 mL/m^3. The latent period was between 6 and 9 years from the beginning of the occupational exposure [3, 6].

Ethylene oxide is formed from ethylene in the human body. An occupational exposure to 1 ppm ethylene oxide is estimated to cause inner stress equivalent to an average exposure to an ethylene concentration of 50 ppm [7].

The Deutsche Forschungsgemeinschaft's Commission for the Investigation of Health Hazards of Chemical Compounds in the Work Area has assigned ethylene oxide to group A2 of the substances shown to be clearly carcinogenic only in animal studies [4]. The TRK value (Technical Exposure Limit) is 1 ppm.

Besides determination of the alveolar air concentration it is advisable to determine the N-2-hydroxyethylvaline adducts on haemoglobin described in this method for the purpose of biological monitoring of persons exposed to ethylene oxide.

On the basis of the results for these biomonitoring parameters the following exposure equivalents have been established [4]:

Air Ethylene oxide		Sampling: during exposure (> 4h) Alveolar air Ethylene oxide		Sampling: no restrictions Erythrocytes N-2-hydroxyethylvaline
[mL/m^3]	[mg/m^3]	[mL/m^3]	[mg/m^3]	[µg/L blood]
0.5	0.92	0.12	0.22	45
1	1.83	0.24	0.44	90
2	3.66	0.48	0.88	180

Methylating agents

Among the numerous organic methylating agents the monohalogenated methane derivatives, bis(chloromethyl) ether, monochlorodimethyl ether and dimethyl sulphate are of particular technical importance.

The stability of the halogen – carbon bond greatly increases in the order I < Br < Cl < F. The fluorine – carbon bond is extremely stable, even in the organism. For this reason aliphatic fluorine compounds exhibit no alkylating activity.

After a short preliminary period accompanied by dizziness, headaches and possible gastro-intestinal disorders, the main consequence of acute poisoning by *monohalogenated methane derivatives* is their neurotoxic effect.

In serious cases of poisoning pronounced sleepiness is observed, after which the victim falls into a deep coma. Eventually, death due to respiratory paralysis can result. After a typical latent period, milder cases of intoxication lead to nervous disorders. Similarly, nervous disorders are of central importance in cases of chronic intoxication. Oedema of the lungs can occur both acutely or after a latent period of several days. Delayed symptoms can appear even after 2–3 weeks, and cases of death have been reported [3, 5].

Chloromethane (molecular mass 50.49 g) is a gas with a slightly sweetish odour. Its solubility in water is low, but it is readily soluble in alcohol.

Chloromethane is produced by thermal chlorination or catalytic oxychlorination of methane. Hydrochlorination of methanol is the most important large-scale process for the production of chloromethane.

Chloromethane is used for the manufacture of more highly chlorinated derivatives of methane. In organic chemistry its uses include the methylation and etherification of phenols, alcohols and cellulose, but also the production of silicones, tetramethyl-lead and quaternary ammonium salts. On account of its high enthalpy of vaporization chloromethane is also used as a cooling agent [1].

The fact that chloromethane has a barely perceptible odour increases the danger of poisoning when it is inhaled over a longer period.

Chloromethane causes sister chromatid exchange and mutations in human lymphocytes in vitro, but does not cause DNA strand breaks [3, 8].

The Deutsche Forschungsgemeinschaft's Commission for the Investigation of Health Hazards of Chemical Compounds in the Work Area has placed chloromethane in category III B with the substances suspected of having carcinogenic potential [4]. Furthermore, chloromethane has been placed in pregnancy risk group B in the list of MAK and BAT values because of the risk of damage to the developing embryo or foetus [4]. This means that such damage cannot be excluded when pregnant women are exposed to the substance, even when the MAK and BAT values are adhered to.

Bromomethane (molecular mass 94.94 g) is a colourless gas with a strong sweetish odour which only becomes noticeable at higher concentrations.

Bromomethane is synthesized from methanol and hydrogen bromide. Its most important use is in organic synthesis, e. g. in Grignard reactions and in the synthesis of pharmaceutical products. In addition, bromomethane is employed as a fungicide, nematocide and insecticide in horticulture and for the preservation of stored products [1].

The Deutsche Forschungsgemeinschaft's Commission for the Investigation of Health Hazards of Chemical Compounds in the Work Area has assigned bromomethane to category III A2 with the substances shown to be clearly carcinogenic only in animal studies [4].

Iodomethane (molecular mass 141.95 g, boiling point 42.5 °C) is a colourless liquid which is barely miscible with water, but readily soluble in alcohol and ether.

Iodomethane is produced from methanol, iodine and red phosphorus and is used, for example, in microscopy, medicine and in organic chemistry as a methylating agent [1].

In contrast to bromomethane and chloromethane, only a few cases of poisoning have been described so far.

Iodomethane methylates deoxyguanosine at the N7 position [3, 9] and induces mutations in cultured mammalian cells. Subcutaneous injection of iodomethane causes local tumours in rats [3, 9].

The Deutsche Forschungsgemeinschaft's Commission for the Investigation of Health Hazards of Chemical Compounds in the Work Area has assigned iodomethane to category III A2 with the substances shown to be clearly carcinogenic only in animal studies [4].

Bis(chloromethyl) ether (molecular mass 114.97 g, boiling point 106 °C) and *mono-chlorodimethyl ether* (molecular mass 80.52 g/mol, boiling point 59 °C) are colourless liquids which are mainly used for alkylations in organic synthesis [1].

Bis(chloromethyl) ether and monochlorodimethyl ether have strongly irritating effects, both on the eyes and on the respiratory tract. Damage to the eyes is noticeable only after some time and can lead to temporary impairment of vision. Bis(chloromethyl) ether has a specific effect on the sense of hearing and the equilibrium centre in the inner ear. In higher concentrations it has narcotic effects. Bis(chloromethyl) ether and monochlorodimethyl ether are strongly carcinogenic. Epidemiological studies have shown that occupational exposure to bis(chloromethyl) ether and monochlorodimethyl ether increases the incidence of lung carcinomas in the exposed persons. The risk of cancer rises with the duration of the exposure [9, 10]. Bis(chloromethyl) ether and monochlorodimethyl ether have been assigned to category III A 1 with the substances shown to induce malignant tumors in humans by the Deutsche Forschungsgemeinschaft's Commission for the Investigation of Health Hazards of Chemical Compounds in the Work Area [4].

Dimethyl sulphate (molecular mass 126.13 g, boiling point 188 °C) is a colourless oil which becomes miscible when it undergoes hydrolysis in water and is also miscible with alcohol, ether, acetone and aromatic hydrocarbons.

Dimethyl sulphate is produced from dimethyl ether and sulphur trioxide in an exothermic reaction. It is used industrially on a large scale for the methylation of carboxylic acids, phenols, thiophenols and amines [1].

Dimethyl sulphate possesses strongly methylating properties even in biological systems. Thus, it is of great importance to realize that inhalation of the vapour cannot be perceived by smelling, even at highly toxic concentrations, and that inhalation causes only relatively mild subjective irritation of the mucous membranes of the eyes and the upper respiratory tract. After a latent period of several hours, however, inflammation of the bronchial tubes and the alveolar walls ensues, which can eventually lead to oedema of the lungs and death by asphyxiation. Acute poisoning in humans has been reported several times. In mild cases the symptoms are limited to the conjunctival and corneal damage (chemical burns) as well as oedematous swelling of the mucous membranes of the upper respiratory tract. Moderate to serious cases are characterized by initially slight, later increasingly serious irritation of the eyes, nose, throat, larynx, trachea and bronchi. Damage to the central nervous system, the liver and the kidneys has also been described [5].

The Deutsche Forschungsgemeinschaft's Commission for the Investigation of Health Hazards of Chemical Compounds in the Work Area has classified dimethyl sulphate as a category III A2 substance shown to be clearly carcinogenic only in animal studies. In addition, it has been labelled with an "H" in the list of MAK and BAT values to indicate the risk of absorption of dimethyl sulphate through the skin [4].

Authors: *N. J. van Sittert*
Examiners: *J. Angerer, M. Bader, M. Blaszkewicz, D. Ellrich, A. Krämer, J.Lewalter*

N-2-Cyanoethylvaline, N-2-Hydroxyethylvaline, N-Methylvaline (as evidence of exposure to acrylonitrile, ethylene oxide as well as methylating agents)

Application	Determination in blood
Analytical principle	Gas chromatography/mass spectrometry
Completed in	February 1996

Contents

Essential Biomonitoring Methods. DFG, Deutsche Forschungsgemeinschaft
Copyright © 2006 WILEY-VCH Verlag GmbH & Co. KGaA, Weinheim
ISBN: 3-527-31478-4

1 General principles

In order to determine the adducts of the N-terminal valine in haemoglobin, the erythrocytes are separated from whole blood and the cells are destroyed through lysis. The globin is precipitated out of the haemoglobin solution, the internal standard N-2-ethoxyethylvaline-alanine-anilide is added and the alkylated N-terminal valines (N-2-cyanoethylvaline (CEV), N-2-hydroxyethylvaline (HEV) and N-methylvaline (MEV)) are converted to their derivatives using pentafluorophenyl isothiocyanate and cleaved from the protein by means of a modified Edman degradation. Alternatively, DL-pipecolinic acid (PCA) can be used as an internal standard.

The derivatives of N-2-cyanoethylvaline, 1-(2-cyanoethyl)-5-isopropyl-3-pentafluoro-phenyl-2-thiohydantoin (N-2-cyanoethylvaline-PFPTH), of N-2-hydroxyethylvaline, 1-(2-hydroxyethyl)-5-isopropyl-3-pentafluorophenyl-2-thiohydantoin (N-2-hydroxyethylvaline-PFPTH), and of N-methylvaline, 1-(2-methyl)-5-isopropyl-3-pentafluorophenyl-2-thiohy-dantoin (N-methylvaline-PFPTH), are extracted, washed several times and separated from the other components of the sample by means of capillary gas chromatography. The quantitative determination is carried out by mass spectrometry using selected ion monitoring based on the masses m/e $= 335$ and m/e $- 377$ (N-2-cyanoethylvaline), m/e $= 308$ and m/e $= 350$ (N-2-hydroxyethylvaline) and m/e $= 310$ and m/e $= 338$ (N-methylvaline). Calibration is carried out using solutions of the dipeptides N-2-cyanoethylvaline-leucine-anilide, N-2-hydroxyethylvaline-leucine-anilide or N-methylvaline-leucine-anilide with added pooled human globin in formamide which are processed in the same manner as the protein samples.

2 Equipment, chemicals and solutions

2.1 Equipment

Gas chromatograph/mass spectrometer with split/splitless injection and the possibility of selected ion monitoring and integration system

Gas chromatographic column:
Fused silica DB-5MS, length 25 m; inner diameter 0.2 mm; film thickness 0.33 μm (e. g. from Hewlett Packard)

10 mL EDTA Monovettes with hypodermic needles (e. g. from Sarstedt, Nümbrecht)

5 or 12 and 20 mL Duran screw-topped jars (e. g. from Schütt)

Roller mixer (e. g. from Denley)

Laboratory shaker (e. g. Vortex from Bender & Hobein)

Vacuum desiccator

Automatic microlitre pipettes, variably adjustable between 10 and 100 μL, and between 100 and 1000 μL (e. g. from Eppendorf)

Centrifuge, at least 3500 g (e. g. from Heraeus)

10 μL syringe for gas chromatography (e. g. from Hamilton)

Thermostatically controlled water bath (e. g. from Julabo)

pH meter (e. g. from Knick)

Evaporation apparatus to remove solvents in a stream of nitrogen (e. g. from Pierce) or vacuum centrifuge (e. g. from Bachhofer)

Ultrasound bath (e. g. from Bandolin Sonorex)

Laboratory balance

10, 20, 100 and 1000 mL volumetric flasks

250 mL glass beaker

2.2 Chemicals

2-Propanol, p. a. (e. g. from Merck)

Ethyl acetate, p. a. (e. g. from Merck)

n-Hexane p. a. (e. g. from Merck)

Sodium chloride p. a. (e. g. from Merck)

Anhydrous sodium carbonate (e. g. from Merck)

37 % Hydrochloric acid p. a. (e. g. from Merck)

Sodium hydroxide pellets p. a. (e. g. from Merck)

Ultrapure water (ASTM type 1) or double-distilled water

Diethyl ether, p. a. (e. g. from Merck)

DL-pipecolinic acid p. a. (e. g. from Merck)

Formamide ultrapure (e. g. from Amersham Life Science) or formamide for molecular biology (e. g. from Merck)

Ethanol p. a. (e. g. from Merck)

N-2-cyanoethylvaline-leucine-anilide (e. g. from Bachem Biochemica)

N-2-ethoxyethylvaline-leucine-anilide (e. g. from Bachem Biochemica)

N-2-hydroxyethylvaline-leucine-anilide (e. g. from Bachem Biochemica)

N-methylvaline-leucine-anilide (e. g. from Bachem Biochemica)

Pentafluorophenyl isothiocyanate (e. g. from Fluka)

Toluene p. a. (e. g. from Merck)

tert.-Butylmethyl ether (e. g. from Merck)

Preparation of the pooled globin:
In order to prepare the pooled globin 2 mL of haemolytic solution from whole blood samples of several non-smokers are processed as described in Section 3. The isolated globins are subsequently pooled and can be stored at -18 °C for up to a year.

2.3 Solutions

50 mmol hydrochloric acid in 2-propanol:
About 500 mL 2-propanol are placed in a 1000 mL volumetric flask and 4.1 mL 37 % hydrochloric acid are pipetted into the flask. Then the flask is filled to the mark with 2-propanol.

0.9 % NaCl solution:
9 g of NaCl are weighed into a 1000 mL volumetric flask. The flask is subsequently filled to the mark with ultrapure water.

1 M NaOH solution:
4 g of NaOH are weighed into a 250 mL glass beaker and dissolved in as little ultrapure water as possible. Then the solution is transferred to a 100 mL volumetric flask and the flask is filled to the mark with ultrapure water.

0.1 M sodium carbonate solution:
1.06 g of sodium carbonate are weighed into 100 mL volumetric flask. The flask is subsequently filled to the mark with ultrapure water.

These solutions can be stored in the refrigerator for 4 weeks.

2.4 Internal standards

N-2-ethoxyethylvaline-alanine-anilide (internal standard: IS):
About 17.7 mg N-2-ethoxyethylvaline-alanine-anilide are weighed exactly into a 100 mL volumetric flask. The flask is subsequently filled to the mark with ethanol (100 mg N-ethoxyethylvaline per litre). 100 µL of this solution are pipetted into a 10 mL volumetric flask. Then the flask is filled to the mark with ethanol (1 mg N-ethoxyethylvaline per litre).

Pipecolinic acid (alternative internal standard: AIS):

About 30 mg of pipecolinic acid are exactly weighed into a 20 mL volumetric flask (starting solution: 1500 mg pipecolinic acid per litre). The volumetric flask is subsequently filled to the mark with formamide.

A solution of the alternative internal standard with a concentration of 150 µg/L (working solution) is prepared by dilution with formamide according to the pipetting scheme in Table 2.

Table 2. Pipetting scheme for the preparation of the alternative internal standard

Designation of the starting solution	Volume of the starting solution	Final volume of the internal standard	Designation of the internal standard	Concentration of the internal standard
Starting solution	100 µL	10 mL	Intermediate dilution	15 mg/L
Intermediate dilution	100 µL	10 mL	Working solution	150 µg/L

These solutions can be stored in the refrigerator at 4 °C for at least 3 months without loss of the internal standards.

2.5 Calibration standards

About 21.1 mg N-2-cyanoethylvaline-leucine-anilide, 21.6 mg N-2-hydroxyethylvaline-leucine-anilide and 24.4 mg N-methylvaline-leucine-anilide are weighed exactly into a 100 mL volumetric flask. The flask is subsequently filled to the mark with ethanol (Content: 100 mg/L expressed in terms of the valine adduct).

Starting solutions in the concentration range from 100 to 10000 µg/L are prepared by dilution of this stock solution with ethanol. Dilution is carried out according to the pipetting scheme in Table 3.

Table 3. Pipetting scheme for the preparation of the starting solutions

Volume of the stock solution [mL]	Final volume of the starting solution [mL]	Conc. of the starting solution [µg/L]	Designation of the starting solution
10	100	10000	Starting solution 1
1	100	1000	Starting solution 2
0.1	100	100	Starting solution 3

In order to prepare the calibration standards the standard globin (100 mg) is dissolved in 3 mL formamide in each of ten 5 mL Duran screw-topped jars and the analytes and the internal standard (N-2-ethoxyethylvaline-alanine-anilide, 125 µL) or the alternative internal standard (pipecolinic acid, 50 µL) are added according to Table 4.

Table 4. Pipetting scheme for the preparation of the calibration standards

Calibration standard	Vol. of the starting solution			Vol. IS	Vol. AIS (PCA)	Conc. of the calibration standards			
No.	[µL]			[µL]	[µL]	µg/L blood	pmol/g globin		
	1	2	3				CEV	HEV	MEV
blank value	–	–	–	125	50	–	–	–	–
1	–	–	10	125	50	1.4	57	60	74
2	–	–	20	125	50	2.9	118	124	154
3	–	–	30	125	50	4.3	176	186	229
4	–	–	75	125	50	10.8	441	466	573
5	–	–	125	125	50	18.0	735	776	954
6	–	30	–	125	· 50	43.2	1765	1863	2290
7	–	75	–	125	50	108.0	4412	4658	5725
8	–	125	–	125	50	180.0	7353	7764	9542
9	25	–	–	125	50	360.0	14706	15528	19084

These calibration standards must be freshly prepared for every analytical series.

3 Specimen collection and sample preparation

3.1 Preparation of the haemolytic solution from the erythrocytes

After disinfection of the puncture point 5 mL of venous whole blood are withdrawn from the test person using an EDTA Monovette. The blood sample is subsequently centrifuged for 10 min at 800 g to separate the erythrocytes from the blood plasma. The supernatant plasma is carefully removed using a pipette, 5 mL of the 0.9 % sodium chloride solution are added to the erythrocyte fraction, swirled around several times and centrifuged once again for 10 min at 800 g. The supernatant is again removed using a pipette and discarded. This washing process is repeated until the supernatant is clear and colourless (experience has shown that three washes are normally required in the case of fresh blood samples). Haemolysis is achieved by subsequently suspending the erythrocytes in ultrapure water and freezing this suspension for at least 60 min at –18 °C. After completion of this sample preparation the erythrocyte solution can be shipped. If immediate shipping is impossible the samples can be kept in the refrigerator until the next day. If longer storage is necessary, the samples must be deep-frozen.

3.2 Isolation of the globin

To prepare 200–250 mg of globin 2 mL of the haemolytic solution are pipetted into a 20 mL Duran screw-topped jar in which 12 mL of the 50 mmol HCl/2-propanol solution has been previously placed. The jar is shaken for 15 min and centrifuged at 3500 g for 10 min. The supernatant is decanted off the precipitated cell components into a 20 mL Duran screw-topped jar and 8 mL ethyl acetate are slowly added. The samples are sealed, cooled to 4 °C (refrigerator) and are left to stand at this temperature for 60 min. The precipitated globin is subsequently centrifuged off (10 min, 800 g) and the supernatant is discarded. The globin is resuspended in 10 mL ethyl acetate and centrifuged once again for 10 min at 800 g. This washing process with ethyl acetate is repeated until the supernatant is clear and colourless (experience has shown that three washes are normally required for the processing of fresh blood samples). Then the globin is resuspended in 5 mL n-hexane and centrifuged once more for 10 min at 800 g. The supernatant is decanted off and the isolated protein is dried in a vacuum desiccator. The globin is stored at –18 °C.

3.3 Derivatization of the globin

100 mg of globin are weighed in a 5 mL Duran screw-topped jar and 3 mL formamide, 30 µL 1 N NaOH as well as 125 µL of the N-2-ethoxyethylvaline-alanine-anilide solution (IS) are added. Alternatively, 50 µL of the pipecolinic acid solution (working solution) can be used as an internal standard instead of the N-2-ethoxyethylvaline-alanine-anilide solution.

In order to ensure that the protein solution is homogeneous, the Duran screw-topped jars are sealed and placed in an ultrasound bath for 30 min. Then 10 µL pentafluorophenyl isothio-cyanate are added for derivatization and the Duran screw-topped jars are sealed again. The samples are first mixed on a roller mixer overnight at room temperature and then tempered for 90 min at 45 °C in a water bath. After cooling the solutions are extracted twice with 3 mL diethyl ether by passing them through a laboratory shaker (e. g. Vortex) for 60 s to ensure that the phases are intensively mixed. In order to improve phase separation the samples are centrifuged for 5 min at 3500 g (10 °C) after each extraction. The combined ether phases (approximately 5 mL) of each sample are evaporated to dryness under a stream of nitrogen, then the residue is dissolved in 1.5 mL toluene. The toluene phases are washed first with 2 mL ultrapure water and subsequently with 2 mL of the 0.1 M sodium carbonate solution (in each case for 60 s on the laboratory shaker). Each wash is followed by centrifugation for 5 min at 3500 g. While being gently warmed to about 30 °C, the toluene phases are evaporated to dryness under a stream of nitrogen. The resulting residue is dissolved in 30 µL toluene and analysed by gas chromatography/mass spectrometry.

4 Operational parameters for gas chromatography

Gas chromatography

Capillary column:	Material:	Quartz (fused silica)
	Stationary phase:	DB-5MS
	Length:	25 m
	Inner diameter:	0.2 mm
	Film thickness:	0.33 μm
Temperatures:	Column:	2 minutes at 100 °C; then increase 15 °C per minute to 220 °C; then increase 5 °C per minute to 280 °C
	Injector:	280 °C
	Transfer line:	300 °C
	Ion source:	200 °C
	Quadrupole:	100 °C
Carrier gas:	Helium at a precolumn pressure of 1 bar	
Sample volume:	1 μL, splitless, after 1 minute split 35 mL/min	

Mass spectrometry

Ionization type:	Electron collision ionization (EI)
Ionization energy:	70 eV
Detection:	Selected ion monitoring (SIM)
Scan time per ion:	100 ms

All other parameters should be optimized as specified in the manufacturer's instructions.

Figure 2 shows an ion chromatogram of a calibration standard.

5 Analytical determination

A 1 μL aliquot of each processed sample is injected splitless into the GC/MS system.

The identification of the amino acid derivatives is based on the retention time and the corresponding characteristic ion sources (cf. Table 5). For quantitative evaluation the ion sources of higher intensity are integrated.

Table 5. Retention times of the analytes as well as the recorded and evaluated ion sources

Substance	Retention time [min]	Recorded ion sources	Evaluated ion sources
N-methylvaline-PFPTH	13.04	310/338	338
N-2-hydroxyethylvaline-PFPTH	16.42	308/350	308
N-2-cyanoethylvaline-PFPTH	16.59	335/377	377
N-2-ethoxyethylvaline-PFPTH (IS)	15.28	367/396	396
Pipecolinic acid-PFPTH (AIS)	15.51	308/336	336

The retention times given in Table 5 should be regarded as an approximate guide.

6 Calibration

The calibration standards prepared as described in Section 2.5 are processed and analysed in the same way as the assay samples (cf. Sections 3 and 4). Calibration curves are obtained by plotting the quotients of the peak areas for the individual alkylvalines with the internal standard as a function of the analyte concentrations in μg per litre blood (cf. Figures 3 to 5). The calibration curves are linear up to 360 μg of the corresponding alkylvalines per litre blood (14706 pmol N-2-cyanoethylvaline, 15528 pmol N-2-hydroxyethylvaline or up to 19084 pmol N-methylvaline per gram globin).

7 Calculation of the analytical result

The average globin content of the blood is 144 g/L [11]. This gives the following approximation:

$$\text{Content [μg/L blood]} = \frac{\text{recorded value [μg/L]} \cdot 0.003 \text{ L} \cdot 144 \text{ g globin}}{0.1 \text{ g globin} \cdot 1 \text{ L blood}}$$

The content in [pmol/g globin] is calculated as follows:

$$\text{Content [pmol/g globin]} = \frac{[\mu g/L \text{ blood}]}{MG \cdot 144 \text{ g/L}}$$

where MG = $131 \cdot 10^{-6} \mu g/mol \; \hat{=}$ molecular weight of N-methylvaline

$161 \cdot 10^{-6} \mu g/mol \; \hat{=}$ molecular weight of N-2-hydroxyethylvaline

$170 \cdot 10^{-6} \mu g/mol \; \hat{=}$ molecular weight of N-2-cyanoethylvaline

In order to determine the content of N-substituted valines the quotient of the area integral of the appropriate analyte with the internal standard is calculated for each sample. The resulting quotients can be used to read off the corresponding concentration in µg per litre blood from the relevant calibration curve.

8 Standardization and quality control

Quality control of the analytical results is carried out as stipulated in TRgA 410 (Regulation 410 of the German Code on Hazardous Working Materials) [12] and in the Special Prelimi-nary Remarks in this series. As no quality control material is commercially available, it must be prepared in the laboratory. For this purpose 2 mL samples of haemolysed blood from the pooled whole blood of smokers are processed. Using blood from smokers ensures that the analytes are present in the globin samples in detectable concentrations. The isolated globins are subsequently combined. About 5 g of this pooled globin are dissolved in 150 mL form-amide (Important: it must be added slowly) and the samples are divided into 3 mL aliquots. The aliquots are stored at –18 °C until analysis is carried out. One of these quality control samples is included in each analytical series. The aliquots can be stored in the deep-freezer for up to one year and used for quality control. The expected value and tolerance range of this quality control material is determined in a pre-analytical period (one analysis of the control material is carried out on each of 20 different days) [13].

9 Reliability of the method

9.1 Precision

In order to check the precision in the series, control material containing mean quantities of the N-alkylvalines between 2.2 and 10 µg per litre blood were processed and analysed as de-scribed in Sections 3 and 4. The relative standard deviations were between 7.0 and 8.8 %, the corresponding diagnostic ranges were between 15.8 and 19.6 %. The individual values are found in Table 6.

Table 6. Precision in the series for the determination of N-alkylvalines

Analyte	Mean N-alkylvaline content of the control material		Precision in the series		
	[µg/L blood]	[µg/g globin]	n	s_w [%]	u [%]
N-2-cyanoethylvaline	10	408	10	7.7	17.2
N-2-hydroxyethylvaline	2.2	95	9	7.0	15.8
N-methylvaline	10	530	10	8.8	19.6

The precision from day to day was determined by analysis of control material containing mean N-alkylvaline concentrations of between 2.3 and 10 µg per litre blood on different days. A sample was processed and analysed daily as described in Sections 3 and 4. The resulting relative standard deviations were between 11.0 and 13 %, the corresponding diagnostic ranges were between 26.9 and 29.3 %. The individual values are shown in Table 7.

Table 7. Precision from day to day for the determination of N-alkylvalines

Analyte	Mean N-alkylvaline content of the control material		Precision from day to day		
	[µg/L blood]	[µg/g globin]	n	s_w [%]	u [%]
N-2-cyanoethylvaline	10	408	10	13.0	29.0
N-2-hydroxyethylvaline	2.3	99	6	12.0	29.3
N-methylvaline	5	265	6	11.0	26.9

9.2 Accuracy

In order to check the accuracy of the method, globin with a defined content of N-substituted valines must be used and compared with a certain weighed amount of a standard globin containing defined concentrations of N-alkylvalines. As such reference globin has been unavailable until now, no particulars about the accuracy of the method can be given.

9.3 Detection limit

The following detection limits (Table 8) were calculated (3 times the signal-background ratio) under the given conditions for sample preparation and the GC/MS analysis.

Table 8. Detection limits

Analyte	Detection limit	
	[μg/L blood]	[pmol/g globin]
N-2-cyanoethylvaline	0.3	12
N-2-hydroxyethylvaline	0.4	19
N-methylvaline	0.2	12

9.4 Sources of error

Globin cannot be selectively precipitated out of whole blood. The applicability of the method is therefore primarily ensured through the separation of intact erythrocytes from the blood plasma which contains further proteins. For this reason separation must be carried out as soon as possible after the blood specimen has been collected. After separation, the erythrocytes must be thoroughly washed with NaCl solution to free them from interfering proteins. Otherwise, undesired proteins, such as serum albumin, can remain in solution during processing and be precipitated only when ethyl acetate is added, which reduces the purity of the isolated globin. Partial haemolysis also produces erroneous results in the adduct analysis. Thus, an 80 % reduction in the adduct content was found for the determination of N-methylvaline from haemolysed whole blood samples compared to samples in which haemolysis had not occurred. Therefore only fresh whole blood samples which have not undergone haemolysis should be used.

During sample processing the isolation of globin should be carried out to completion in order to prevent possible artefacts. Törnqvist [15] was able to prove that the N-2-hydroxyethylvaline content of globin in haemolysed samples, even when they are kept deep-frozen, can increase up to 8 times within six months in certain cases. The mechanisms responsible could not be clarified.

Therefore haemolysed erythrocytes should not be deep-frozen for long periods. In contrast, the adduct content of isolated globin is stable at −20 °C for at least 3 months.

The purity of the formamide used in this method is of critical importance. Formamide from various manufacturers and of different degrees of purity was used during the development of this method. It was established that, due to production or storage, the formamide can be contaminated by amines. The alkaline pH value due to the amines can cause a reduced yield in the Edman cyclization reaction. In addition, the presence of free amines can lead to unwanted side reactions. While ammonia can be relatively easily driven out of the formamide by purging it with nitrogen, removal of other amines is a laborious process. Experience with this method has shown that the use of formamide for molecular biological purposes (e.g. from Amersham Life Science or Merck) is most successful.

The users of this method must therefore ensure that the formamide they employ is pure. Volatile alkaline contaminants are indicated by the discoloration of litmus paper in the vapour above the formamide.

After evaporation of the diethyl ether extracts, precipitates have been known to occur which are presumably generated by reactions of the derivatization reagent with contaminants in the chemicals used. In this case the sample preparation can be continued as long as the subsequent washing of the toluene phase with water and 0.1 M sodium carbonate solution is carefully carried out (cf. Section 3.3).

Contamination of the chemicals by N-substituted valines or pipecolinic acid was not detected. In the case of N-2-hydroxyethylvaline an unidentified interfering peak sometimes occurred at the ion source m/e = 350 used as a qualifier. However, the ion source used for the evaluation m/e = 308 was not influenced by it.

As the investigations of one examiner have shown, the globin can also be centrifuged at 1800 g (cf. Section 3.2) without obvious interference occurring.

In the meantime, there is evidence that the analytical results can be influenced by the globin matrix in certain cases. In order to minimize the resulting error, as many different blood samples from non-smokers as possible should be used to prepare the pooled globin for calibration.

10 Discussion of the method

The preparation of human globin described in this method is based on a process developed by Mowrer et al. [16]. Besides minimizing the use of solvents, the main benefit of this method lies in the high quality of the isolated globin which has a direct effect on the reliability of the subsequent determination of the adducts. In contrast to Mowrer's method, no dialysis of the haemolytic blood was carried out. Alternatively, smaller molecules can be removed from the globin by the considerably simpler, shorter and equally effective procedure of gel permeation chromatography. In principle, however, such a step is not necessary for the subsequent analysis of the adducts of the N-terminal valine. When the standard operating procedure (SOP) described here is followed, determination of the adduct content of the N-terminal valine can be carried out without interference [17].

The original version of the multi-step Edman degradation requires both a drastic elevation of the pH value and a temperature increase to about 80 °C. In addition, almost all the N-terminal valines are cleaved from the globin and converted to their derivatives so that analytical problems can be expected for the determination of N-substituted amino acids which occur in the ratio of about $1 : 10^5$ compared to the unsubstituted amino acids.

Ehrenberg and co-workers thus developed a variation of the reaction which is generally known as the "modified Edman degradation" (cf. Figure 6) [18] and which enables a selective cleavage of the alkylated valines. A simplification of the reaction ("one-pot reaction") and the use of perfluorinated phenyl isothiocyanate were introduced to improve the detection sensitivity in the ECD or NCI mass spectrometer. The neutral pH value in the reaction solution primarily ensures a limited selectivity of the cyclization reaction for the adduct-carrying amino acids. Thus cyclization to the thiazole-5-one cation is facilitated. The reaction at pH 7 favours the cyclization of alkylated valines. However, the Edman degradation reactions are equilibrium reactions, with the exception of the electrophilic addition to the thiocarbamyl derivative. Adherence to the optimum pH range of 6–7 is therefore of central importance for the reproducibility of the method.

Calibration of the method is carried out using standard substances in which an appropriately modified valine is bound to another amino acid by way of a peptide bond. Only in this case are the pK_a values of the standards and thus the derivatization yields equivalent to those of the N-alkylvaline from the globin.

Until recently, in order to quantify e. g. the N-2-hydroxyethylvaline adducts, globin has been treated with ^{14}C-ethylene oxide and its N-2-hydroxyethylvaline content has been determined by scintillation counting after hydrolysis of the protein and liquid chromatographic separation of the amino acids. An analogue incubation with non-radioactive ethylene oxide then generates globin with the same content of modified amino acids which can be used as a standard for calibration. The generation of such a standard is relatively complicated and is possible only in laboratories in which radioactive labelling and measurement can be carried out. In addition, the techniques of protein hydrolysis and amino acid separation must be mastered. Therefore, the widespread use of the Edman method for biochemical effect monitoring is limited if the type of calibration described above is necessary.

The use of free modified amino acids offers an alternative calibration method. The disadvantage of this calibration method for the modified Edman degradation lies in the differing derivatization yield for free and protein-bound amino acids. As a result of the low reactivity of the free carboxyl group compared with the peptide bond, the cyclization reaction to the pentafluoroanilino thiazolinone cation proceeds with higher yields for the protein-bound valines. In practice the use of free amino acids for calibration results in N-2-hydroxyethylvaline being over-determined by a factor of about 8 [18]. Round-robin experiments have verified the systematic differences which arise from the choice of various standards [19].

An alternative to calibration with protein standards or free amino acids was established and applied in the course of examining this method [17]. The dipeptides N-2-cyanoethylvaline-leucine-anilide, N-2-hydroxyethylvaline-leucine-anilide and N-methylvaline-leucine-anilide were used for the determination of the N-alkylvalines. These standards are equivalent to the N-terminus of the alkylated α-globin chain. The amino group of the valine carries the alkyl adduct, while the carboxyl group of the leucine is protected by the anilide. Therefore the standards exhibit a pK_a value equivalent to that of the protein-bound amino acid in the α-chain of the globin and are assumed to lead to identical results. The results of the alkylvaline analyses are in good agreement with the results achieved using the protein standard in the literature. The calibration method presented here can be regarded as a suitable and advantageous alternative to the radioactively labelled protein standard.

In the described method it is possible to use either 2-N-ethoxyethylvaline-alanine-anilide or DL-pipecolinic acid as an internal standard. As the chemical behaviour of 2-N-ethoxyethyl-valine-alanine-anilide is distinctly closer to that of the analytes, however, it is preferable as an internal standard.

The precision both in the series and from day to day of this procedure to determine the alkyl haemoglobin adducts can be regarded as satisfactory. It was impossible to determine the recovery rate and thus the accuracy of the method, as neither globin with a defined content of N-alkyl-substituted valines nor the equivalent thiohydantoin derivative of these compounds were available. In the literature recovery rates of 35–40 % are given [20]. It can be assumed that the recovery of the analytes using this method would yield results of a similar order of magnitude.

The detection limits of the method are adequate for the range relevant to occupational and environmental medicine and their agreement with those given in the current literature is good. All three parameters are therefore suitable for the detection of the content of endogenous adducts or those caused by individual lifestyles. Table 9 shows the mean concentrations and ranges for N-2-cyanoethylvaline, N-2-hydroxyethylvaline and N-methylvaline measured with the present method in people who were not occupationally exposed to these substances.

Instruments used:
HP Engine System consisting of:
Gas chromatograph 5890 Series II from Hewlett Packard
Mass spectrometer 5989 A from Hewlett Packard

Table 9. Mean Concentrations and ranges for N-2-cyanoethylvaline, N-2-hydroxyethylvaline and N-methylvaline for people who have not been occupationally exposed to acyrlonitrile, ethylene oxide and methylating substances

Adduct	Smoker or non-smoker	Concentration (range)		Literature data (range) [21]	
		[pmol/g globin]	[µg/L blood]	[pmol/g globin]	[µg/L blood]
N-2-cyanoethylvaline	NS	< 20	< 0.5		
	S	150 (30 - 365)	4.0 (0.8 - 9.2)	16 (9 - 29)	0.5 (0.2 - 0.8)
N-2-hydroxyethylvaline	NS	< 75	< 2		
	S	150 (< 75 - 550)	4.0 (< 2 - 14)	143 (50 - 355)	3.9 (1.4 - 9.6)
N-methylvaline	NS	320 (170 - 480)	7.0 (3.5 - 10.4)	225 (166 - 284)	5.0 (3.7 - 6.3)
	S	390 (248 - 530)	8.5 (5.1 - 11.4)	268 (217 - 372)	6.0 (4.8 - 8.2)

Addendum

While testing this method one of the examiners deviated from the operational specifications described above in several points. The following changes were made:

In each case 100 mg of globin are dissolved in 1.5 mL formamide. After addition of NaOH and the IS, 20 µL pentafluorophenyl isothiocyanate are added to convert the N-terminus to its derivative. The derivatization is carried out at a temperature of 80 °C for three hours. Tertiary butyl ether is used for the subsequent extraction of the solutions. The volume of the combined ether phases is reduced by vacuum centrifugation. The final volume of the samples is 50 µL.

Furthermore, a different combination of GC/MS instruments (consisting of: gas chromatograph 5890 Series II from Hewlett Packard, mass selective detector 5970 A from Hewlett Packard and autosampler 7673 A from Hewlett Packard) was used.

The detection limits (calculated as 3 times the signal/background ratio) obtained under the modified conditions for the sample preparation GC/MS determination are given in Table 10.

Table 10. Detection limits

Analyte	Detection limit	
	[µg/L blood]	[pmol/g globin]
N-2-cyanoethylvaline	0.5	20
N-2-hydroxyethylvaline	1.7	75
N-methylvaline	0.9	46

Otherwise the results obtained using the modified procedure are essentially comparable with those attained using the described test specifications.

11 References

[1] *Römpp Chemie Lexikon*. 9th extended and revised edition. Georg Thieme Verlag, Stuttgart – New York 1989.

[2] *D. Fanini, N. M. Trieff, V. M. Sadagopa-Ramanujam, A. E. Ahmed and P. M. Adams*: Effect of acute acrylonitrile exposure on metrazol induced seizures in the rat. Neurotoxicology *6(1)*, 29 – 34 (1985).

[3] *H. Greim* (ed.): Gesundheitsschädliche Arbeitsstoffe. Toxikologisch-arbeitsmedizinische Begründungen von MAK-Werten. Deutsche Forschungsgemeinschaft. VCH Verlagsgesellschaft, Weinheim, 20th issue (1994).

[4] *Deutsche Forschungsgemeinschaft*: List of MAK and BAT Values 1995. Maximum Concentrations at the Workplace and Biological Tolerance Values for Working Materials. Commission for the Investigation of Health Hazards of Chemical Compounds in the Work Area. Report No. 31. VCH Verlagsgesellschaft, Weinheim 1995.

[5] *S. Moeschlin* (ed.): Klinik und Therapie von Vergiftungen. Georg Thieme Verlag, Stuttgart 1986.

[6] International Agency for Research on Cancer: Monographs on the evaluation of the carcinogenic risk of chemicals to humans: Allyl compounds, aldehydes, epoxides and peroxides. IARC Monographs Vol. 36, Lyon, IARC 1974

[7] *H. Greim* and *G. Lehnert* (eds.): Biologische Arbeitsstoff-Toleranz-Werte (BAT-Werte) und Expositionsäquivalente für krebsezeugende Arbeitsstoffe (EKA). Arbeitsmedizinisch-toxikologische Begründungen. Deutsche Forschungsgemeinschaft. VCH Verlagsgesellschaft, Weinheim, 7th issue(1995).

[8] International Agency for Research on Cancer: Monographs on the evaluation of the carcinogenic risk of chemicals to humans: some halogenated hydrocarbons and pesticide exposures. IARC Monographs Vol. 41. Lyon, IARC 1986.

[9] International Agency for Research on Cancer: Monographs on the evaluation of the carcinogenic risk of chemicals to humans: Supplement 7: Overall evaluation of carcinogenicity: An updating of IARC Monographs Vols. 1 to 42. Lyon, IARC 1987.

[10] International Agency for Research on Cancer: Monographs on the evaluation of the carcinogenic risk of chemicals to humans: some aromatic amines, hydrazines and miscellaneous alkylating agents. IARC Monographs Vol. 4. Lyon, IARC 1974.

[11] *H. F. Bunn:* Hemoglobin. In: A. Haeberli (ed.): Human Protein Data. VCH Verlagsgesellschaft, Weinheim, (1992).

[12] *Bundesministerium für Arbeit und Sozialordnung*: TRgA 410. Statistische Qualitätssicherung. In: Technische Regeln und Richtlinien des BMA zur Verordnung über gefährliche Stoffe. Bek. des BMA vom 9.4.1979, Bundesarbeitsblatt 5/1979, p. 88 ff.

[13] *J. Angerer* and *K. H. Schaller*: Erfahrungen mit der statistischen Qualitätskontrolle im arbeitsmedizinisch-toxikologischen Laboratorium. Arbeitsmed. Sozialmed. Präventivmed. *1*, 33–35 (1977).

[14] *M. Bader*: Gaschromatographisch-massenspektrometrische Analyse von Proteinaddukten als Beitrag zum Biochemischen Effekt-Monitoring kanzerogener Arbeitsstoffe. Dissertation. Universität Erlangen-Nürnberg, 1996.

[15] *M. Törnqvist*: Formation of reactive species that lead to hemoglobin adducts during storage of blood samples. Carcinogenesis *11*, 51–54 (1990).

[16] *J. Mowrer, M. Törnqvist, S. Jensen* and *L. Ehrenberg*: Modified Edman degradation applied to hemoglobin for monitoring occupational exposure to alkylating agents. Toxikol. Environ. Chem. *11*, 215–231 (1986).

[17] *M. Bader, J. Lewalter* and *J. Angerer*: Analysis of N-alkylated amino acids in human hemoglobin: Evidence for elevated N-methylvaline levels in smokers. Int. Arch. Occup. Environ. Health *67*, 237–242 (1995).

[18] *L. Ehrenberg, S. Osterman-Golkar, D. Segerbäck, K. Svensson* and *C. J. Calleman*: Evaluation of genetic risks of alkylating agents. III. Alkylation of haemoglobin after metabolic conversion of ethene to ethene oxide in vivo. Mutat. Res. *45(2)*, 175–184 (1977).

[19] *M. Törnqvist, A. L. Magnusson, P. B. Farmer, Y. S. Tang, A. M. Jeffrey, L. Wazneh, G. D. Beulink, H. van-der-Waal* and *N. J. van Sittert*: Ring test for low levels of N-2-hydroxyethylvaline in human hemoglobin. Anal. Biochem. *203(2)*, 357–360 (1992).

[20] *M. Törnqvist, J. Mowrer, S. Jensen* and *L. Ehrenberg*: Monitoring of environmental cancer initiators through hemoglobin adducts by a modified Edman degradation method. Anal. Biochem. *154*, 255–266 (1986).

[21] *M. Törnqvist, M. Svartengren* and *Ch. Ericsson*: Methylations in hemoglobin from monozygotic twins discordant for cigarette smoking: Hereditary and tobacco-related factors. Chem. Biol. Interact. *82*, 91–98 (1992).

Author: *N. J. van Sittert*

Examiners: *J. Angerer, M. Bader, M. Blaszkewicz, D. Ellrich, A. Krämer, J. Lewalter*

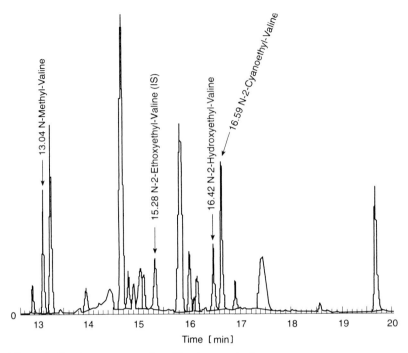

Figure 2. Gas chromatogram of a calibration standard

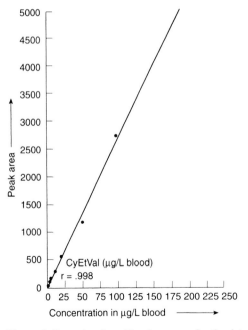

Figure 3. Example of a calibration curve for the determination of N-2-cyanoethylvaline

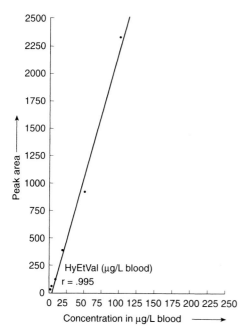

Figure 4. Example of a calibration curve for the determination of N-2-hydroxyethylvaline

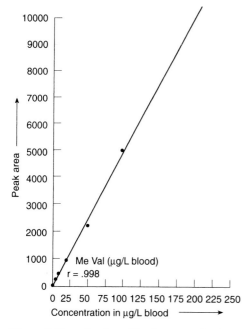

Figure 5. Example of a calibration curve for the determination of N-methylvaline

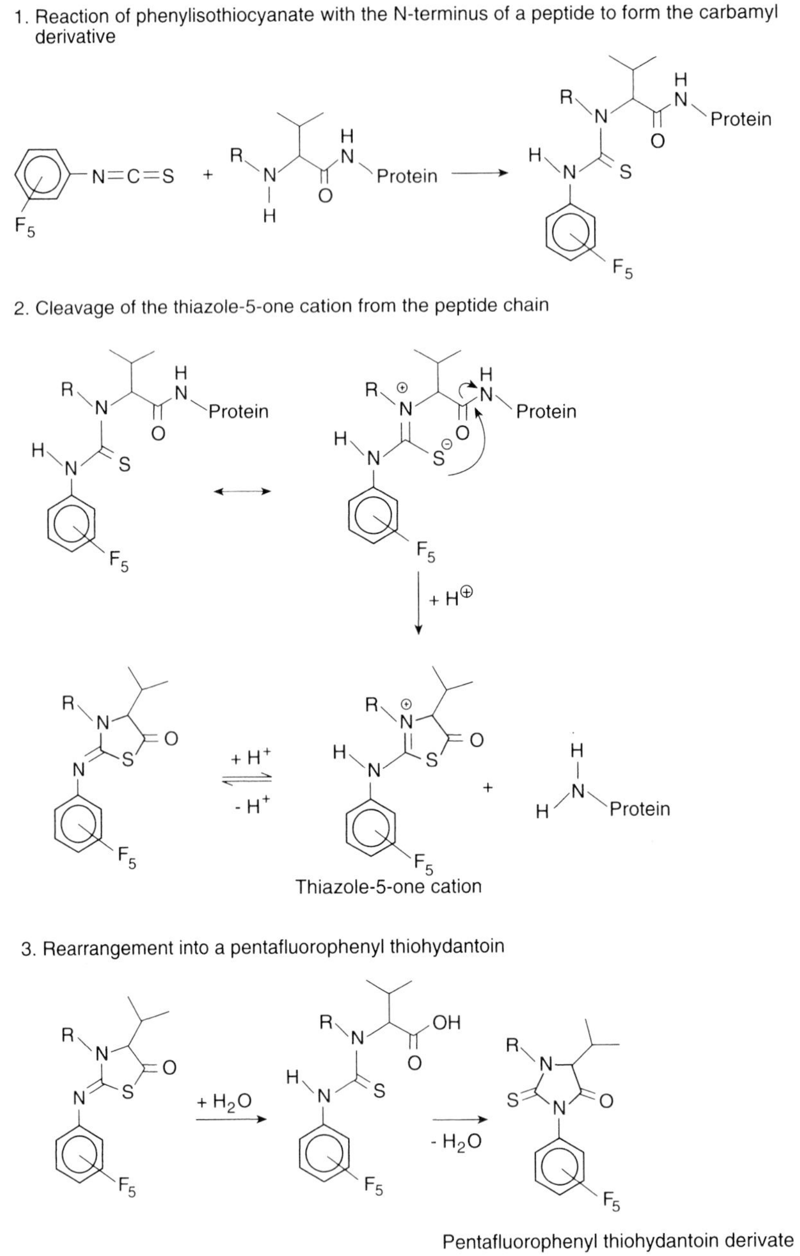

1. Reaction of phenylisothiocyanate with the N-terminus of a peptide to form the carbamyl derivative

2. Cleavage of the thiazole-5-one cation from the peptide chain

Thiazole-5-one cation

3. Rearrangement into a pentafluorophenyl thiohydantoin

Pentafluorophenyl thiohydantoin derivate

Figure 6. Mechanism of the modified Edman degradation [16]

Halogenated hydrocarbons (dichloromethane, 1,2-dichloro-ethylene, 2-bromo-2-chloro-1,1,1-trifluoroethane (halothane), trichloromethane, 1,1,1-trichloro-ethane, tetrachloromethane, trichloroethylene, tetrachloro-ethylene)

Application	Determination in blood
Analytical principle	Capillary gas chromatographic headspace technique
Completed in	June 1990

Summary

This is a sensitive method for the determination in one analytical operation of eight halogenated hydrocarbons in blood. It is suitable for occupational health surveillance. Moreover, the procedure can be adapted to allow sensitive determination of tetrachloro-ethylene, trichloroethylene and trichloroethane. However, considerably more extensive precautions must be taken to prevent contamination in the latter case.

The volatile halogenated hydrocarbons present in blood are determined by gas chromatography using the headspace analysis. Blood samples are warmed in sealed, airtight crimp top vials. After distribution of the halogenated hydrocarbons between the liquid and vapour phases has reached equilibrium, an aliquot is taken from the headspace and analysed by gas chromatography with the aid of an electron capture detector (ECD).

Calibration curves are obtained by analysing blood samples to which known amounts of halogenated hydrocarbons have been added. The resulting peak areas are plotted as a function of the concentrations used. An aqueous solution of the halogenated hydrocarbons can be used as an external standard. As the relationship between the results obtained

Essential Biomonitoring Methods. DFG, Deutsche Forschungsgemeinschaft
Copyright © 2006 WILEY-VCH Verlag GmbH & Co. KGaA, Weinheim
ISBN: 3-527-31478-4

for this aqueous standard and those achieved for the calibration standard solutions in blood is constant, they can be used for calibration for about six months.

Dichloromethane

Within-series imprecision: Standard deviation (rel.) $s_w = 6.9\%$
Prognostic range $u = 15.7\%$
At a concentration of 742 µg dichloromethane per litre blood and where $n = 10$ determinations

Inaccuracy: Recovery rate $r = 94\%$

Detection limit: 25 µg Dichloromethane per litre blood

2-Bromo-2-chloro-1,1,1-trifluoroethane

Within-series imprecision: Standard deviation (rel.) $s_w = 10.1\%$
Prognostic range $u = 22.7\%$
At a concentration of 204 µg halothane per litre blood and where $n = 10$ determinations

Inaccuracy: Recovery rate $r = 79-112\%$

Detection limit: 0.2 µg Halothane per litre blood

1,2-Dichloroethylene

Within-series imprecision: Standard deviation (rel.) $s_w = 7.9\%$
Prognostic range $u = 17.8\%$
At a concentration of 699 µg 1,2-dichloroethylene per litre blood and where $n = 10$ determinations

Inaccuracy: Recovery rate $r = 89-114\%$

Detection limit: 55 µg 1,2-Dichloroethylene per litre blood

Trichloromethane

Within-series imprecision: Standard deviation (rel.) $s_w = 5.9\%$
Prognostic range $u = 13.4\%$
At a concentration of 101 µg trichloromethane per litre blood and where $n = 10$ determinations

Inaccuracy: Recovery rate $r = 82-109\%$

Detection limit: 0.8 µg Trichloromethane per litre blood

1,1,1-Trichloroethane

Within-series imprecision: Standard deviation (rel.) $s_w = 7.8\%$
Prognostic range $u = 17.7\%$
At a concentration of 144 µg 1,1,1-trichloroethane per litre blood and where $n = 10$ determinations

Inaccuracy: Recovery rate $r = 78-112\%$

Detection limit: 1.0 µg 1,1,1-Trichloroethane per litre blood

Tetrachloromethane

Within-series imprecision: Standard deviation (rel.) $s_w = 3.8\%$
                            Prognostic range $u = 8.5\%$
                            At a concentration of 168 µg carbon tetrachloride per litre
                            blood and where $n = 10$ determinations
Inaccuracy: Recovery rate $r = 86{-}117\%$
Detection limit: 0.3 µg Carbon tetrachloride per litre blood

Trichloroethylene

Within-series imprecision: Standard deviation (rel.) $s_w = 6.3\%$
                            Prognostic range $u = 14.3\%$
                            At a concentration of 148 µg trichloroethylene per litre blood
                            and where n = 10 determinations
Inaccuracy: Recovery rate $r = 77{-}101\%$
Detection limit: 1.1 µg Trichloroethylene per litre blood

Tetrachloroethylene

Within-series imprecision: Standard deviation (rel.) $s_w = 7.1\%$
                            Prognostic range $u = 16.1\%$
                            At a concentration of 178 µg tetrachloroethylene per litre
                            blood and where n = 10 determinations
Inaccuracy: Recovery rate $r = 82{-}99\%$
Detection limit: 0.5 µg Tetrachloroethylene per litre blood

Halogenated hydrocarbons

With the exception of 2-bromo-2-chloro-1,1,1-trifluoroethane, all the halogenated hydrocarbons in question have industrial and household uses, primarily as solvents. In particular, on account of their capacity to readily dissolve fats and oils they are widely used for surface treatment, for dry-cleaning of textiles as well as for the extraction of fats and aroma substances. As a rule, the halogenated hydrocarbons are very stable. Their degradation in the environment, e.g. as a result of photochemical reactions, generally proceeds very slowly.

Approximately 25 aliphatic halogenated hydrocarbons are included in the list of German MAK/BAT values [1]. In addition to some of their physico-chemical properties, Table 1 shows the MAK and BAT values for a number of compounds of significance to occupational medicine, the peak limitation categories to which they have been assigned, as well as their carcinogenic properties. Comprehensive data about the properties, industrial importance and distribution of this heterogeneous group of compounds are given in recent reviews [2, 3] as well as the General Introduction to the loose-leaf collection "Analysen in biologischem Material" which contains descriptions of analytical methods for determination of the individual substances.

Detailed descriptions of the intake, metabolism, excretion and kinetics of the various aliphatic halogenated hydrocarbons are included in the special "Arbeitsmedizinisch-toxikologische Begründungen von MAK- und BAT-Werten" [4, 5]. Relevant enzymatic metabolic pathways are presented in an overview by *Anders* and *Jacobson* [6].

The toxicity of most halogenated solvents is primarily dependent on their biotransformation in the human organism. Either stable toxic metabolites or reactive, electrophilic intermediates are formed. Thus, occupational medicine is not only concerned with the typical acute and chronic toxic symptoms of excessive exposure to certain solvents containing halogenated hydrocarbons, but also with the carcinogenic properties of these substances. This is especially true of the halogenated ethylenes [7]. Cases of illness caused by halogenated hydrocarbons are considered in the German list of occupational diseases, published in April 1988 [8], with the No. 1302. As a rule, about ten cases are awarded compensation for the first time every year [3].

Professional Association's Guidelines for Occupational Health Examinations have laid down principles concerning preventative examinations for the medical surveillance of persons occupationally exposed to halogenated hydrocarbons [9]. The solvent or metabolite concentrations in blood or urine measured as part of the biomonitoring of certain halogenated hydrocarbons are judged on the basis of the BAT values (Table 1).

The wide industrial use, especially of trichloroethylene and tetrachloroethylene, and the great biological persistence of these solvents account for their occurrence in the environment. The fine sensitivity of the analytical detection systems used here permits determination of some of these halogenated hydrocarbons in the blood of persons who do not come in contact with these substances at work ("normal persons"). However, contamination which often occurs during specimen collection or from the vessels or their caps must be taken into account for the "normal values" given in the literature. As a rule, the mean concentrations of the solvents trichloroethylene and tetrachloroethylene in blood samples of the general population are less than 1 µg/L. *Hajimiragha* et al. [10] found the following levels of halogenated hydrocarbons in the blood of 39 persons who were not exposed to these substances at work: trichloromethane: median 0.2 µg/L (range < 0.1–1.7 µg/L); 1,1,1-trichloroethane: median 0.2 µg/L (range < 0.1–3.4 µg/L); tetrachloroethylene: median 0.4 µg/L (range < 0.1–3.7 µg/L); trichloroethylene: median 0.1 µg/L (range < 0.1–1.3 µg/L). The concentration of halogenated hydrocarbons in blood ($n = 218$) shown in Table 2 were determined in a current cross-sectional study.

Table 1: Characteristic data for certain halogenated hydrocarbons which are relevant to occupational medicine.

Substance	MAK value mL/m^3	Peak limit. category	Carcinogenic group	Vapour pressure in mbar at 20°C	Boiling point in °C	BAT value µg/L blood	Time of sampling
Chlorethane	1000	n. b.	–	–	13.1	–	–
1,1-Dichloroethane	100	II, 1	–	240	57.3	–	–
1,2-Dichloroethane	–	–	III A 2	87	83.7	–	–
1,2-Dichloroethylene	200	II ,1	–	220	60 (cis-) 47 (trans-)	–	–
Dichloromethane	100	II ,2	III B	475	40.1	1000	b
1,2-Dichloro-1,2,2-tetrafluoroethane (R 114)	1000	IV	–	–	3.55	–	–
2-Bromo-2-chloro-1,1,1-trifluoroethane (Halothane)	5	II ,1	–	242	50.2	2500**	b, c
Hexachloroethane	1	n. b.	–	–	185 (sublimated)	–	–
1,1,2,2-Tetrachloroethane	1	–	III B	7	146.35	–	–
Tetrachloroethylene	50	II ,1	III B	19	121.1	1000	d
Tetrachloromethane	10	II ,1	III B	120	76.7	–	–
Trichloroethylene	50	II, 2	III B	77	87	5000*	b, c
1,1,2-Trichloroethane	10	II, 2	III B	25	113.5	–	–
1,1,1-Trichloroethane	200	II, 2	–	133	74.1	550	c, d
1,1,2-Trichloro-1,2,2-trifluoroethane (R 113)	500	IV	–	360	47.57	–	–
Trichloromethane	10	II, 1	III B	210	61.7	–	–

a) = not fixed

* measured as trichloroethanol

c) = for long-term exposures: after several shifts

Table 2: Tetrachloroethylene, 1,1,1-trichloroethane and trichloroethylene in blood samples of unexposed persons.

	Tetrachloro-ethylene	1,1,1-Trichloro-ethane	Trichloro-ethylene
Proportion of the results below the DL [%]	42.7	49.1	78
Median [ng/L]	39.5	38.0	< DL
95-percentile [ng/L]	381.0	590.5	185.5
Mean value [ng/L]	107.5	155.5	32.0
Range [ng/L]	< DL – 2392	< DL – 3091	< DL – 468

DL = Detection limit

Author: *J. Angerer*
Examiners: *H. Muffler, R. Eisenmann*

Halogenated hydrocarbons (dichloromethane, 1,2-dichloroethylene, 2-bromo-2-chloro-1,1,1-trifluoroethane (halothane), trichloromethane, 1,1,1-trichloroethane, tetrachloromethane, trichloroethylene, tetrachloroethylene)

Application   Determination in blood

Analytical principle   Capillary gas chromatographic headspace technique

Completed in   June 1990

Contents

Essential Biomonitoring Methods. DFG, Deutsche Forschungsgemeinschaft
Copyright © 2006 WILEY-VCH Verlag GmbH & Co. KGaA, Weinheim
ISBN: 3-527-31478-4

1 General principles

The volatile halogenated hydrocarbons present in blood are determined by gas chromatography using the headspace analysis. Blood samples are warmed in sealed, airtight crimp top vials. After distribution of the halogenated hydrocarbons between the liquid and vapour phases has reached equilibrium, an aliquot is taken from the headspace and analysed by gas chromatography with the aid of an electron capture detector (ECD).

Calibration curves are obtained by analysing blood samples to which known amounts of halogenated hydrocarbons have been added. The resulting peak areas are plotted as a function of the concentrations used. An aqueous solution of the halogenated hydrocarbons can be used as an external standard. As the relationship between the results obtained for this aqueous standard and those achieved for the calibration standard solutions in blood is constant, they can be used for calibration for about six months.

2 Equipment, chemicals and solutions

2.1 Equipment

Gas chromatography with split-splitless injector, electron capture detector, analogous recorder as well as possibly with integrator and plotter

Dry thermostat, adjustable to 60 °C with moulds into which the crimp top vials can be fitted

250 µL Syringe for gas chromatography, airtight

Alternatively:

Instead of the dry thermostat and the airtight syringe, automatic equipment for injecting headspace samples can be used to advantage.

Microlitre pipettes, adjustable between 10 and 100 µL as well as 100 and 1000 µL (e.g. from Eppendorf)

Magnetic stirrer

1, 2, 4, 5, 8, 10 and 20 mL Transfer pipettes

20, 50,100, 250 and 1000 mL Volumetric flasks

20 mL Crimp top vials with teflon-coated butyl rubber stoppers and aluminium caps as well as crimping tongs for sealing and opening them. The crimp top vials and especially the stoppers must be heated in the drying cupboard at 110 °C for at least two days. It is advisable to heat the stoppers for a week.

Disposable syringes containing an anticoagulant (e.g. Monovetten® from Sarstedt containing potassium EDTA)

2.2 Chemicals

Chemicals of the highest available purity should be used.
Dichloromethane (e.g. Pestanal from Riedel-de Haën)
2-Bromo-2-chloro-1,1,1-trifluoroethane (halothane), 99 % (e.g. from Aldrich)
1,2-Dichloroethylene for gas chromatography (e.g. from Riedel-de Haën)
Trichloromethane (chloroform) (e.g. Pestanal from Riedel-de Haën)
1,1,1-Trichloroethane for gas chromatography (e.g. from Riedel-de Haën)
Tetrachloromethane (carbon tetrachloride) for gas chromatography (e.g. from Riedel-de Haën)
Trichloroethylene for gas chromatography (e.g. from Riedel-de Haën)
Tetrachloroethylene for gas chromatography (e.g. from Riedel-de Haën)
2-Ethoxyethanol
Ammonium oxalate, p.a. (e.g. from Merck)
Defibrinated sheep's blood (e.g. from GLD, Gesellschaft für Labordiagnostica, Essen, FRG)
Ultrapure water (ASTM type 1) or double distilled water
Physiological saline (154.0 mmol/L) (e.g. isotonic saline solution from Fresenius)
K_2-EDTA (cf. Section 3)
Purified nitrogen (99.999 %)

2.3 Solutions

Ammonium oxalate solution (10 g/L):
10 g ammonium oxalate are dissolved in ultrapure water in a 1000 mL volumetric flask. The flask is filled up to the mark with ultrapure water. The solution can be stored for about 12 weeks at room temperature.

2.4 Preparation of the crimp top vials

Crimp top vials (20 mL) each containing about 20 mg ammonium oxalate as an anticoagulant are prepared in advance to receive the blood samples. 2 mL of an aqueous solution of ammonium oxalate (see Section 2.3) are placed in the 20 mL crimp top vials. The solution is evaporated in a drying cupboard at 40 °C. The anticoagulant remains as a finely dispersed precipitate on the walls of the vessels. After heating and cooling, the teflon-coated rubber stoppers and the aluminium caps, are used to seal the crimp top vials. Alternatively, crimp top vials each containing approximately 5 mg K_2-EDTA in solid form can be also used.

2.5 Calibration standards

a) Solution of the halogenated hydrocarbons in 2-ethoxyethanol and water

Starting solution in 2-ethoxyethanol:
After placing about 15 mL 2-ethoxyethanol in a 20 mL volumetric flask, 100 µL dichloromethane, 1,2-dichloroethylene, as well as 20 µL 2-bromo-2-chloro-1,1,1-trifluoroethane, trichloromethane, 1,1,1-trichloroethane, tetrachloromethane, trichloroethylene and tetrachloroethylene are each pipetted into the flask one after another. The mass of the halogenated hydrocarbons added is determined gravimetrically. The volumetric flask is filled up to its nominal volume to give a solution containing the individual halogenated hydrocarbons at concentrations ranging from 1000 to 6000 mg/L. Table 3 shows the volumes of the individual halogenated hydrocarbons added, their corresponding masses as well as their concentrations in this starting solution.

Stock solution in water:
1 mL of the starting solution is pipetted into a 1000 mL volumetric flask, 5 mL 2-ethoxyethanol are added and the flask is subsequently filled up to the mark with ultrapure water. The concentrations of the individual halogenated hydrocarbons in the stock solution are shown in Table 3.

Aqueous external standard:
100 mL of the stock solution in water are pipetted into a 1000 mL volumetric flask which is filled up to the mark with ultrapure water. The concentrations of the individual halogenated hydrocarbons are given in Table 3.

2 mL of this external standard solution are pipetted into each 20 mL crimp top vial using an automatic pipette. The vials are subsequently tightly sealed with teflon-coated rubber septa and crimped with aluminium caps. These standards are kept in the deep freezer at –16 to –20 °C. In each analytical series one or two standard samples are processed and analysed.

b) Solution of the halogenated hydrocarbons in physiological saline

10 mL of the starting solution of the halogenated hydrocarbons in 2-ethoxyethanol are diluted to the 50 mL mark with physiological saline in a 50 mL volumetric flask. The concentrations of the various halogenated hydrocarbons are shown in the first column of Table 4.

c) Solutions of the halogenated hydrocarbons in blood

Starting solution:
20 mL animal blood are placed in a 250 mL volumetric flask. 5 mL of the solution containing the halogenated hydrocarbons in physiological saline are added. The flask is

filled to the mark with animal blood. The resulting solution is mixed in the sealed volumetric flask at room temperature with a magnetic stirrer. The concentrations of the various halogenated hydrocarbons are shown in the second column of Table 4.

Stock solution:
10 mL of the starting solution are transferred to a 100 mL volumetric flask and filled up to the mark with blood. The stock solution is mixed in the sealed flask at room temperature for 2 h. The solution serves as the stock solution for the preparation of the blood calibration standards. The concentrations of the various halogenated hydrocarbons are shown in the third column of Table 4.

Calibration standard solutions of the halogenated hydrocarbons in blood:
Between 0.1 and 4 mL of the stock solution of the halogenated hydrocarbons in blood are pipetted into a 10 mL volumetric flask, which already contains 2 mL blood (see Table 5). In addition, a reagent blank is included. This consists of the blood (10 mL) used to prepare the halogenated hydrocarbon solution in blood. Then the flask is filled to the mark with blood. After the calibration standards are thoroughly mixed, they are divided into aliquots (each 2 mL) in crimp top vials. These crimp top vials are immediately sealed with teflon-coated rubber septa and crimped with aluminium caps. The calibration standards can be stored in this state in the deep freezer at -16 to -20 °C. No changes in the concentrations were detectable after a storage period of six months.

3 Specimen collection and sample preparation

For occupational health surveillance the blood specimen for determination of the chlorinated hydrocarbons is taken at the end of the exposure. If the substance in question has been assigned a BAT value the sampling time can be found in the list of BAT values [1]. In order to prevent contamination of the blood specimen by solvents, the arm of the test person is cleaned with soap and water rather than the usual disinfectant. The blood specimen is withdrawn using a disposable syringe containing an anticoagulant (e.g. Monovette from Sarstedt). Approximately 2 mL of the blood sample are injected into a crimp top vial immediately after sampling. The liquid is vigorously swirled around the vessel to dissolve the anticoagulant in the crimp top vial. The sample can be transported in this state, no further precautions are necessary. Samples can be stored in a deep freezer for at least six months (–16 to –20 °C) until analysis without fear of distortion of the analytical results. Before the determination is carried out the samples are allowed to reach room temperature.

The crimp top vials are placed in the moulds of the thermostat or in the automatic equipment for the headspace technique. The samples are incubated at a temperature of 50 °C until equilibrium between the liquid and vapour phase has been reached. The incubation period is 90 min.

4 Operational parameters for gas chromatography

Column:	Material:	Fused silica
	Length:	60 m
	Inner diameter:	0.33 mm
	Stationary phase:	100 % Dimethylpolysiloxane, chemically bonded and cross-linked, DB, film thickness 1.0 μm
Detector:		Electron capture detector with ^{63}Ni as a source of radiation
Temperatures:	Column:	60 °C, isotherm
	Injection block:	230 °C
	Detector:	300 °C
Column pressure:		16 psi
Carrier gas:		Purified nitrogen
Split:		1 : 30
Make up gas:		Purified nitrogen, 40 mL/min
Sample volume:		100 μL

The following retention times of the individual halogenated hydrocarbons can be regarded as a guide:

Dichloromethane	5.18 min
2-Bromo-2-chloro-1,1,1-trifluoroethane	5.58 min
1,2-Dichloroethylene	5.79 min
Trichloromethane	6.84 min
1,1,1-Trichloroethane	7.95 min
Tetrachloromethane	8.76 min
Trichloroethylene	10.21 min
Tetrachloroethylene	20.04 min

Figure 1 shows the gas chromatogram of a calibration standard in blood. The gas chromatogram for the analysis of a whole blood sample of an exposed person is illustrated in Figure 2.

5 Analytical determination

After incubating the crimp top vial for 90 min, 100 μL are taken from the vapour in the headspace with a pre-warmed airtight syringe and injected into the gas chromatograph. Alternatively this can be carried out by an autosampler for the headspace technique.

6 Calibration

A sample of each of the calibration standard solutions shown in Table 5 is analysed as described. A blood sample which has not been spiked with halogenated hydrocarbons serves to determine the reagent blank value. These samples are kept deep frozen until analysis in sealed airtight crimp top vials, as described in Section 2.5. The resulting peak areas for the individual halogenated hydrocarbons are plotted against the corresponding concentrations used. If reagent blank values are found they must be previously subtracted.

Figure 3 shows examples of calibration curves. The calibration curves are linear in the investigated range (cf. Table 5).

The basic necessity of plotting a new calibration curve for each analytical series can be avoided by external standardization. The aqueous solution of the external standard is determined in addition to the calibration standards in blood. A sufficient number of aliquots of this aqueous standard are kept deep frozen in crimp top vials (cf. Section 2.5). One or more of these aqueous standards is included in each analytical series. The relationship is calculated between the resulting value for the aqueous standard and the value obtained for this standard when the calibration curve in blood was plotted. Each concentration obtained from the calibration curve is corrected by this factor.

7 Calculation of the analytical result

a) Evaluation with the aid of calibration curves
When the peak areas have been analytically determined the corresponding concentration of each substance in µg/L can be read off from the calibration curve.

b) Evaluation with the aid of an aqueous external standard
When an external standard is used, the concentration obtained for a halogenated hydrocarbon from the calibration curve is corrected by multiplying it with the quotient of the peak areas measured for the aqueous external standard according to the following formula:

$$c_p^i = c_{cal}^i \cdot \frac{s_a^i}{s_{a+n}^i}$$

c_p^i concentration of substance i in the sample
c_{cal}^i concentration of substance i read off from the calibration curve
s_a^i value for substance i from the aqueous external standard on the day the calibration curve was plotted (day a)
s_{a+n}^i value for substance i from the aqueous external standard on the day of the sample analysis (day a + n)

8 Reliability of the method

8.1 Precision

To determine the within-series imprecision, animal blood was spiked with known quantities of halogenated hydrocarbons to give solutions containing the various halogenated hydrocarbons at concentrations between 101 and 742 µg/L. The blood samples were analysed ten times. The relative standard deviations lay between 3.8 and 10.1%, the corresponding prognostic ranges were between 16.1 and 22.7% (see Table 6).

Blood samples of 18 persons exposed to tetrachloroethylene were determined in duplicate analyses. At tetrachloroethylene concentrations ranging between 85 and 1005 µg/L ($\bar{x}$ = 580 µg/L), a standard deviation of 16 µg/L was calculated from the duplicate analyses. This is equivalent to a relative standard deviation of 2.7%.

To test the precision of the method the samples were injected manually.

8.2 Accuracy

Recovery experiments were carried out to check the accuracy of the method. For this purpose animal blood was spiked with different quantities of each halogenated hydrocarbon. At the lower concentration the recovery rates ranged from 98.7 to 116.9% depending on the halogenated hydrocarbon (cf. Table 7). Dichloromethane was not considered in the evaluation because of a disproportionately high reagent blank value of the animal blood (Table 7).

At the higher concentration recovery rates between 77.2 and 94.4% were calculated (Table 7).

8.3 Detection limit

The detection limits were calculated as three times the signal/background ratio:

Dichloromethane	25 µg/L
2-Bromo-2-chloro-1,1,1-trifluoroethane	0.2 µg/L
1,2-Dichloroethylene	55 µg/L
Trichloromethane	0.8 µg/L
1,1,1-Trichloroethane	1.0 µg/L
Tetrachloromethane	0.3 µg/L
Trichloroethylene	1.1 µg/L
Tetrachloroethylene	0.5 µg/L

8.4 Sources of error

a) Pre-analytical phase

The time at which the blood sample is taken has a considerable effect on the analytical result. The half-life for the elimination of the solvents from the blood is very short in certain cases, so that the solvent level sinks rapidly after exposure. Therefore it is absolutely essential to adhere to the sampling time stipulated in the list of BAT values [1], usually at the end of exposure.

Coagulation of the blood sample alters the distribution of the halogenated hydrocarbons between the biological matrix and the vapour phase at equilibrium and the reproducibility of the results suffers. Thus, the blood sample should be vigorously swirled round the crimp top vial to ensure it is properly mixed with the anticoagulant. The samples should be placed into the crimp top vials immediately after sample collection to prevent loss of the readily volatile halogenated hydrocarbons during transport and storage. It is especially important to ensure that the seals are airtight. It should be impossible to turn the seals by hand. The crimping tongs must be appropriately adjusted.

If Vacutainers® are used for withdrawing the blood sample, losses of solvents occur when the vessels are opened. Headspace analysis is impossible directly from the Vacutainers®. Thus, Vacutainers® are neither suitable for sample collection nor as transport vessels for the determination of halogenated hydrocarbons in blood.

Exogenous contamination can influence the results of this method to determine halogenated hydrocarbons in blood. This applies particularly to the examination of blood samples from the general population. The use of an electron capture detector permits an extremely sensitive detection of the majority of halogenated hydrocarbons in blood. Therefore, e.g., even the slightest traces of halogenated hydrocarbons present in the air of the laboratory can cause erroneous results. In order to prevent the contamination of the samples it is necessary to dispense with the usual disinfectant for cleaning the arm before sampling. Furthermore, the crimp top vials and especially the teflon-coated butyl rubber stoppers must be heated in the drying cupboard at 110 °C for at least 2 days. It is advisable to heat the stoppers for a week. After heating it is absolutely essential to test the sealed crimp top vials for a blank value.

b) Analytical phase

The blood samples should be incubated at 50 °C long enough to ensure that the distribution equilibrium between the liquid and the vapour phase has definitely been reached. If adjustment is incomplete the reliability and precision of the results suffer. An incubation period of 90 min has proved sufficient. However, the user of this method should check how long it takes to reach the distribution equilibrium under the working conditions of his apparatus. If an autosampler is used it is important to allow enough pressurization before injection. As a rule, six minutes is sufficient.

Various possible errors can occur during the withdrawal of an aliquot from the headspace and during its transfer to the separation capillary. It is important to ensure that the syringe is airtight and condensation of the sample components in the syringe must be avoided. The latter is achieved by warming the airtight syringe at 60 °C before the sample vapour is withdrawn. To prevent halogenated hydrocarbons being carried over from one sample to another the syringe is carefully flushed with nitrogen after every sample injection. When an autosampler is used the temperature of the transfer line should be 20 °C higher than the incubation temperature of the samples.

After each analytical series a stream of carrier gas at a high temperature is passed through the gas chromatographic column, preferably overnight. Less volatile contaminants which have settled onto the column are flushed from the stationary phase. This creates active absorptive sites in the gas chromatographic separation system which can lead to lower findings in the subsequent analyses. Therefore an aqueous calibration standard of the highest concentration should be injected into the system several times before the next assay series to saturate the absorptive sites.

The preparation of the calibration standards is of great importance. The procedure chosen here, which involves several dilution steps and the use of a solutizing agent, has proved successful in practice. It ensures the homogeneous dispersion of the halogenated hydrocarbons in blood, although they are only sparingly soluble in aqueous media. At the same time, the preparation of these standards present problems, especially for the inexperienced user. It is particularly important that the standard solutions are prepared in a room in which no halogenated hydrocarbons are detectable. The starting solution in ethoxyethanol and the stock solution in water are prepared in separate rooms.

Tests were carried out to investigate whether interference occurs between various industrially important solvents when they are determined by headspace gas chromatography. · Table 8 shows the retention times of these substances. No interference occurred between the substances investigated here. However, the examiner found that solvent R 113 interfered with the determination of dichloromethane. If exposure to 2-bromo-2-chloro-1,1,1-trifluoroethane as well as 1,2-dichloroethane occurs then it is important to ensure that they are adequately separated or that the separation performance is optimized.

9 Discussion of the method

Eight different halogenated hydrocarbons can be determined in one analytical operation by means of this procedure. For the substances, which have already been assigned a BAT value in blood (dichloromethane 1000 μg/L, tetrachloroethylene 1000 μg/L, 1,1,1-trichloroethane 550 μg/L), concentrations of these solvents between 1/40 and 1/200 of the BAT value can be determined. Although the detection limits for substances with two chlorine atoms are relatively high compared with the substances with three or four chlorine atoms, it is still possible to detect a concentration of 1/40 of the BAT value for dichloromethane.

The reliability criteria given here were determined using manual injection into the gas chromatograph. When an autosampler for headspace injection is used, relative standard deviations of less than 3 % can be achieved at the given concentrations for the substances with three or four chlorine atoms (10 determinations). Relative standard deviations of less than 5 % were calculated for dichloromethane and 1,2-dichloroethylene when the samples were automatically injected.

Numerous errors can occur during the pre-analytical and analytical phases of the method described here (cf. Section 8.4). It is important to take all the above-described precautionary measures to minimize errors and to ensure reliable determination of the halogenated hydrocarbons in blood. Experience with headspace analysis of blood samples is necessary for successful use of this method. Under these conditions the reliability criteria given here can be achieved without difficulty.

The German loose-leaf collection of "Analysen in biologischem Material" [12] gives descriptions of the headspace determination of some of the halogenated hydrocarbons included here. These methods are based on the use of packed columns. As the separation achieved by packed columns is, in some cases, considerably poorer than that of the capillary columns used here, interference between the substances must be expected. The use of a capillary column is especially advantageous not only because it permits the determination of the various halogenated hydrocarbons in one analytical operation, but they are also separated from most of the readily volatile components which are naturally present in blood or which also occur at the workplace. This is of particular importance because workers are frequently exposed to a mixture of solvents at their workplace. In addition to the described halogenated hydrocarbons, even more substances can be detected with this method.

It is possible to lower the detection limits obtained here by at least a factor of 10 when larger sample volumes and other split ratios are used. In this way the blood level of tetrachloroethylene, trichloroethylene and trichloroethane resulting from environmental exposure can be detected. However, considerably more extensive precautions must be taken against exogenous contamination in this case.

As this is the first description of capillary gas chromatographic analysis using the headspace technique published in this series, the following remarks seem appropriate. As many years of experience have shown, separation capillaries have made headspace analysis of organic solvents in blood samples more efficient and practicable. The whole range of exposure to industrial solvents can be covered using only a few separation columns. The so-called fused silica columns with chemically bounded phases have proved to be extraordinarily successful in long-term service. The separation efficiency of the column described here remains extremely satisfactory for routine use over a period of months.

Instruments used:
Gas chromatograph 3300 with ECD and data system DS 650 from Varian

10 References

[1] *Deutsche Forschungsgemeinschaft:* Maximum Concentrations at the Workplace and Biological Tolerance Values for Working Materials 1990. Report XXVI of the Commission for the Investigation of Health Hazards of Chemical Compounds in the Work Area. VCH, Weinheim 1990.

[2] VDI: Halogenierte organische Verbindungen in der Umwelt. VDI, 1989.

[3] *G. Triebig:* Gesundheitsrisiken durch organische Lösemittel in Arbeits- und Umwelt – eine kritische Analyse aus arbeitsmedizinischer Sicht. Hamburger Ärzteblatt *43*, 181–185 (1989).

[4] *D. Henschler* and *G. Lehnert* (eds.): Biologische Arbeitsstoff-Toleranz-Werte (BAT-Werte) und Expositionsäquivalente für krebserzeugende Arbeitsstoffe (EKA). Arbeitsmedizinisch-toxikologische Begründungen. Deutsche Forschungsgemeinschaft, VCH, Weinheim, 1st–4th issue 1983–1989.

[5] *D. Henschler* (ed.): Gesundheitsschädliche Arbeitsstoffe. Toxikologisch-arbeitsmedizinische Begründung von MAK-Werten. Deutsche Forschungsgemeinschaft, VCH, 1st–15th issue 1972–1989.

[6] *M. W. Anders* and *I. Jacobson:* Biotransformation of halogenated solvents. Scand. J. Work Environ. Health *11*, Suppl. 1, 23–32 (1985).

[7] *H. M. Bolt:* Die toxikologische Beurteilung halogenierter Äthylene. Arbeitsmed. Sozialmed. Präventivmed. *15*, 49–53 (1980).

[8] Berufskrankheitenliste: Erkrankungen durch Halogenkohlenwasserstoffe. Bek. des BMA vom 29. März 1988, Bundesarbeitsblatt 6/1988.

[9] *Hauptverband der gewerblichen Berufsgenossenschaften* (ed.): Berufsgenossenschaftliche Grundsätze für arbeitsmedizinische Vorsorgeuntersuchungen. Gentner, Stuttgart.

[10] *H. Hajimiragha, U. Ewers, R. Jansen-Rosseck,* and *A. Brockhaus:* Human exposure to volatile halogenated hydrocarbons from the general environment. Int. Arch. Occup. Environ. Health *58*, 141–150 (1986).

[11] *A. Böttger, I. Schäfer, U. Ewers, R. Engelke,* and *J. Majer:* Belastung der Anwohner von Chemisch-Reinigungen durch Tetrachlorethen. Hygiene-Tagung, Kiel, 29./30. 9. 88.

[12] *D. Henschler* (ed.): Analytische Methoden zur Prüfung gesundheitsschädlicher Arbeitsstoffe. Band 2, Analysen in biologischem Material. VCH, Weinheim.

Author: *J. Angerer*
Examiners: *H. Muffler, R. Eisenmann*

Table 3: Preparation and concentrations of the halogenated hydrocarbons in 2-ethoxyethanol as well as in the stock solution and in the aqueous external standard.

Substance	Volume of the individual halogenated hydrocarbons in 20 mL ethoxyethanol	Corresponds to a mass in 20 mL 2-ethoxyethanol	Starting solution in 2-ethoxyethanol	Stock solution in water	Aqueous external standard
	µL	mg	mg/L	µg/L	µg/L
Dichloromethane	100	105.70	5282.0	5285.0	528.5
1,2-Dichloroethylene	100	111.00	5550.0	5550.0	555.0
2-Bromo-2-chloro-1,1,1-trifluoroethane	20	32.48	1624.0	1624.0	162.4
Trichloromethane	20	24.28	1214.0	1214.0	121.4
1,1,1-Trichloroethane	20	23.87	1193.5	1193.5	119.4
Tetrachloromethane	20	29.47	1473.5	1473.5	147.4
Trichloroethylene	20	29.75	1487.5	1487.5	148.8
Tetrachloroethylene	20	31.90	1595.0	1595.0	159.5

Table 4: Concentrations of the halogenated hydrocarbons in the physiological saline as well as in the starting solution and stock solution in blood.

Substance	Solution in physiological saline mg/L	Starting solution in blood mg/L	Stock solution in blood mg/L
Dichloromethane	1057.0	21.14	2.114
1,2-Dichloroethylene	1110.0	22.20	2.220
2-Bromo-2-chloro-1,1,1-trifluoroethane	324.8	6.50	0.650
Trichloromethane	242.8	4.86	0.486
1,1,1-Trichloroethane	238.7	4.77	0.477
Tetrachloromethane	294.7	5.89	0.589
Trichloroethylene	297.5	5.95	0.595
Tetrachloroethylene	319.0	6.38	0.638

Table 5: Pipetting scheme for the preparation of the calibration standards in blood.

Standard No.	Volume of the stock solution in blood mL	Final volume of the calibration standard in blood mL	Concentration of the calibration standard							
			Dichloro-methane µg/L	1,2-Di-chloro-ethylene µg/L	2-Bromo-2-chloro-1,1,1-trifluoro-ethane µg/L	Trichloro-methane µg/L	1,1,1-Tri-chloro-ethane µg/L	Tetrachloro-methane µg/L	Trichloro-ethylene µg/L	Tetra-chloro-ethylene µg/L
0	–	10	–	–	–	–	–	–	–	–
1	0.1	10	21.14	22.20	6.50	4.86	4.77	5.89	5.95	6.38
2	0.4	10	84.56	88.80	26.00	19.44	19.08	23.56	23.80	25.52
3	1.0	10	211.40	222.00	65.00	48.60	47.70	58.90	59.50	63.80
4	2.0	10	422.80	444.00	130.00	97.20	95.40	117.80	119.00	127.60
5	4.0	10	845.60	888.00	260.00	194.40	190.80	235.60	238.00	255.20

Table 6: Within-series imprecision ($n = 10$) for the determination of eight halogenated hydrocarbons in a whole blood sample.

	Mean value µg/L	Standard deviation µg/L	Standard deviation (rel.) %	Prognostic range %
Dichloromethane	741.7	51.4	6.9	15.7
2-Bromo-2-chloro-1,1,1-trifluoroethane	203.7	20.5	10.1	22.7
1,2-Dichloroethylene	698.5	55.0	7.9	17.8
Trichloromethane	100.6	6.0	5.9	13.4
1,1,1-Trichloroethane	144.4	11.3	7.8	17.7
Tetrachloromethane	167.8	6.3	3.8	8.5
Trichloroethylene	147.9	9.4	6.3	14.3
Tetrachloroethylene	178.1	12.7	7.1	16.1

Table 7: Results of recovery experiments for eight halogenated hydrocarbons in whole blood.

	Added µg/L		Recovered µg/L		Recovery rate %	
	conc. 1	conc. 2	conc. 1	conc. 2	conc. 1	conc. 2
Dichloromethane	252.9	404.7	340.0	381.9	–	94.4
2-Bromo-2-chloro-1,1,1-trifluoroethane	66.6	106.5	74.7	84.3	112.2	79.1
1,2-Dichloroethylene	254.6	407.0	289.7	363.1	113.8	89.2
Trichloromethane	33.0	52.8	35.8	43.1	108.5	81.7
1,1,1-Trichloroethane	49.7	79.5	55.6	62.0	111.8	78.0
Tetrachloromethane	50.8	81.2	59.3	70.0	116.9	86.2
Trichloroethylene	55.5	88.7	55.7	68.6	100.5	77.2
Tetrachloroethylene	61.8	98.9	61.0	81.1	98.7	81.9

Table 8: Retention times of some industrially important solvents determined by means of headspace gas chromatography according to the methodical instructions presented here.

Temperature program: 60 °C isotherm

Frigen 114	3.911 min
Methanol	3.939 min
Acetonitrile	4.407 min
Acetone	4.587 min
Propanol-2	4.654 min
Cyclohexanone	4.680 min
Ethanol	4.698 min
Dichloromethane	*5.180 min*
Propanol-1	5.378 min
2-Bromo-2-chloro-1,1,1-trifluoroethane	*5.580 min*
Acetaldehyde	5.709 min
1,2-Dichloroethylene	*5.790 min*
Hexane	5.905 min
Methylethylketone	6.085 min
Ethyl acetate	6.739 min
Trichloromethane	*6.840 min*
p-Xylene	6.899 min
Butanol-2	6.920 min
1,1,1-Trichloroethane	*7.950 min*
Butanol-1	8.050 min
Benzene	8.572 min
Toluene	8.590 min
m-Xylene	8.635 min
Tetrachloromethane	*8.760 min*
o-Xylene	8.827 min
Dibromomethane	9.554 min
Trichloroethylene	*10.210 min*
Methylisobutylketone	12.000 min
Dioxane	12.786 min
Cumene	13.870 min
n-Amyl alcohol	14.468 min
Styrene	19.686 min
Tetrachloroethylene	*20.040 min*
1,1,2,2-Tetrabromomethane	> 22 min
Pentachloroethane	> 22 min

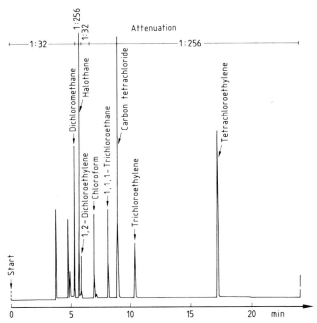

Fig. 1: Gas chromatogram of a calibration standard in blood.
Concentrations:
Dichloromethane 252.9 μg/L, 2-Bromo-2-chloro-1,1,1-trifluoromethane (Halothane) 66.6 μg/L, 1,2-Dichloroethylene 254.6 μg/L, Trichloromethane (Chloroform) 33.0 μg/L, 1,1,1-Trichloroethane 49.7 μg/L, Tetrachloromethane 50.8 μg/L, Trichloroethylene 55.5 μg/L, Tetrachloroethylene μg/L, 61.8 μg/L

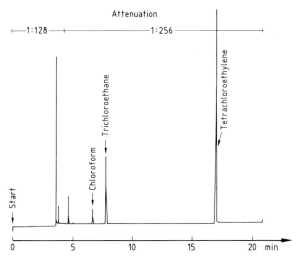

Fig. 2: Gas chromatogram of the blood sample of an exposed person.
Concentrations:
Trichloromethane (Chloroform) 6.0 μg/L, Trichloroethane 37.0 μg/L, Tetrachloroethylene 79.0 μg/L

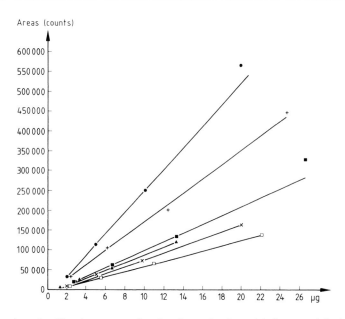

Fig. 3: Examples of calibration curves for the determination of halogenated hydrocarbons in blood.

- ● Tetrachloromethane
- + Tetrachloroethylene
- ■ 2-Bromo-2-chloro-1,1,1-trifluoroethane
- ▲ Trichloromethane
- × 1,1,1-Trichloroethane
- □ Trichloroethylene

Hexane metabolites (2,5-Hexanedione, 2-Hexanone)

Application   Determination in urine

Analytical principle   Capillary gas chromatography

Completed in   June 1993

Summary

The hexane metabolites (2,5-hexanedione, 2-hexanone) present in urine as a result of occupational exposure can be simply, sensitively and specifically determined by the capillary gas chromatographic method described here.

The hexane metabolites contained in urine are determined by capillary gas chromatography. For this purpose the urine is subjected to acid hydrolysis. The hexane metabolites are subsequently separated from the other constituents of the urine and, at the same time, enriched using a strongly acidic cation exchanger and a macroreticular resin. The qualitative and quantitative analysis of the hexane metabolites contained in the eluate is carried out by capillary gas chromatography with a flame ionisation detector (FID).

Aqueous standards which are processed like the samples and analysed by capillary gas chromatography are used for calibration. The peak areas of the individual hexane metabolites are calculated in relation to the peak area of the internal standard and the resulting values plotted as a function of the concentrations of the hexane metabolites used to give a calibration curve.

Within-series imprecision: Standard deviation (rel.) s_w = 2.5 and 1.1 %
                                  Prognostic range u = 6.1 and 2.7 %
                                  At concentrations of 4.49 and 5.74 mg 2-hexanone per litre urine and where n = 6 determinations

                                  Standard deviation (rel.) s_w = 2.2 and 2.5 %
                                  Prognostic range u = 5.3 and 6.1 %

Essential Biomonitoring Methods. DFG, Deutsche Forschungsgemeinschaft
Copyright © 2006 WILEY-VCH Verlag GmbH & Co. KGaA, Weinheim
ISBN: 3-527-31478-4

	At concentrations of 5.81 and 6.48 mg 2,5-hexanedione per litre urine and where $n = 6$ determinations	
Inaccuracy:	Recovery rate	$r = 90.5$ and 96.5% 2-hexanone in urine
	Recovery rate	$r = 89.1$ and 85.9% 2,5-hexanedione in urine
Detection limit:	0.2 mg 2-hexanone per litre urine 0.2 mg 2,5-hexanedione per litre urine	

Note: The given reliability criteria are based on the concentration of 2,5-hexanedione *after hydrolysis* (the sum of free 2,5-hexanedione and 4,5-dihydroxy-2-hexanone).

n-Hexane and its metabolites

n-Hexane (MW 86.18 g, b.p. 69 °C) is a colourless, highly volatile and readily flammable aliphatic hydrocarbon with a typical, weak odour. Its vapour is heavier than air and can form an explosive mixture with air. n-Hexane burns in air to give carbon dioxide and water.

n-Hexane is barely soluble in water. In the presence of salts its solubility in water increases slightly [1].

n-Hexane is miscible with many organic solvents, for example, aliphatic and aromatic solvents, ethyl alcohol, ketones, diethyl ether and halogenated hydrocarbons.

Like all paraffins n-hexane is relatively inert at room temperature. It reacts with neither concentrated sulfuric nor concentrated nitric acid. However, suitable oxidizing agents can react with n-hexane under certain conditions to form products which can contain one or more hydroxy and/or keto groups [2].

n-Hexane is contained in low boiling mineral oil fractions, from which the commercial solvent is obtained using special refining processes.

Pure n-hexane is produced only in very small quantities for special purposes, such as for use in the laboratory (elution agent and solvent in thin layer chromatography and in spectroscopy).

Solvent mixtures containing other hydrocarbons as well as n-hexane are of industrial importance. Technical hexane (approximately 55 % weight of n-hexane), special gasoline and motor fuel are of particular importance because of the high consumption of these substances [2].

Technical hexane is primarily used in the chemical industry, but also in the food, adhesive, rubber, varnish and paint industries. Moreover, technical hexane is employed either alone or in combination with other solvents in fine mechanical clock-making and the metal industry, as well as in printing works.

As a result of the numerous uses of hydrocarbons containing n-hexane many people come in contact with n-hexane. In U.S.A. it is estimated that some 2.5 million workers come into contact with working materials containing n-hexane [3].

The occurrence of n-hexane in the environment is primarily attributed to the emission of exhaust fumes from vehicles with internal combustion engines and the emission of industrial solvents. In Germany (i.e. the former West German states) an annual total emission of about 15 500 tonnes of n-hexane can be estimated from the data available [4–8].

The main mode of intake in humans is absorption in the lungs after inhalation with respiratory air [9]. This initially causes inspecific symptoms, such as dizziness, headache, nausea and vomiting. High concentrations of n-hexane lead to a narcosis-like condition due to depression of the CNS [10, 11]. This effect is abused by glue sniffers [1], as n-hexane is a frequently used component of thinners for adhesives.

n-Hexane has an irritating effect on the eyes and mucous membranes of the nose and throat or windpipe. A slight irritation of the eyes was reported at 20 ppm n-hexane – mixed with 1 ppm nitrogen dioxide, however – [12], while no irritation (eyes, nose, throat) could allegedly be ascertained even at a concentration of 500 ppm n-hexane in an older investigation [13]. Skin contact with n-hexane primarily leads to the removal of fat and to irritation of the skin.

When evaluating the critical toxicity of n-hexane or its metabolites the neurotoxic effects observed after chronic exposure are of foremost interest [14]. As published, not only n-hexane [15, 16], but also its metabolites, particularly 2,5-hexanedione [17–20], cause peripheral polyneuropathy after chronic exposure. On account of the structure of hexane and its metabolites (C6 basic structure) the neurological syndrome is known as hexacarbon polyneuropathy both in workers who have been exposed to n-hexane and in glue sniffers.

The exact relationship between dosage and effect in humans cannot be found in the literature, as the reported exposure concentrations fluctuate greatly and data on duration of exposure, composition of the atmosphere and analysis are usually unavailable [2]. The current MAK value of 50 ppm for n-hexane was ascertained from experiments on the relationship between dosage and effect in animals, whereby known quantitative differences in the metabolism from n-hexane to 2,5-hexanedione between humans and experimental animals were taken into account in the documentation [14].

The widespread use of n-hexane has recently been greatly restricted largely because of the risk of damage to the peripheral nervous system.

n-Hexane absorbed by the human organism after inhalation is eliminated unchanged via the respiratory air and excreted in the form of its metabolites in the urine.

The biotransformation of n-hexane takes place mainly in the liver, whereby 2-hexanol is initially formed in several oxidation steps. This compound is further oxidized to diols, hydroxyketones and to caproic acids.

Thus, lipophilic n-hexane is converted to hydrophilic products which can be directly eliminated in the urine or renally excreted after conjugation with glucuronic and sulfuric acid [9].

Perbellini et al. [21] identified 2,5-hexanedione (approx. 55%), 2-hexanol (approx. 3%), 2,5-dimethylfuran (approx. 29%) and γ-valerolactone (approx. 13%) in the urine of 10 workers in a shoe factory who were exposed to technical hexane and other solvents (mean concentration of n-hexane 117 ppm, at some workplaces peak values of 285–483 ppm).

A recent publication [22] dealing with relevant investigations on 41 workers from 5 shoe factories who were similarly exposed (mean n-hexane concentration 52 ppm) gave the mean concentrations of the following substances in urine as: 36.5% 2,5-hexanedione, 1.1% 2-hexanol, 30.5% 2,5-dimethylfuran and 31.9% γ-valerolactone. In most cases 2,5-hexanedione was the main catabolic product.

Furthermore, it was shown that a considerable proportion of the intermediate 5-hydroxy-2-hexanone (see Fig. 1) is further hydroxylated to give dihydroxy-2-hexanone which is excreted in the urine in conjugated form (glucuronide). Under the mildly acidic hydrolytic conditions (acetic acid) normally used for sample preparation up till now this compound is converted to 2,5-dimethylfuran, while drastic acidic hydrolysis converts it to 2,5-hexanedione. These relationships have not been taken into account in the studies on the excretion of 2,5-hexanedione in exposed groups published so far. Thus, published data on the excretion of 2,5-hexanedione is based on the sum of the actual 2,5-hexanedione excretion plus the excretion of 4,5-dihydroxy-2-hexanone which is converted to 2,5-hexanedione during sample preparation. 2,5-dimethylfuran measured at the same time is also formed from 4,5-dihydroxy-2-hexanone during sample preparation. It is important to note that the excretion of 4,5-dihydroxy-2-hexanone is several times higher than that of 2,5-hexanedione [23].

The method described here permits a differentiation between the total hexanedione (sum of the 4,5-dihydroxy-2-hexanone and 2,5-hexanedione) and the so-called freely extractable 2,5-hexanedione. Investigations carried out by the examiners of the method on 40 persons occupationally exposed to n-hexane showed that an average of only about 10.5% of the total hexanedione is present as free 2,5-hexanedione.

The metabolism of n-hexane is schematically shown in Figure 1.

Using the method described here the examiners found a normal total content of hexanedione of less than 1.0 mg/L in urine.
The BAT value (1987) is 9 mg/L urine [9].

Author: *J. Angerer, J. Gündel*
Examiners: *R. Heinrich-Ramm, U. Knecht*

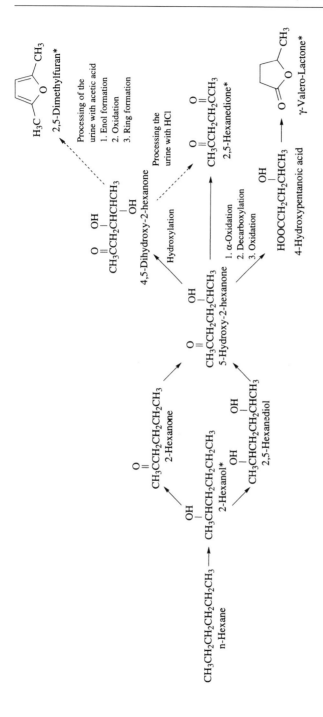

Fig. 1: Biotransformation of n-hexane in humans according to [21, 22].
*Metabolites of n-hexane found in the urine of occupationally exposed persons according to [21, 22].

Hexane metabolites (2,5-Hexanedione, 2-Hexanone)

Application   Determination in urine

Analytical principle   Capillary gas chromatography

Completed in   June 1993

Contents

Essential Biomonitoring Methods. DFG, Deutsche Forschungsgemeinschaft
Copyright © 2006 WILEY-VCH Verlag GmbH & Co. KGaA, Weinheim
ISBN: 3-527-31478-4

1 General principles

The urine is subjected to hydrolysis with hydrochloric acid. The hexane metabolites are subsequently separated from the other components of the urine and, at the same time, enriched using a strongly acidic cation exchanger and a macroreticular resin. The qualitative and quantitative determination of the hexane metabolites contained in the eluate is carried out by means of capillary gas chromatography with a flame ionisation detector.

Aqueous standards are used for calibration. They are processed and determined like the samples to be analysed. Cyclohexanone serves as an internal standard.

2 Equipment, chemicals and solutions

2.1 Equipment

Gas chromatograph with capillaries, flame ionisation detector, compensation recorder or integrator

Gas chromatographic column: Length 60 m; inner diameter 0.33 mm; stationary phase silicone oil (dimethylpolysiloxane DB-1), film thickness 1 μm (e.g. from J & W)

5 μL syringe for gas chromatography, preferably an autosampler

Water bath

Soxhlet apparatus

5, 50, 100 and 1000 mL volumetric flasks

20 mL crimp top vials with PTFE-coated crimp caps as well as crimping tongs

1, 2, 4, 5 and 10 mL pipettes

Dry thermostat

3 mL cation exchange columns, benzenesulfonic acid (e.g. from Baker)

Disposable glass tubes with flat bottoms

Apparatus for evaporation under a stream of nitrogen

25 mL empty cartridges with frits (e.g. from Baker)

3 mL empty cartridges with frits which can be packed in the laboratory (e.g. from Baker)

Work station for vacuum extraction

2.2 Chemicals

2,5-Hexanedione p.a. (acetonylacetone) (e.g. from Fluka)

2-Hexanone p.a. (e.g. from Aldrich)

Cyclohexanone p.a. (e.g. from Fluka)

36% Hydrochloric acid

Sodium hydroxide p.a.

XAD-4, purest (e.g. from Serva) particle size 0.2–0.4 mm

Acetone p.a. (e.g. from Merck)

Methanol p.a.

Ultrapure water (ASTM type 1) or double distilled water

Dichloromethane (e.g. Pestanal from Riedel-de Haën)

Anhydrous sodium sulfate

Purified nitrogen (99.999%)

Hydrogen (99.90%)

Synthetic air (80% purified nitrogen, 20% oxygen)

2.3 Solutions

3.6 M hydrochloric acid:
About 10 mL ultrapure water are placed in a 100 mL volumetric flask. Then 30 mL concentrated hydrochloric acid (36%) are added. After swirling the contents the flask is filled up to the mark with ultrapure water.

0.1 M hydrochloric acid:
About 500 mL ultrapure water are placed in a 1000 mL volumetric flask. Then 8.3 mL concentrated hydrochloric acid (36%) are added. After swirling the contents the flask is filled up to the mark with ultrapure water.

0.1 M sodium hydroxide:
4.0 g sodium hydroxide are weighed, transferred to a 1000 mL volumetric flask and dissolved in ultrapure water. The flask is then filled to the mark with ultrapure water.

Preparation of the solution containing the internal standard:

Stock solution:
25 mg cyclohexane are weighed in a 50 mL volumetric flask and dissolved in ultrapure water. The flask is then filled to the mark with ultrapure water (500 mg/L).

Solution of the internal standard:
5 mL of the stock solution are transferred to a 50 mL volumetric flask. The flask is filled to the mark with ultrapure water (50 mg/L).

2.4 Calibration standards

Stock solution:
25 mg each of 2,5-hexanedione and 2-hexanone are placed in a 50 mL volumetric flask and dissolved in ultrapure water. The flask is then filled to the mark with ultrapure water (content of each: 500 mg/L).

Calibration standards containing between 1 and 20 mg of the individual hexane metabolites per litre (cf. Table 1) are prepared from this stock solution by diluting with ultrapure water.

Table 1: Pipetting scheme for the preparation of the calibration standards.

Volume of the stock solution	Final volume of the calibration standard	Concentration of calibration standard
mL	mL	mg/L
0.2	100	1
1	100	5
2	100	10
3	100	15
4	100	20

2.5 Preparation of the XAD extraction columns

50 g of the XAD-4 resin are washed with 250 mL acetone in the Soxhlet extractor three times for eight hours each time and subsequently dried. The purified resin is filled into brown glass bottles for storage. It can be kept there indefinitely.

The XAD extraction columns are prepared by packing 3 mL XAD-4 into each of the empty 3 mL cartridges with frits. They are conditioned by vacuum suction of 10 mL acetone and three times 10 mL of 0.1 M hydrochloric acid through the columns. The cartridges are discarded after use.

2.6 Preparation of the cation exchange columns

The cation exchange columns are conditioned by vacuum suction of 5 mL methanol and 10 mL ultrapure water through the columns. The cartridges are discarded after use.

3 Specimen collection and sample preparation

Specimens are collected at the end of a working week.

The urine is collected in sealable plastic bottles and stored in the deep freezer until they are processed.

3.1 Sample preparation with hydrolysis (to determine the total content of 2,5-hexanedione as well as 2-hexanone)

The urine is thawed in the water bath at about 60 °C. After cooling to room temperature and intensive shaking, 5 mL urine are withdrawn and transferred to a crimp top vial. 1 mL of the internal standard solution and 2 mL 3.6 M hydrochloric acid are added. The crimp top vial is then sealed. Hydrolysis is carried out at 90 °C in the dry thermostat for an hour.

A work station for vacuum extraction which permits injection of the samples onto the chromatography columns and elution under vacuum is practical for sample preparation by means of ion exchange chromatography and liquid/solid extraction.

After cooling, the prepared sample is sucked through the series of chromatographic columns. The middle column contains the cation exchanger, the lower the macroreticular resin XAD-4. The upper column consists of an empty column (25 mL) with frit to separate the solid products of hydrolysis (Fig. 2). After the sample volume has been sucked through, the upper column and the cation exchange column are removed. The cartridges are discarded after use.

The macroreticular resin which absorbs the hexane metabolites is washed with 10 mL ultrapure water and then sucked dry in a stream of nitrogen (approx. 30 min). 5 mL dichloromethane are used for the elution of the hexane metabolites. The eluate is collected in a 5 mL volumetric flask which is then filled to the mark with dichloromethane. The eluate is dried with anhydrous sodium sulfate. An aliquot of 4 mL is pipetted into a disposable tube with a flat bottom and reduced to a volume of 1 mL by passing a stream of nitrogen over it.

3.2 Sample preparation without hydrolysis (to determine the freely extractable 2,5-hexanedione as well as 2-hexanone)

In principle, the free 2,5-hexanedione can also be detected with the process outlined here. It is carried out as follows:

25 mL of the urine sample are centrifuged, 0.5 mL of the internal standard solution are added and this is injected onto the columns arranged in series. Otherwise the processing and analysis are carried out as previously described.

4 Operational parameters for gas chromatography

Capillary column:	Material:	Fused silica
	Length:	60 m
	Inner diameter:	0.33 mm
	Stationary phase:	Dimethylpolysiloxane DB-1 Film thickness 1 μm
Detector:	Flame ionisation detector	
Temperatures:	Column:	7 min. at 80 °C, then increase 20 °C per min. until 200 °C; 2 min. at final temperature
	Injection block:	250 °C
	Detector:	300 °C
Carrier gas:	Purified nitrogen with a column pressure of 1380 hPa	
Split:	30 mL/min	
Sample volume:	1 μL	

The retention times observed under these conditions are:

 8.4 min 2-hexanone
11.0 min cyclohexanone (internal standard)
11.3 min 2,5-hexanedione

5 Analytical determination

For the gas chromatographic analysis 1 µl of each of the prepared samples are injected into the gas chromatograph.

6 Calibration

Like the urine samples, the aqueous calibration standards (cf. Section 2.4) are processed according to Section 3.1 and analysed by gas chromatography as described in Sections 4 and 5. A calibration curve is obtained by plotting the quotients of the peak areas of the individual hexane metabolites and the internal standard as a function of the concentrations used (cf. Fig. 3). The linearity of the calibration curves for each of the individual hexane metabolites was checked up to a concentration of 20 mg/L.

In order to determine the free 2,5-hexanedione, calibration should be carried out in the lower concentration range (about 2 mg/L).

7 Calculation of the analytical result

The peak area of each hexane to be determined is divided by the peak area of the internal standard. The quotients thus obtained are used to ascertain the concentration of the relevant hexane metabolite in mg/L urine. It can be read off the appropriate calibration curve.

8 Standardization and quality control

Quality control of the analytical results is carried out as stipulated in TRG A 410 of the German Arbeitsstoffverordnung (Regulation 410 of the German Code on Hazardous Working Materials) [29] and in the Spezial Preliminary Remarks. At present no standard material with a specified 2,5-hexanedione and 2-hexanone content is commercially available.

9 Reliability of the method

The reliability criteria given here are based on the concentration of 2,5-hexanedione after hydrolysis (sum of the free 2,5-hexanedione and 4,5-dihydroxy-2-hexanone).

9.1 Precision

In order to assess the precision in the series pooled urine of people who were not exposed to n-hexane at the workplace was spiked with two concentrations of 2,5-hexanedione or 2-hexanone respectively. The relative standard deviation for 2,5-hexanedione was 2.2 or 2.5%, the prognostic ranges were 5.3 and 6.1% for 6 determinations and average 2,5-hexanedione concentrations in urine of 5.81 or 6.48 mg/L respectively. For 2-hexanone the relative standard deviations were 2.5 or 1.1%, the prognostic ranges were 6.1 or 2.7%, when 6 determinations were carried out and the average concentrations of 2-hexanone in urine were 4.49 or 5.74 mg/L respectively.

9.2 Accuracy

The accuracy of the method was checked by carrying out recovery experiments and evaluated with the help of the aqueous calibration curves for 2,5-hexanedione and 2-hexanone. The samples described in 9.1 were used for this purpose. The total hexanedione content in urine before spiking was 1.43 mg/L. 2-Hexanone was not detectable in the unspiked urine. The recovery rate for 2,5-hexanedione was 89.1 or 85.9% (spiked concentrations of 5.09 or 6.11 mg/L respectively), for 2-hexanone the percentage recovery rate was 90.5 or 96.5 (spiked concentrations of 4.96 or 5.95 mg/L respectively).

9.3 Detection limit

As no reagent blank values were detected, the detection limit was estimated from the signal of the aqueous standard of lowest concentration which was subjected to complete sample processing. Thus, the detection limit in urine is given as 0.2 mg/L for both 2,5-hexanedione and 2-hexanone.

9.4 Sources of error

After familiarization with the principles of extraction of the solid phase the method showed little susceptibility to interference. However, two points should be noted: in order to ensure reproducible and complete elution of the absorbed hexane metabolites from the XAD resin, the packed resin cartridges must be very thoroughly sucked dry (about 30 min, laboratory vacuum pump). This should be carried out under nitrogen to prevent contaminants from the laboratory atmosphere being absorbed and subsequently eluted with the hexane metabolites.

The absorbent resin should undergo a Soxhlet extraction with acetone before use to lower the background interference. In our experience better results are thus achieved than with ready-packed, commercially available XAD cartridges.

Before cyclohexanone can be used as an internal standard it must be ensured that this substance is not present at the workplace.

10 Discussion of the method

The procedure described here is based on a method from *Fedtke* and *Bolt* [24, 25], in which hydrolysis of the urine is also carried out under strongly mineral acidic conditions. The n-hexane metabolite 4,5-dihydroxy-2-hexanone is also converted to 2,5-hexanedione and simultaneously determined with it.

The clean-up described here has the advantage over the original method that the interfering accompanying substances, such as proteins etc. are retained on the strongly acidic cation exchanger. Thus, the analytical interference is reduced and the specificity of the method has been improved. Moreover, the more efficient clean-up allows an improvement in the sensitivity of the method, as the extract can be evaporated. The addition of the internal standard at the beginning of the sample preparation has proved a further advantage of this method.

The method described here permits the determination of the hexane metabolites 2-hexanone and 2,5-hexanedione. In addition, the metabolite 4,5-dihydroxy-2-hexanone can be determined. This metabolite is converted to 2,5-hexanedione by means of the drastic acidic hydrolysis prescribed here. It is then analysed by gas chromatography together with the metabolite 2,5-hexanedione itself [25]. The 4,5-dihydroxy-2-hexanone is quantitatively determined by ascertaining the 2,5-hexanedione content after complete acid hydrolysis and subtracting the content of freely extractable 2,5-hexanedione without previous acid hydrolysis. The BAT value [9] is based on the concentration of the hexane metabolites 2,5-hexanedione plus 4,5-dihydroxy-2-hexanone. Thus, the method fulfils the conditions for the surveillance of, and compliance with the BAT value.

This method for the total hexanedione determination was compared with the procedure already published in the collection of methods [26] (extraction with dichloromethane after acid hydrolysis and subsequent gas chromatographic determination). Urine samples of 41 persons who were exposed to n-hexane at the workplace were used for this purpose (see Fig. 4).

This previously published gas chromatographic method is valid and simply carried out. It is especially suitable for high level exposure to n-hexane.

In addition, a high performance liquid chromatographic [27] and another gas chromatographic [28] method with derivatization are described in the literature. The detection limit given for the high performance liquid chromatographic method of about 3 µg/L seems scarcely credible, as the specificity of the procedure at low concentrations is very limited. In contrast, the derivatization method with subsequent gas chromatographic determination has the advantage of both high specificity and a low detection limit. However, the fact that three isomers are formed from the 2,5-hexanedione is a drawback for the quantitative evaluation.

Intruments used:

Gas chromatograph 3700 with FID, Varian, and integrator SP 4290, Spectra Physics

11 References

[1] Römpps Chemie-Lexikon. Franckh'sche Verlagshandlung, Stuttgart, 8th edition, 1979.

[2] DGMK (Deutsche Gesellschaft für Mineralölwissenschaft und Kohlechemie e.V.): DGMK-Projekt 174–2. Wirkung von n-Hexan auf Mensch und Tier. Hamburg 1982.

[3] NIOSH (National Institute for Occupational Safety and Health): Occupational exposure to alkanes (C_5-C_8), March 1977, p. 21.

[4] W. Dulcson: Schriftenreihe des Vereins für Wasser-, Boden- und Lufthygiene No. 47. Organisch-chemische Fremdstoffe in atmosphärischer Luft. Gustav Fischer Verlag, Stuttgart 1978, pp. 50, 60, 67.

[5] W. Weigert, E. Koberstein and E. Lakatos: Katalysatoren zur Reinigung von Autoabgasen. Chemiker-Ztg. 97, 469–478 (1973).

[6] H. May: Abschätzung des derzeitigen Anteils der Verdampfungsemmission an der Kohlenwasserstoffemmission von Fahrzeugen in der Bundesrepublik Deutschland. Vortrag vor der VDI-Kommission „Reinhaltung der Luft" am 2.5.1979.

[7] II. Schröder: Kohlenwasserstoffemmissionen von Kraftfahrzeugen mit Ottomotor. Umwelt 5, 370–371 (1979).

[8] MWV (Mineralölwirtschaftsverband): Jahresbericht 1980, pp. T 29, T 55.

[9] Deutsche Forschungsgemeinschaft: Maximum Concentrations at the Workplace and Biological Tolerance Values for Working Materials 1989. Report No. XXV of the Commission for the Health Hazards of Chemical Compounds in the Work Area. VCH Verlagsgesellschaft, Weinheim 1989.

[10] CIIT (Chemical Industry Institute of Toxicology): CIIT Current Status Reports No. 1: n-Hexane, Feb. 1977.

[11] W. Braun and A. Dönhardt: Vergiftungsregister. Georg Thieme Verlag, Stuttgart, 2nd edition, 1975, p. 196.

[12] L.G Wayne and J.A. Orcutt: The relative potentials of common organic solvents as precursors of eye irritants in urban atmospheres. J. Occup. Med. 2, 383–387 (1960).

[13] K.W. Nelson, J.F. Ege jun., M. Ross, L.E. Woodman and L. Silverman: Sensory response to certain industrial solvent vapors. J. Industr. Hyg. Toxicol. 25, 282–285 (1943).

[14] D. Henschler (ed.): „n-Hexan". Gesundheitsschädliche Arbeitsstoffe. Toxikologisch-arbeitsmedizinische Begründung von MAK-Werten. VCH Verlagsgesellschaft, Weinheim 1983, 9th issue.

[15] A. Herskowitz, N. Ishii and H.H. Schaumburg: n-Hexane neuropathy. A syndrome occuring as a result of industrial exposure. New Engl. J. Med. 285, 82–85 (1971).

[16] H.H. Schaumburg and P.S. Spencer: Degeneration in central and peripheral nervous systems produced by pure n-hexane: an experimental study. Brain. 99, 183–192 (1976).

[17] *J.L. O'Donoghue, W.J. Krasavage* and *C.J. Terhaar:* The relative neurotoxicity of methyl n-butyl ketone and its metabolites. Toxicol. Appl. Pharmacol. *45*, 269 (1978).

[18] *M.J. Politis, R.G. Pelligrino* and *P.S. Spencer:* Ultra structural studies of the dying-back process. V. Axonal neurofilaments accumulate at sites of 2,5-hexanedione application: evidence for nerve fibre disfunction in experimental hexacarbon neuropathy. J. Neurocytol. *9*, 505–516 (1980).

[19] *P.S. Spencer* and *H.H. Schaumburg:* Experimental neuropathy produced by 2,5-hexanedione. A major metabolite of the neurotoxic industrial solvent methyl n-butylketone. J. Neurol. Neurosurg. Psychiat. *38*, 771–775 (1975).

[20] *A.B. Sterman* and *R.C. Sheppard:* The role of the neuronal cell body in neurotoxic injury. Neurobehav. Toxicol. Teratol. *4*, 567–572 (1982).

[21] *L. Perbellini, F. Brugnone* and *I. Pavian:* Identification of the metabolites of n-hexane, cyclohexane and their isomers in men's urine. Toxicol. Appl. Pharmacol. *53*, 220–229 (1980).

[22] *L. Perbellini, F. Brugnone* and *G. Faggionato:* Urinary excretion of the metabolites of n-hexane and its isomers during occupational exposure. Brit. J. Industr. Med. *38*, 20–26 (1981).

[23] *N. Fedtke* and *H.M. Bolt:* The relevance of 4,5-dihydroxy-2-hexanone in the excretion kinetics of n-hexane metabolites in rat and man. Arch. Toxicol. *61*, 130–136 (1987).

[24] *N. Fedtke* and *H. Bolt:* Methodological investigations on the determinations of n-hexane metabolites in urine. Int. Arch. Occup. Environ. Health *57*, 149–158 (1986).

[25] *N. Fedtke* and *H.M. Bolt:* The relevance of 4,5-dihydroxy-2-hexanone in the excretion kinetics of n-hexane metabolites in rat and man. Arch. Toxicol. *61*, 131–137 (1987)

[26] *A. Eben:* 2,5-Hexandion. In: *J. Angerer, K.H. Schaller* (eds.): Analysen in biologischem Material. Deutsche Forschungsgemeinschaft, VCH Verlagsgesellschaft, Weinheim 1980, 4th issue.

[27] *I. Marchiseppe, M. Valentino* and *M. Governa:* Determination of total 2,5-hexanedione by reversed-phase high performance liquid chromatography. J. Chromatogr. *495*, 288–294 (1989).

[28] *S. Kezic* and *A.C. Monster:* Determination of 2,5-hexanedione in urine and serum by gas chromatography after derivatization with O-(pentafluorobenzyl)-hydroxylamine and solid-phase extraction. J. Chromatogr. *563*, 199–204 (1991).

[29] TRg A 410: Statistische Qualitätssicherung. In: Technische Regeln und Richtlinien des BMA zur Verordnung über gefährliche Stoffe. Bek. des BMA vom 3.4.1979, Bundesarbeitsblatt 6/1979, ppß. 88 ff.

Authors: *J. Angerer, J. Gündel*
Examiners: *R. Heinrich-Ramm, U. Knecht*

Fig. 2: Illustration of the columns used for solid phase extraction of the hexane metabolites (above: empty cartridge with frit; middle: 3 mL cation exchange column; below: 3 mL cartridge packed with XAD-4).

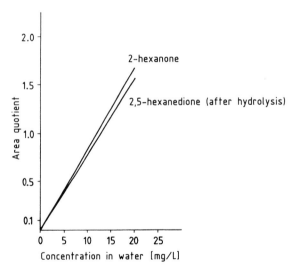

Fig. 3: Example of a calibration curve for the determination of 2-hexanone and 2,5-hexanedione (after hydrolysis).

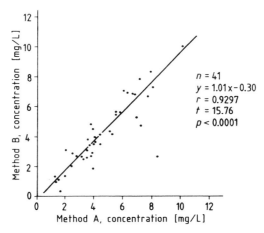

Fig. 4: Correlation diagram for the parallel analysis of 41 urine samples of persons exposed to n-hexane using the method described here (A) and the method according to *Eben* [26] (B).

t,t-Muconic acid

Application   Determination in urine

Analytical principle   High pressure liquid chromatography

Completed in   June 1995

Summary

The concentration of t,t-muconic acid in urine can be sensitively and reliably determined using the high pressure liquid chromatography method described here. On account of its sensitivity this method is suitable for the detection of the concentration range due to ecological benzene exposure and those benzene concentrations relevant to occupational medicine.

In order to determine t,t-muconic acid, it is separated from the acidified urine by ion chromatography. The t,t-muconic acid is subsequently eluted with 10 % acetic acid and separated from its concomitants by means of high pressure liquid chromatography. The detection is carried out by a UV detector at 259 nm. Aqueous standards which are processed and analysed by high pressure liquid chromatography in the same way as the urine samples are used for calibration.

Within-series imprecision:	Standard deviation (rel.)	$s_w = 2\%$
	Prognostic range	$u = 4.5\%$
	At a concentration of 2.5 mg t,t-muconic acid per litre urine and where $n = 10$ determinations	
	Standard deviation (rel.)	$s_w = 1.1\%$
	Prognostic range	$u = 2.5\%$
	At a concentration of 10 mg t,t-muconic acid per litre urine and where $n = 10$ determinations	
Between-day imprecision:	Standard deviation (rel.)	$s = 6.8\%$
	Prognostic range	$u = 14.2\%$
	At a concentration of 2.5 mg t,t-muconic acid per litre urine and where $n = 20$ days	
Inaccuracy:	Recovery rate	$r = 100-110\%$
	At concentrations of between 0.1 and 20 mg t,t-muconic acid per litre urine	
Detection limit:	0.1 mg t,t-muconic acid per litre urine	

t,t-Muconic acid

Benzene is a colourless, volatile, flammable liquid with a characteristic odour. Benzene, together with the so-called BTX aromatics (benzene, toluene and the three xylene isomers), used to be obtained mainly by distillation from coal and by washing out of coke gas. As a result of increasing demand, new production processes were developed using mineral oil as their source. Thus, in the oil refining process during the upgrading of petroleum by reforming and during the cracking process for the manufacture of olefins, fractions occur which are rich in aromatic hydrocarbons. These are reformate, hydrolysis or cracked gasoline which represent valuable sources for the production of BTX aromatics [1, 2].

Benzene is added to motor fuels (as an antiknock agent). It is also used as the starting material for the synthesis of many benzene derivatives, such as aniline, nitrobenzene, styrene, synthetic rubber, plastics, phenol and dyestuffs. However, benzene is no longer used as a solvent because of its carcinogenic effect [1].

Benzene has been assigned to category III A1 of carcinogenic substances by the Deutsche Forschungsgemeinschaft's Commission for the Investigation of Health Hazards of Chemical Compounds in the Work Area and classified as a substance "shown to induce malignant tumours in humans" [3].

It is regarded as proven that benzene can cause acute myeloic leukaemia. At present there is increasing discussion whether the chronic intake of benzene could cause other forms of leukaemia, e.g. lymphatic leukaemia.

In contrast to its homologues, benzene manifests carcinogenic effects: this is due to the aromatic ring being oxidized in human metabolism. Benzene oxide, which is formed as an intermediate, is considered to be the ultimate carcinogenic agent, as it can form covalent bonds with DNA. In contrast, the aliphatic side chain of the alkylbenzenes is oxidized to form the relatively innoxious aromatic carboxylic acids which are renally excreted in their free form or bound to glycine.

Only a small proportion of the alkylbenzenes is oxidized at the aromatic ring to alkylphenols [3].

In occupational medicine benzene exposure at the workplace can be monitored by determination of benzene in biological samples such as blood, exhaled air and urine. In addition, exposure can be detected by analysis of benzene metabolites. Besides the main metabolite phenol, small quantities of benzene are also metabolized to t,t-muconic acid and excreted renally by the human organism [4] (cf. Figure 1).

In 1993 the German Committee for hazardous substances (AGS) reduced the technical exposure limit (TRK) for benzene as a carcinogenic substance from the previous limit of 5 ppm to 1 ppm or 2.5 ppm. Thus, it was necessary to reconsider the diagnostic validity of the investigation parameter used for biological monitoring. The specificity and sensitivity of phenol determination in urine proved insufficient for this purpose. The upper norm limit for the phenol concentration in urine is given as 17 mg/L for the general public [5]. It is impossible to distinguish between normal excretion and the phenol excretion of a person

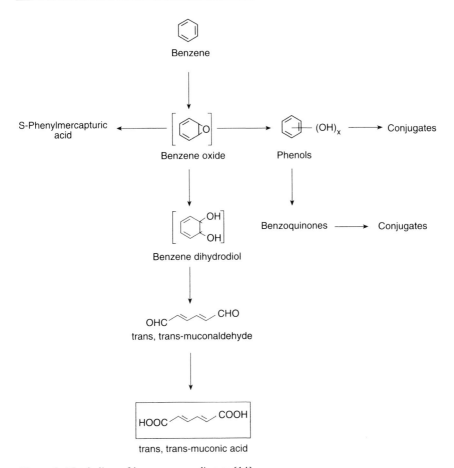

Figure 1. Metabolism of benzene according to [14]

exposed to benzene under TRK conditions at the workplace. For this reason the Commission for the Investigation of Health Hazards of Chemical Compounds in the Work Area abandoned the exposure equivalent for phenol in 1993 in favour of t,t-muconic acid and S-phenylmercapturic acid in urine [3].

The concentrations of t,t-muconic acid or S-phenylmercapturic acid in urine represent diagnostically more sensitive parameters [6]. However, from the analytical point of view the quantitative determination of t,t-muconic acid is more practicable, as it can be carried out by high pressure liquid chromatography, in contrast to the GC/MS analysis of S-phenylmercapturic acid.

The determination of t,t-muconic acid in urine as an indicator of occupational and ecological exposure to benzene has been recommended by various working groups. The working groups headed by Inoue [7] and Ducos [8] carried out the quantitative analysis of t,t-muconic acid by means of HPLC and UV detection. Bechtold [9] used the GC/MS technique for this purpose. The first two working groups give normal values for t,t-muconic acid excretion between 0.1 mg/L and < 0.5 mg/L respectively. Results of our own investigations show t,t-muconic acid concentrations of up to 3 mg/L in the normal population [10]. However, the possibility that these elevated values are due to limited exposure to benzene during leisure activities cannot be excluded. Another influencing factor under discussion is the intake of sorbic acid from preserved food. Apart from benzene, sorbic acid is the only currently known precursor of t,t-muconic acid [8].

Simultaneous exposure to toluene can interfere with the benzene dependent t,t-muconic acid excretion in urine [11]. This should be taken into account under the conditions at the workplace, as benzene is frequently encountered in a mixture with other alkylbenzenes. Furthermore, motor fuels contain toluene [12].

In order to evaluate the EKA correlation, experience with the relationship between the benzene concentration in air and the t,t-muconic acid excretion for occupationally exposed persons as well as the normal level of t,t-muconic acid excretion in the general population was taken into account [13].

The results of comparative studies from different laboratories and those from quality control show that t,t-muconic acid can be reliably quantified. An upper t,t-muconic acid norm limit of 1 mg/L was found in the general population. On the basis of this value and results for people who have been occupationally exposed to benzene the following exposure equivalents were obtained:

Benzene in air	0.6	1	2	4	6	ppm
t,t-muconic acid in urine	1.6	2	3	5	7	mg/L

Further information about the toxicity and metabolism of benzene and the other BTX aromatic compounds can be found in recent monographs [14].

Authors: *J. Angerer, D. Rauscher, W. Will*
Examiner: *M. Blaszkewicz*

t,t-Muconic acid

Application   Determination in urine

Analytical principle High pressure liquid chromatography

Completed in   June 1995

Contents

Essential Biomonitoring Methods. DFG, Deutsche Forschungsgemeinschaft
Copyright © 2006 WILEY-VCH Verlag GmbH & Co. KGaA, Weinheim
ISBN: 3-527-31478-4

1 General principles

t,t-Muconic acid is separated from the acidified urine by ion chromatography. The t,t-muconic acid is subsequently eluted with 10 % acetic acid and separated from its concomitants by means of high pressure liquid chromatography. The detection is carried out by a UV detector at 259 nm. Aqueous standards, to which a defined amount of t,t-muconic acid has been added, are used for calibration.

2 Equipment, chemicals and solutions

2.1 Equipment

High pressure liquid chromatograph, UV detector capable of reading at 259 nm, column thermostat, integrator.

High pressure liquid chromatographic column:
Column: steel, 250×4 mm
Precolumn: steel, 21×2.1 mm
Column packing: LiChrosorb RP 18; 5 μm (e.g. from Merck)

Syringe for sample injection (e.g. from Hamilton) or autosampler (e.g. from Merck-Hitachi)

Vacuum station (e.g. VAC ELUT SPS 24, ICT Analytichem International)

Anion exchange columns 500 mg SAX commercially available columns Bond Elut (e.g. from ICT Analytichem International)

10 mL Disposable polyethylene tubes

100 mL Volumetric flasks

100 mL and 1000 mL graduated cylinders

1 mL Eppendorf pipettes

Hand dispensers for various, exactly adjustable dosage volumes between 1 mL and 5 mL (e.g. Multipette)

Variable pipettes from 100 to 1000 μL

Ultrasonic bath

2.2 Chemicals

t,t-Muconic acid, 98 % (e.g. from Aldrich)

Methanol, HPLC quality (e.g. from Baker)

Acetic acid, 100 % p.a. (e.g. from Merck)

Absolute ethanol (e.g. from Merck)

Ultrapure water (ASTM type 1) or double-distilled water

2.3 Solutions

1 % and 10 % acetic acid:
10 mL and 100 mL 100 % acetic acid are each filled into separate 1000 mL volumetric flasks which are filled to the mark with ultrapure water and then shaken vigorously.

HPLC-flow agent (1 % acetic acid/methanol (10:1)):
900 mL 1 % acetic acid are placed into a 1000 mL volumetric flask. 100 mL methanol are filled into a 100 mL volumetric flask. The solutions are mixed and degassed for about 15 minutes in an ultrasonic bath or in other customary commercially available devices.

2.4 Calibration standards

Stock solution (100 mg/L):
About 10 mg t,t-muconic acid are weighed exactly, dissolved in about 5 mL 100 % ethanol in a 100 mL volumetric flask (ultrasonic bath) and the flask is then filled to the mark with 1 % acetic acid.

Calibration standards in the concentration range from 0.1 to 20 mg/L are prepared by diluting this stock solution with 1 % acetic acid according to the following pipetting scheme:

Volume of the stock solution [mL]	Final volume of the calibration standards [mL]	Concentration of the calibration standards [mg/L]
0.1	100	0.1
0.2	100	0.2
1.0	100	1.0
2.0	100	2.0
2.5	100	2.5
5.0	100	5.0
10.0	100	10.0
20.0	100	20.0

These calibration standards can be stored for at least 6 months in a deep-freezer without affecting the t,t-muconic acid content.

3 Specimen collection and sample preparation

Urine is collected in plastic bottles at the end of a working shift. The urine samples are acidified with acetic acid (1 mL glacial acetic acid/100 mL urine) and kept in the refrigerator for up to 5 days if they cannot be immediately processed. If the interval between collection and processing is longer the urine samples are stored in the deep-freezer until analysis. Then the samples are thawed in a water bath at 40 °C and subsequently brought to room temperature. Before an aliquot is taken for analysis the samples are thoroughly shaken.

For the analytical investigation the anion exchange column (quaternary amine) on the vacuum station is primed with 3 mL methanol followed by 3 mL ultrapure water. Then 1 mL of the urine sample is injected onto the column and drawn slowly (10 to 15 min) through it by means of a weak vacuum. The column is rapidly rinsed with 3 mL 1 % acetic acid and sucked out until it is almost dry.

Then solutes are eluted twice with 2 mL 10 % acetic acid and collected in a 10 mL centrifugation tube. After the latter elution the column is sucked dry. The ion exchange column is discarded after use. Figure 2 shows the sample preparation schedule with the solid phase extraction.

4 Operational parameters for high pressure liquid chromatography

Column:	Material:	Steel
	Length:	250 mm
	Inner diameter:	4 mm
Precolumn:	Material:	Steel
	Length:	21 mm
	Inner diameter:	2.1 mm
Column packing:	LiChrosorb RP-18; 5 μm	
Separation mode:	Reversed phase	
Column temperature:	Room temperature (or column thermostat at 35 °C)	
Detection:	UV detector which can record at 259 nm	
Mobile phase:	1 % Acetic acid/methanol (10:1) (filtration under vacuum)	
Pressure:	approx. 50 bar	
Flow rate:	0.2 mL/min	
Sample volume:	10 μL	
Flow time:	20 min	

Under the chromatographic conditions described above the retention time for t,t-muconic acid was approximately 6.4 min (cf. Figure 3).

After about 200 injections the precolumn must be exchanged.

5 Analytical determination

10 μL of the sample eluted from the solid phase column are injected into the high pressure liquid chromatograph with a syringe through a loop valve (or by means of an autosampler).

If the readings are not within the linear range of the calibration curve the urine samples are diluted and processed anew. A quality control sample is included in each analytical series.

6 Calibration

A calibration curve is obtained using aqueous standard solutions which are processed and analysed in the same way as the assay material.

The concentrations used are plotted as a function of the peak areas to give the calibration curve. The linearity of the process was checked up to a concentration of 20 mg/L t,t-muconic acid (cf. Figure 4).

Each analytical series includes an aqueous solution to check the validity of the calibration curve (sensitivity check).

The entire calibration curve must be plotted anew if the analytical conditions change or if the quality control values indicate that this is necessary.

Investigations have shown that aqueous standard solutions and standard solutions in pooled urine generate calibration curves which are congruent within the error limits.

7 Calculation of the analytical result

Calculation of the t,t-muconic acid concentration of the samples is carried out using the peak areas recorded by the integrator. Using these values the results in mg/L are read off the calibration curve (calculated using the gradient of the calibration curve). If necessary, dilution of the urine samples must be taken into account.

To date no reagent blank values have been observed when using this method.

8 Standardization and quality control

Quality control of the analytical results is carried out according to TRgA 410 (Regulation 410 of the German Code on Hazardous Working Materials) [15] and the special remarks on quality control in this series. In order to check the precision a control sample containing a specific concentration of t,t-muconic acid is included in the analysis. As commercially prepared material is not available, these control samples must be prepared in the laboratory. Urine is spiked with a defined amount of t,t-muconic acid. Aliquots of this solution can be stored in a deep-freezer for up to a year and used for quality control. The mean expected value and the tolerance range of this quality material is obtained in a pre-analytical period (one determination of the control material on 20 different days) [20].

9 Reliability of the method

The reliability of the method was checked using spiked urine and/or aqueous standard solutions from concentrations of 0.1 to 20 mg/L.

9.1 Precision

In order to determine the precision in the series a urine sample to which a defined quantity of t,t-muconic acid had been added was analysed 10 times (creatinine < 1 g/L). At a mean concentration of 2.5 mg/L the standard deviation was 0.05 mg/L (relative standard deviation 2 %). When the mean concentration of t,t-muconic acid was 10 mg/L a standard deviation of 0.11 mg/L was calculated (relative standard deviation 1.1 %).

The examiner found a standard deviation of 0.12 mg/L (relative standard deviation 4.7 %) with a more highly concentrated urine sample (creatinine >1 g/L) containing a t,t-muconic acid concentration of 2.5 mg per litre of urine.

The between-day precision was also determined using a spiked urine sample. The samples were analysed on 20 days. With a mean concentration of 2.5 mg/L a standard deviation of 0.17 mg/L (relative standard deviation 6.8 %) was calculated.

9.2 Accuracy

Recovery experiments were carried out to determine the accuracy of the method. Spiked urine samples containing concentrations of t,t-muconic acid from 0.1 to 20 mg/L were used for this purpose. These samples were processed as described above, analysed and compared with the analogously processed aqueous standard solutions.

Spiked concentration in urine [mg/L]	Recovery [mg/L]	Recovery rate [%]
0.1	0.10	100.0
0.2	0.21	105.0
2.0	2.20	110.0
10.0	10.48	104.8
20.0	20.32	101.6

The mean recovery rate was 104.2 %.

Parallel to the recovery experiments, in order to estimate the processing losses, especially due to solid phase extraction, aqueous solutions of t,t-muconic acid ranging in concentration from 0.1 to 20 mg/L were analysed by high pressure liquid chromatography without further processing. Samples were also processed and analysed according to the above instructions. Comparison of the results showed a mean loss of t,t-muconic acid of 6.58 % as a result of processing.

Spiked concentration [mg/L]	Recovery [mg/L]	Recovery rate [%]	Loss [%]
2.0	1.91	95.50	4.50
10.0	9.45	94.50	5.50
20.0	18.75	93.80	6.20

9.3 Detection limit

Under the conditions given for sample processing and for HPLC the detection limit was below 0.1 mg/L in urine and in water (3 times the signal-noise ratio).

9.4 Sources of error

Concomitants which cause interference are effectively removed from the urine by means of ion exchange chromatography so that no matrix effects were observed in the HPLC chromatogram. The calibration curves in water and urine were congruent. It was discovered that people who are not exposed to benzene at the workplace also excrete small amounts of t,t-muconic acid (up to 1 mg/L). Consumption of food preserved with sorbic acid may be responsible for a slight excretion of t,t-muconic acid. Sorbic acid is the only presently known precursor of t,t-muconic acid apart from benzene. However, exposure to benzene during leisure activities and due to environmental pollution are also conceivable causes.

Instrument-specific validation of the method should be carried out in each individual laboratory and, if necessary, the method should be optimized. Setting the HPLC column thermostat at 35 °C proved advantageous.

The integration markers should be checked, as the chromatogram can be complex.

10 Discussion of the method

The method described here is based on the work of Ducos [8]. The solid phase extraction presented here offers one possibility of separating the matrix components and simultaneously enriching the t,t-muconic acid on the ion exchange column. This not only reduces the analytical background interference, but also enhances the sensitivity of the method. Matrix effects can be largely eliminated, as shown by the identical calibration curves obtained using the standard solutions in urine and in water.

The methods of Inoue and Gad-El Karim do not offer these advantages. In Inoue's method [7] the urine sample is diluted with methanol, centrifuged and then transferred to the HPLC. According to Gad-El Karim's method [16] a liquid-liquid extraction of the urine sample is carried out using ether before the sample is analysed by high pressure liquid chromatography.

The reliability criteria of the described method, such as imprecision in the series, between-day imprecision (relative standard deviation with values between 1.1 % and 6.8 %), recovery rate and detection limit can be described as good. Losses of t,t-muconic acid of < 10 % due to processing are very small. The detection limit of less than 0.1 mg/L is comparable with the values given in the literature.

The detection limit is sufficiently low to reach the range due to ecological exposure. On the basis of data obtained from a group of persons it was possible to differentiate between smokers and non-smokers [17, 18].

The method described here proved to be practicable and suitable for routine analytical use, as a high sample throughput is possible. Sample processing requires little equipment and few staff and the instructions are easily followed. It saves a complicated liquid-liquid extraction with organic solvents as well as expensive waste disposal and thus conforms with the current trend to process samples using solid phase extraction. If necessary, the solid phase column can be prepared in the laboratory by loosely filling the column with packing material.

Absorption of the t,t-muconic acid on the ion exchange column leads to enhancement of the specificity of this method, as no background interference occurs as a result of the absorption and no interference due to the substances present in the urine is discernible.

Intake of sorbic acid restricts the diagnostic specificity of this determination of t,t-muconic acid in urine. This substance is used as a food preservative. After intake of 1 g of sorbic acid the renal excretion of t,t-muconic acid reaches brief peak values (half-life 0.5 h) of

15 mg/L [19]. Therefore, the time of collection of the urine sample (after the working shift) is of great importance with regard to food intake.

Instruments used:
HPLC pump: Series 1050, Hewlett Packard
UV Detector: L4000, Merck-Hitachi
Column thermostat: T-6300, Merck-Hitachi
Autosampler: AS-2000, Merck-Hitachi
Integrator: SP 4290, Spectra Physics

11 References

[1] Römpps Lexikon der Chemie. 9th edition, Franckh'sche Verlagshandlung, Stuttgart, 1992.
[2] *K. Weisermehl* and *H.-J. Arpe (eds.)*: Industrielle Organische Chemie. 2nd. revised and extended edition. VCH Verlagsgesellschaft, Weinheim 1978.
[3] *Deutsche Forschungsgemeinschaft*: List of MAK and BAT Values 1994. Maximum Concentrations and Biological Tolerance Values at the Workplace. Report No. 30 of the Commission for the Investigation of Health Hazards of Chemical Compounds in the Work Area. VCH Verlagsgesellschaft, Weinheim 1994.
[4] *A. Yardley-Jones, D. Anderson* and *D. V. Parke*: The toxicity of benzene and its metabolism and molecular pathology in human risk assessment. Br. J. Ind. Med. *48*, 437–444 (1991).
[5] *J. Angerer*: Überwachung von Arbeitsbereichen durch Biological Monitoring: Benzol und seine Homologen. In: Berufsgenossenschaft der chemischen Industrie (Hrsg.). Bericht über das 8. internationale Kolloquium über die Verhütung von Arbeitsunfällen und Berufskrankheiten in der chemischen Industrie, Frankfurt/Main, pp. 423–436 (1982).
[6] *W. Popp, D. Rauscher, G. Müller, J. Angerer,* and *K. Norpoth*: Concentrations of benzene in blood and S-phenylmercapturic and t,t-muconic acid in urine in car mechanics. Int. Arch. Occup. Environ. Health *66*, 1–6 (1994).
[7] *O. Inoue, K. Seiji, H. Nakatsuka, T. Watanabe, S. N. Yin, G. L. Li, S. X. Cai, C. Jin* and *M. Ikeda*: Urinary t,t-muconic acid as an indicator of exposure to benzene. Br. J. Ind. Med. *46*, 122–127 (1989).
[8] *P. Ducos, R. Gaudin, A. Robert, J.M. Francin* and *C. Maire*: Improvement in HPLC analysis of urinary trans,trans-muconic acid, a promising substitute for phenol in the assessment of benzene exposure. Int. Arch. Occup. Environ. Health *62*, 529–534 (1990).
[9] *W. E. Bechtold, G. Lucier, L. S. Birnbaum, S. N. Yin, G. L. Li* and *R. F. Henderson*: Muconic acid determinations in urine as a biological exposure index for workers occupationally exposed to benzene. Am. Ind. Hyg. Ass. J. *52*, 473–478 (1991).
[10] *J. Angerer, D. Rauscher, W. Will, K. H. Schaller* and *D. Weltle*: Biomonitoring einer Benzolbelastung anhand der t,t-Muconsäure-Ausscheidung im Harn. In: R. Kreutz and C. Piekarskie (eds.). Verhandlungen der Deutschen Gesellschaft für Arbeitsmedizin e.V., Gentner Verlag, Stuttgart 1992.
[11] *O. Inoue, K. Seiji, T. Watanabe, M. Kasahara, H. Nakatsuka, S. Yin, G. Li, S. Cai, C. Jin* and *M. Ikeda*: Mutual metabolic suppression between benzene and toluene in man. Int. Arch. Occup. Environ. Health *60*, 15–20 (1988).
[12] *DGMK (Deutsche Gesellschaft für Mineralölwissenschaft und Kohlechemie e.V.)*: DGMK-Projekt 174–6. Wirkung von Toluol auf Mensch und Tier. Hamburg 1985.

[13] *P. Ducos, R. Gaudin, J. Bel, C. Maire, J.M. Francin, A. Robert* and *P. Wild*: trans,trans-muconic acid, a reliable biological indicator for the detection of individual benzene exposure down to the ppm level. Int. Arch. Occup. Environ. Health *64*, 309–313 (1992).

[14] *J. Angerer* and *B. Hörsch*: Determination of aromatic hydrocarbons and their metabolites in human blood and urine. J. Chromatogr. *580*, 229–255 (1992).

[15] *Bundesministerium für Arbeit und Sozialordnung*: TRgA 410. Statistische Qualitätssicherung. In: Technische Regeln und Richtlinien des BMA zur Verordnung über gefährliche Stoffe. Bek. des BMA vom 9.4.1979, Bundesarbeitsblatt 5/1979, p. 88 ff.

[16] *M.M. Gad-El Karim, V.M. Sadagopa Ramanujam* and *M.S. Legator*: trans,trans-muconic acid, an open-chain urinary metabolite of benzene in mice. Quantification by high pressure liquid chromatography. Xenobiotica *15*, 211–220 (1985).

[17] *D. Rauscher, G. Lehnert* and *J. Angerer*: Biomonitoring of occupational and environmental exposures to benzene by measuring t,t-muconic acid in urine. Clin. Chem. *40/7*, 1468–1470 (1994).

[18] *A.A. Melikian, A.K. Prahalad* and *D. Hoffmann*: Urinary trans, trans-muconic acid as an indicator of exposure to benzene in cigarette smokers. Cancer Epidemiology, Biomarkers & Prevention 2, 47–51 (1993).

[19] *W. Will*: Personal Communication (1994).

[20] *J. Angerer* and *K.H. Schaller*: Erfahrungen mit der statistischen Qualitätskontrolle im arbeitsmedizinisch-toxikologischen Laboratorium. Arbeitsmed. Sozialmed. Präventivmed. *1*, 33 (1977).

Authors: *J. Angerer, D. Rauscher, W. Will*
Examiner: *M. Blaszkewicz*

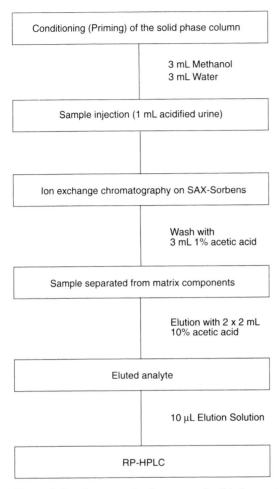

Conditioning (Priming) of the solid phase column

3 mL Methanol
3 mL Water

Sample injection (1 mL acidified urine)

Ion exchange chromatography on SAX-Sorbens

Wash with
3 mL 1% acetic acid

Sample separated from matrix components

Elution with 2 x 2 mL
10% acetic acid

Eluted analyte

10 µL Elution Solution

RP-HPLC

Figure 2. Sample preparation by means of solid phase extraction

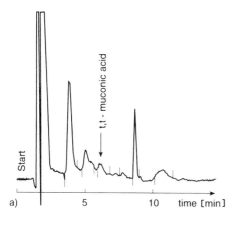

a)

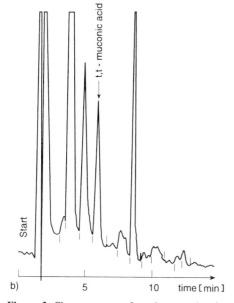

b)

Figure 3. Chromatogram of a urine sample of a person who was not exposed to benzene (a) and a chromatogram of a urine sample containing t,t-muconic acid (b)

Area [units]

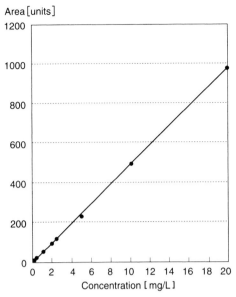

Figure 4. Example of a calibration curve for the determination of t,t-muconic acid

Organochlorine compounds in whole blood and plasma

Application   Determination in whole blood and plasma

Analytical principle   Capillary gas chromatography/
mass spectrometric detection (MS)

Completed in   May 2001

Summary

The method presented here permits the sensitive and specific quantification of poly-chlorinated biphenyls, hexachlorobenzene, DDT and DDE, as well as α-, β- and γ-hexachlorocyclohexane. This method can determine the inner exposure of persons who have absorbed these substances from the environment.

Blood and plasma are shaken with formic acid and thus homogenised. The analytes are subsequently extracted in a mixture of hexane and toluene. The extracts are puri-fied and enriched using silica gel, and after capillary gas chromatographic separation they are quantified by means of mass selective detection in the SIM mode. Calibra-tion standards prepared in bovine blood are processed and measured in the same manner as the samples to be analysed. 4,4'-Dibrombiphenyl, δ-hexachlorocyclohex-ane and Mirex serve as internal standards.

p,p'-DDT

Within-series imprecision: Standard deviation (rel.) s_w = 12.0% or 6.4%
                                       Prognostic range u = 26.7% or 15.8%
                                         at a concentration of 0.25 µg or 1.0 µg p,p'-DDT per litre blood
                                         and where n = 10 or 6 determinations

Between-day imprecision: Standard deviation (rel.) s_w = 14%
                                         Prognostic range u = 35%
                                         at a concentration of 0.4 µg p,p'-DDT per litre blood and where n = 6 determinations

Accuracy: Recovery rate r = 80% at a concentration of 0.5 µg/L

Detection limit: 0.05 µg p,p'-DDT per litre blood

Essential Biomonitoring Methods. DFG, Deutsche Forschungsgemeinschaft
Copyright © 2006 WILEY-VCH Verlag GmbH & Co. KGaA, Weinheim
ISBN: 3-527-31478-4

p,p′-DDE

Within-series imprecision: Standard deviation (rel.) s_w = 9.0% or 6.7%
Prognostic range u = 20.1% or 16.6%
at a concentration of 0.25 µg or 1.0 µg p,p′-DDE per litre blood
and where n = 10 or 6 determinations

Between-day imprecision: Standard deviation (rel.) s_w = 7.0%
Prognostic range u = 17.4%
at a concentration of 1.8 µg p,p′-DDE per litre blood and where n = 6 determinations

Accuracy: Recovery rate r = 90% at a concentration of 0.5 µg/L

Detection limit: 0.02 µg p,p′-DDE per litre blood

Hexachlorobenzene (HCB)

Within-series imprecision: Standard deviation (rel.) s_w = 5.3% or 2.7%
Prognostic range u = 11.8% or 6.7%
at a concentration of 0.25 µg or 1.0 µg HCB per litre blood
and where n = 10 or 6 determinations

Between-day imprecision: Standard deviation (rel.) s_w = 8.1%
Prognostic range u = 20.1%
at a concentration of 3.1 µg HCB per litre blood
and n = 6 determinations

Accuracy: Recovery rate r = 88% at a concentration of 0.5 µg/L

Detection limit: 0.02 µg HCB per litre blood

α-Hexachlorocyclohexane (α-HCH)

Within-series imprecision: Standard deviation (rel.) s_w = 10.0% or 2.8%
Prognostic range u = 22.3% or 6.9%
at a concentration of 0.25 µg or 1.0 µg α-HCH per litre blood
and where n = 10 or 6 determinations

Between-day imprecision: Standard deviation (rel.) s_w = 7.3%
Prognostic range u = 18.1%

at a concentration of 0.7 µg α-HCH per litre blood and
where n = 6 determinations

Accuracy: Recovery rate r = 86% at a concentration of
                                                     0.5 µg/L

Detection limit: 0.05 µg α-HCH per litre blood

β-Hexachlorocyclohexane (β-HCH)

Within-series imprecision: Standard deviation (rel.) s_w = 11.0% or 8.1%
                          Prognostic range u = 24.5% or 20.1%
                          at a concentration of 0.25 µg or 1.5 µg β-HCH per litre
                          blood
                          and where n = 10 or 6 determinations

Between-day imprecision: Standard deviation (rel.) s_w = 8.3%
                         Prognostic range u = 20.6%
                         at a concentration of 2.2 µg β-HCH per litre blood
                         and where n = 6 determinations

Accuracy: Recovery rate r = 80% at a concentration of
                                                     0.5 µg/L

Detection limit: 0.05 µg β-HCH per litre blood

γ-Hexachlorocyclohexane (γ-HCH)

Within-series imprecision: Standard deviation (rel.) s_w = 11.0% or 7.4%
                          Prognostic range u = 18.3% or 21.8%
                          at a concentration of 0.25 µg or 1.0 µg γ-HCH per litre
                          blood
                          and where n = 10 or 6 determinations

Between-day imprecision: Standard deviation (rel.) s_w = 9.0%
                         Prognostic range u = 22.3%
                         at a concentration of 1.5 µg γ-HCH per litre blood and
                         where n = 6 determinations

Accuracy: Recovery rate r = 85% at a concentration of
                                                     0.5 µg/L

Detection limit: 0.05 µg γ-HCH per litre blood

PCB 28

Within-series imprecision: Standard deviation (rel.) s_w = 9.8% or 4.3%
Prognostic range u = 21.8% or 10.7%
at a concentration of 0.25 µg or 1.0 µg PCB 28 per litre blood
and where n = 10 or 6 determinations

Between-day imprecision: Standard deviation (rel.) s_w = 7.2%
Prognostic range u = 17.9%
at a concentration of 3.4 µg PCB 28 per litre blood and
where n = 6 determinations

Accuracy: Recovery rate r = 97% at a concentration of 0.5 µg/L

Detection limit: 0.02 µg PCB 28 per litre blood

PCB 52

Within-series imprecision: Standard deviation (rel.) s_w = 9.2% or 4.4%
Prognostic range u = 20.5% or 10.9%
at a concentration of 0.25 µg or 1.0 µg PCB 52 per litre blood
and where n = 10 or 6 determinations

Between-day imprecision: Standard deviation (rel.) s_w = 6.3%
Prognostic range u = 15.6%
at a concentration of 2.4 µg PCB 52 per litre blood and
where n = 6 determinations

Accuracy: Recovery rate r = 97% at a concentration of 0.5 µg/L

Detection limit: 0.02 µg PCB 52 per litre blood

PCB 101

Within-series imprecision: Standard deviation (rel.) s_w = 10.1% or 4.7%
Prognostic range u = 22.5% or 11.6%
at a concentration of 0.25 µg or 1.0 µg PCB 101 per litre blood
and where n = 10 or 6 determinations

Between-day imprecision: Standard deviation (rel.) s_w = 5.4%
Prognostic range u = 13.4%

at a concentration of 2.1 µg PCB 101 per litre blood and where $n = 6$ determinations

Accuracy: Recovery rate $r = 95\%$ at a concentration of 0.5 µg/L

Detection limit: 0.02 µg PCB 101 per litre blood

PCB 138

Within-series imprecision: Standard deviation (rel.) $s_w = 9.7\%$ or 6.4%
Prognostic range $u = 21.6\%$ or 15.9%
at a concentration of 0.25 µg or 1.0 µg PCB 138 per litre blood
and where $n = 10$ or 6 determinations

Between-day imprecision: Standard deviation (rel.) $s_w = 7.0\%$
Prognostic range $u = 17.4\%$
at a concentration of 2.7 µg PCB 138 per litre blood and where $n = 6$ determinations

Accuracy: Recovery rate $r = 91\%$ at a concentration of 0.5 µg/L

Detection limit: 0.03 µg PCB 138 per litre blood

PCB 153

Within-series imprecision: Standard deviation (rel.) $s_w = 8.4\%$ or 4.9%
Prognostic range $u = 18.7\%$ or 12.1%
at a concentration of 0.25 µg or 1.0 µg PCB 153 per litre blood
and where $n = 10$ or 6 determinations

Between-day imprecision: Standard deviation (rel.) $s_w = 5.9\%$
Prognostic range $u = 14.6\%$
at a concentration of 2.6 µg PCB 153 per litre blood and where $n = 6$ determinations

Accuracy: Recovery rate $r = 93\%$ at a concentration of 0.5 µg/L

Detection limit: 0.03 µg PCB 138 per litre blood

PCB 180

Within-series imprecision: Standard deviation (rel.) s_w = 9.9% or 6.7%
Prognostic range u = 22.1% or 16.6%
at a concentration of 0.25 µg or 1 µg PCB 180 per litre
blood
and where n = 10 or 6 determinations

Between-day imprecision: Standard deviation (rel.) s_w = 6.4%
Prognostic range u = 15.8%
at a concentration of 3.2 µg PCB 180 per litre blood and
where n = 6 determinations

Accuracy: Recovery rate r = 100% at a concentration of
0.5 µg/L

Detection limit: 0.03 µg PCB 180 per litre blood

Polychlorinated biphenyls (PCB)

All the biphenyls bearing substituted chlorine atoms belong to the group of the poly-chlorinated biphenyls. A total of 209 so-called congeners are possible. Depending on the degree of chlorination (1 to 10 chlorine atoms), there are 3 monochloro-, 12 di-chloro-, 24 trichloro-, 42 tetrachloro-, 46 pentachloro-, 42 hexachloro-, 12 octa-chloro-, 3 nonachlorobiphenyls and 1 decachlorobiphenyl. The nomenclature of the PCBs was simplified by Ballschmiter and Zell [1] who arranged and numbered them according to the number and position of the chlorine atoms.

The PCB substances were widely used on account of their chemical and physical properties (e.g. inertness, thermal stability, low electrical conductivity) and the fact that they can be relatively simply and inexpensively manufactured. Thus among other applications PCBs are used as hydraulic fluids, industrial fats and oils, heat-ex-change media, impregnating agents, non-conductors in the electronics industry, sea-lants, organic solvents, paper-coating agents, flame-proofing agents for paper, woven materials and wood, and also as plasticisers in paints.

All the commercially available PCB products are composed of a mixture of various chlorinated biphenyls. As a rule, the chlorine content is between 20 and 60%. Infor-mation on the chemical composition of the technical mixtures may be derived from their names (e.g. Aroclor 1242, Chlophen A60, etc.). PCB congeners were generally selected at random for analytical determination in scientific studies until the end of the 1980s. In 1988 six congeners (Figure 1) were selected by agreement in Germany as indicators in routine analyses. These particular congeners have been selected largely for practical purposes to minimise the problems of separating all the conge-ners from each other by chromatography in one analytical run and thus of quantify-ing them without interference. The concept of indicator congeners [2] is generally ac-cepted and also used worldwide.

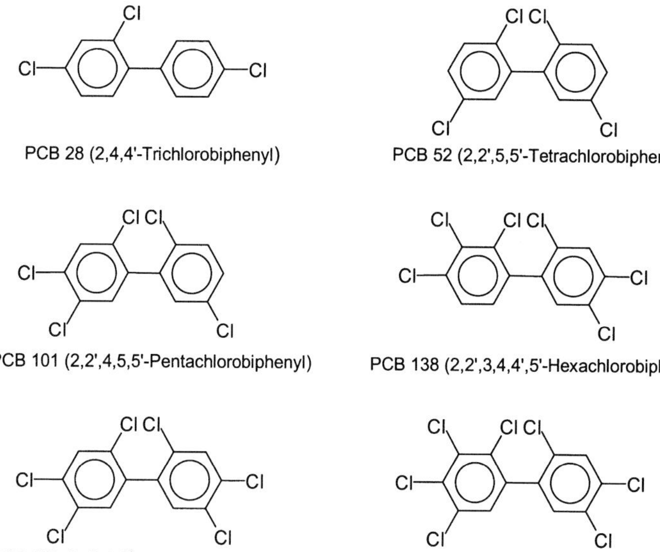

PCB 28 (2,4,4'-Trichlorobiphenyl)

PCB 52 (2,2',5,5'-Tetrachlorobiphenyl)

PCB 101 (2,2',4,5,5'-Pentachlorobiphenyl)

PCB 138 (2,2',3,4,4',5'-Hexachlorobiphenyl)

PCB 153 (2,2',4,4'5,5'-Hexachlorobiphenyl)

PCB 180 (2,2',3,4,4',5,5'-Heptachlorobiphenyl)

Fig. 1. PCB indicator congeners

Considerable amounts of PCBs are released into the environment due to improper handling (leaking transformers and condensers, disposal in non-sealed refuse sites, accidents, etc.). This causes great problems, as the stable PCBs are not readily decomposed and therefore remain for long periods in the ecosystem where they are not eliminated in many cases but merely relocated. PCBs accumulate in the food chain on account of their lipophilic nature. After the persistence of PCBs in the environment had been recognised, Monsanto and Bayer stopped their production in 1977 and 1983 respectively. The PCB, PCT and VC Ordinance passed in 1989 and the Chemikalien-Verbotsverordnung [Chemical Prohibition Ordinance] that became effective in 1993 ensured that the production, marketing and use of PCBs, polychlorinated terphenyls (PCTs) and vinyl chloride (VC) were forbidden in Germany.

A comprehensive overview of the metabolism of the polychlorinated biphenyls, which has not been completely clarified in detail to date, cannot be presented in this publication. Therefore the reader should refer to the relevant specialised literature (e.g. [3–7]). However, some basic aspects of the absorption, distribution, metabolism and excretion of PCBs in humans are discussed below.

PCBs accumulate in the organism because they are persistent and strongly lipophilic substances. More than 90% of the PCBs found in human tissue have been absorbed with fatty food. The absorption, distribution, metabolism and excretion of PCBs are strongly dependent on the degree of chlorination and the position of the chlorine atoms. Introduction of phenolic OH groups and subsequent conjugation reactions with glucuronic acid or sulphuric acid and glutathione represents the main metabolic pathway in humans, and this leads to accelerated excretion via the kidneys and intestine compared with the non-metabolised PCBs. The low-chlorinated PCBs (e.g. PCB 28, PCB 52, PCB 101) are relatively rapidly metabolised and eliminated, whereas the highly chlorinated congeners (e.g. PCB 138, PCB 153, PCB 180) accumulate in the body.

This also has a considerable effect on the PCB concentrations measured in the blood and fatty tissue of the general population. There is an almost linear relationship between the age of a person and the blood level of the hexachlorinated and heptachlorinated congeners [8], which have mainly been absorbed from the environment over a long period of time. In contrast, if low-chlorinated PCBs (dichlorinated, trichlorinated and tetrachlorinated biphenyls) are detected in the blood, this indicates recent exposure to these substances. Therefore if elevated blood levels of low-chlorinated biphenyls (e.g. PCB 28, PCB 52) compared to the general population are measured, it can be assumed that the exposure to PCBs is current or very recent. Table 1 shows the concentrations of the indicator PCBs found in blood samples from the general population. In general, a distinct drop in the PCB levels in blood samples in Germany has been observed in recent years [8].

Although PCBs generally exhibit a low acute toxicity, chronic exposure to these substances has diverse toxic effects. This applies to the liver in particular. PCBs induce a series of cytochrome P450-dependent enzymes which metabolise extraneous substances. Disorders of the immune systems have been observed in various species of animals after administration of PCBs. In animal studies PCBs are not genotoxic, but they are carcinogenic. The Deutsche Forschungsgemeinschaft's Commission for the Investigation of Health Hazards of Chemical Compounds in the Work Area has assigned technical mixtures of PCBs to category 3B of the carcinogenic working materials [21].

Table 1. Concentrations of polychlorinated biphenyls in blood, plasma or serum samples from persons who have not handled these substances in the course of their work

Substance	n	Group (average age)	Median [ng/L]	Reference
PCB 28	96	Teachers from schools contaminated with PCBs (48)	45 to 98[a]	Gabrio et al. 2000 [9]
	55	Controls (49)	35[a]	
PCB 28	35	General population in Spain (51)	24.5[a, b]	Wingfors et al. 2000 [10]
	26	General population in Sweden (68)	20.5[a, b]	
PCB 52	35	General population in Spain (51)	4.5[a, b]	
	26	General population in Sweden (68)	7[a, b]	
PCB 101	35	General population in Spain (51)	10[a, b]	
	26	General population in Sweden (68)	11.5[a, b]	
PCB 138	35	General population in Spain (51)	1100[a, b]	
	26	General population in Sweden (68)	1150[a, b]	
PCB 153	35	General population in Spain (51)	1500[a, b]	
	26	General population in Sweden (68)	1500[a, b]	
PCB 180	35	General population in Spain (51)	1400[a, b]	
	26	General population in Sweden (68)	1000[a, b]	

Table 1 (continued)

Substance	n	Group (average age)	Median [ng/L]	Reference
PCB 28	120	General population in Sweden (63)	19[b]	Glynn et al. 2000 [11]
PCB 52			<10[b]	
PCB 101			18[b]	
PCB 138			670[b]	
PCB 153			1475[b]	
PCB 180			1030[b]	
PCB 138	13	Immigrants from the former Yugoslavia (27)	200	Schmid et al. 1997 [12]
	29	Immigrants from the former USSR (26)	400	
	28	Immigrants from Asia (28)	<100	
	33	Immigrants from Africa (27)	<100	
	34	General population in Germany (26)	530	
PCB 153	13	Immigrants from the former Yugoslavia (27)	250	
	29	Immigrants from the former USSR (26)	400	
	28	Immigrants from Asia (28)	<100	
	33	Immigrants from Africa (27)	200	
	34	General population in Germany (26)	830	
PCB 180	13	Immigrants from the former Yugoslavia (27)	130	
	29	Immigrants from the former USSR (26)	<100	
	28	Immigrants from Asia (28)	<100	
	33	Immigrants from Africa (27)	<100	
	34	General population in Germany (26)	700	

[a] mean value, [b] calculated as ng/g fat in serum, assuming a constant blood fat content of 0.5%.

PCBs can also cause chloracne, oedema of the eyelids, atrophy of the thymus and the spleen, pathological changes to the kidneys and swelling of the meibomian glands. These effects are mainly observed after exposure to technical PCB mixtures. Therefore the possibility that these symptoms are at least partly caused by poly-chlorinated dibenzofuran impurities formed in the production process cannot be completely ruled out [14].

See Table 7 for the MAK values that have been assigned to PCB mixtures.

Table 2. Reference values (Germany) for polychlorinated biphenyls in stabilised whole blood and blood plasma in µg/L (Source: Umweltbundesamt [German Federal Environmental Agency] [13])

Age	PCB 138		PCB 153		PCB 180	
	Whole blood	Plasma	Whole blood	Plasma	Whole blood	Plasma
7–10	0.5	–	0.5	–	0.3	–
18–25	0.8	0.8	1.0	1.0	0.7	0.8
26–35	1.0	1.5	1.5	1.9	1.0	1.5
36–45	1.3	2.2	2.0	2.8	1.4	2.2
46–55	1.6	3.0	2.5	3.7	1.9	2.9
55–65	1.8	3.7	3.0	4.6	2.2	3.5

1,2,3,4,5,6-Hexachlorocyclohexane (HCH)

The hexachlorocyclohexane group consists of 8 stereoisomers (Figure 2). Only the a,a,a,e,e,e-isomer (3 neighbouring chlorine atoms in the axial position and the remaining chlorine atoms in the equatorial position), which is also known as γ-HCH or lindane, acts as an insecticide.

While a general prohibition of the use of technical HCH mixtures has been in force in the European Union since 1978, the use of lindane as a biocide is still permitted in agriculture and forestry, as a preservative for wood and textiles (moth protection), in veterinary medicine and for external application in human medicine (e.g. Jacutin a medication against lice). Lindane is also added to the insulation for electric cables to protect them from attack by termites (in Germany only for export).

About nine tonnes of HCH waste isomers are formed as by-products for every tonne of lindane that is manufactured. Until about 1972 some of this waste was simply dumped on open disposal sites in the Federal Republic of Germany. From there mainly α-HCH and β-HCH were dispersed in the environment by the effects of wind and evaporation [15]. As a result of the relatively high persistence of the hexachlorocyclohexanes this still poses an environmental problem today. If it is considered that β-HCH with a half-life of 8 to 10 years and α-HCH and γ-HCH with a half-life of over one year remain in the soil, then it becomes clear that even now this path of introduction still represents a considerable exposure source for animals and plants. As a consequence, food contaminated by HCHs is the main exposure route for man. More than 90% of the HCHs absorbed by humans are taken in with their food.

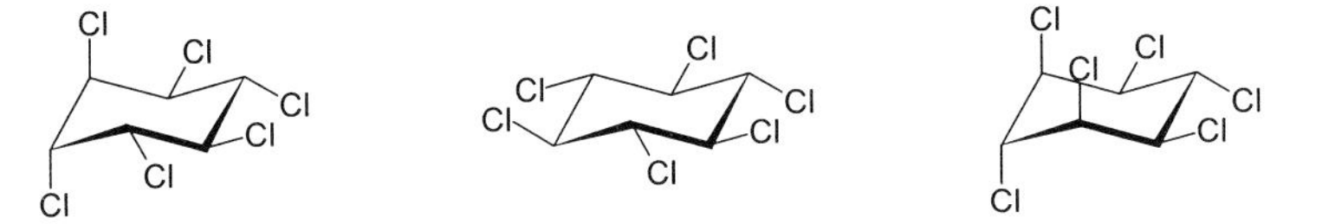

α-Hexachlorocyclohexane
(a,a,e,e,e,e-HCH)

β-Hexachlorocyclohexane
(e,e,e,e,e,e-HCH)

γ-Hexachlorocyclohexane
(a,a,a,e,e,e-HCH)

Fig. 2. Important HCH isomers

β-HCH accumulates in the organism on account of its relatively long biological half-life. This is also reflected in an age-dependency of the blood levels. Table 3 shows the data on the β-HCH exposure of different groups (immigrants shortly after their arrival in Germany), the age-dependent reference values for β-HCH are found in Table 4.

Due to the short half-life in humans ($t_{1/2} = 1$ d) values of less than 0.1 µg/L are generally measured for γ-HCH (lindane) in the general population. As in the case of α-HCH, recent or still existing exposure can be assumed if a level of 0.1 µg per litre blood is exceeded.

Table 3. β-HCH concentrations in plasma samples of the general population

Substance	n	Group (average age)	Median [ng/L]	Reference
β-HCH	13	Immigrants from the former Yugoslavia (27)	<500	Schmid et al. 1997 [12]
	29	Immigrants from the former USSR (26)	1600	
	28	Immigrants from Asia (28)	990	
	33	Immigrants from Africa (27)	<500	
	34	General population in Germany (26)	<500	

Table 4. Reference values (Germany) for β-HCH in whole blood in µg/L (Source: Umweltbundesamt [German Federal Environmental Agency] [19])

Age	β-HCH Whole blood
7–10	0.3
18–25	0.2
26–35	0.4
36–45	0.7
46–55	1.3
55–65	1.3
>65	2.0

The configuration of the HCH isomers determines their metabolism and excretion. β-HCH is metabolised most slowly and is predominantly excreted unchanged with the faeces. In contrast, α-hexachlorocyclohexane and γ-hexachlorocyclohexane are mainly excreted as metabolites via the urine. In man a major part of the α-HCH and γ-HCH is finally converted to various chlorophenols by dehydrochlorination, dechlorination and dehydrogenation [16]. The resulting metabolites are partly eliminated through the kidneys as conjugates of glutathione or glucuronic acid, as sulphate conjugates or as mercapturic acids.

Hexachlorocyclohexanes are not mutagenic. Carcinogenesis experiments with lindane resulted in neoplastic foci in the liver of rats, mice showed an elevated incidence of liver tumours. No irritation was observed in initiation/promotion studies on rats, but a tumour-promoting effect was reported for lindane [17].

α- and β-HCH also exhibited tumour-promoting properties in rats and mice [18].

A MAK value exists for the hexachlorocyclohexanes, lindane has been assigned a MAK and a BAT value (see Table 7).

Hexachlorobenzene (HCB)

Hexachlorobenzene (Figure 3) was mainly used as a pesticide for treating cereal seeds until its use was prohibited in Germany in 1974. In addition, it is formed as a by-product in industrial chemical production, such as the manufacture of trichloro-ethylene and tetrachloroethylene, during the chlorination of hydrocarbons and in the course of the production of other biocides such as pentachlorophenol or HCH. In the environment HCB mainly accumulates in the soil and water sediments. This is primarily due to the use of hexachlorobenzene as a fungicide for seeds and is also caused by industrial emissions. As a result of its half-life in the soil of 3 to 6 years HCB accumulates in the food chain. Once again fatty foods are the most important source of exposure for humans. However, a continuous reduction in hexachloroben-zene contamination in food has been ascertained since its use was forbidden.

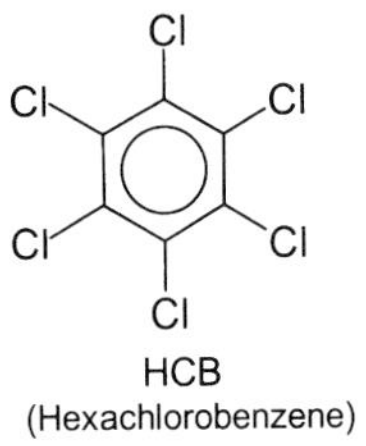

HCB
(Hexachlorobenzene)

Fig. 3. Hexachlorobenzene

Hexachlorobenzene accumulates in the human body. The corresponding HCB concentrations measured in the blood are again dependent on age. Table 5 shows the age-dependent reference values given by the Umweltbundesamt [German Ministry for the Environment] for Germany.

Table 5. Reference values (Germany) for HCB in whole blood in µg/L (Source: Umweltbundesamt [German Federal Environmental Agency] [19])

Age	HCB Whole blood
7–10	0.4
18–25	0.4
26–35	1.2
36–45	2.1
46–55	2.9
55–65	4.0
>65	4.6

A mass intoxication of humans with fungicides containing HCB over a period of 4 to 5 years led to a high fatality rate. General weakness, skin damage, porphyria, hyposomia, osteoporosis, arthritis and neuritis were described as symptoms.

In animal studies the acute toxicity is low. In rats it is in the order of 3500 to 10,000 mg/kg body weight. The symptoms of intoxication include tremors, convulsions and ataxia. If animals are given HCB orally over a long period of time, they suffer from dermal lesions such as loss of hair, blistering, formation of scabs and hyperpigmentation. Animal studies have shown that the main target organ is the liver. Porphyria is observed, especially in female rats and pigs. Degenerative effects on the liver, pre-neoplastic foci and liver tumours are observed in rats as a result of chronic administration of hexachlorobenzene. Liver tumours are also manifested by mice and hamsters. In addition, tumours of the bile duct, adenomas of the kidneys and phaeochromocytomas of the adrenal gland occur in rats.

The thyroid gland is a further target organ. Hexachlorobenzene also has an immunomodulating influence.

No mutagenic effect of HCB has been observed, but there is evidence of a slight clastogenic effect [20].

HCB has been assigned a BAT value (see Table 7).

p,p′-DDT and p,p′-DDE

The production, import, export, purchase and use of DDT have been prohibited in the Federal Republic of Germany since 1972. DDT is still in use in tropical countries for the purpose of controlling epidemics. DDT has an extraordinarily high persistence and can still be detected in all environmental compartments today as a consequence of its massive worldwide use over many years. DDT is poorly degradable, and on account of its lipophilic character it also accumulates in the human body after intake via the food chain.

DDT is extremely lipophilic and it is readily absorbed via the gastrointestinal tract and the lungs. As a solid DDT is hardly absorbed through the skin, but in solution it can easily penetrate the skin.

The main metabolic pathway of DDT is the enzymatic elimination of HCl to p,p′-dichlorodiphenyldichloroethylene (DDE). DDE is very resistant to further metabolisation and, like DDT, it is also strongly lipophilic. The half-life for the excretion of DDT is approx. 1 year for humans. However, the metabolite DDE is more suitable for detecting long-term exposure to DDT.

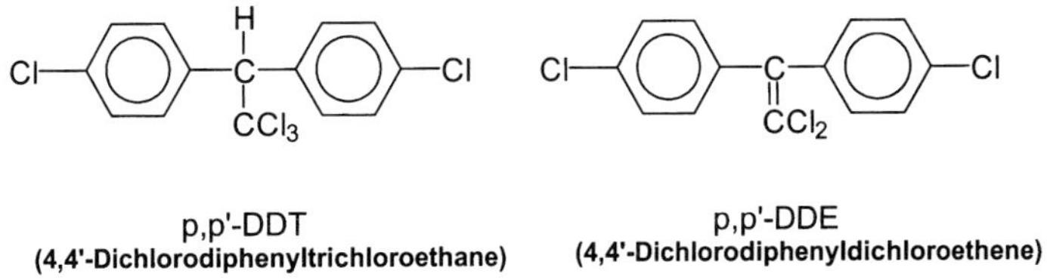

p,p′-DDT
(4,4′-Dichlorodiphenyltrichloroethane)

p,p′-DDE
(4,4′-Dichlorodiphenyldichloroethene)

Fig. 4. DDT and DDE

Table 6 shows DDE concentrations measured in persons with no occupational contact with DDT. In the case of DDE there is also a marked dependence of the blood concentration on age. The mean blood levels in persons over 60 years of age are higher by a factor of 4 to 7 than the levels found in young adults [19].

Table 6. DDE concentrations found in plasma samples of persons who had no occupational contact with DDT

Substance	n	Group (average age)	Median [ng/L]	Reference
DDE	13	Immigrants from the former Yugoslavia (27)	2300	Schmid et al. 1997 [12]
	29	Immigrants from the former USSR (26)	11900	
	28	Immigrants from Asia (28)	16900	
	33	Immigrants from Africa (27)	10850	
	34	General population in Germany (26)	1400	

DDT is a non-systemic contact and intestinal insecticide with a wide application range. DDT is effective against insects by interference with their nerve conduction. The same neurotoxic effect has been observed in humans and other warm-blooded animals.

The lethal oral dose of DDT for humans is approximately 0.1 to 1 g/kg body weight, or in the case of dermal absorption about one order of magnitude higher.

The target organ for chronic DDT intoxication is the liver. The effects of chronic exposure to high doses of DDT range from changes caused to the liver to necrosis of the liver. The NOAEL for man is about 1.5 mg DDT per kg body weight and day.

DDT and its metabolites induce cytochrome P450-dependent enzymes, in particular the P450-B sub-family.

No evidence of a toxic effect on human reproduction has been found for DDT or its metabolites. There is no indication that DDT is mutagenic, except for an impairment of the intercellular communication (fibrobasts of the skin) induced by DDT.

However, numerous animal studies on rats, mice and hamsters have shown that chronic exposure to DDT or DDE leads to liver tumours, lymphomas and tumours of the adrenal cortex.

A MAK value has been assigned to p,p'-DDT (see Table 7).

Table 7. Limit values, H = absorbed through the skin

Substance	Absorbed through the skin	Classification category of carcinogenic substances (DFG [21])	Exposure limit
p,p'-DDT	H	–	MAK: 1 mg/m^3
HCB	H	4	BAT: 150 µg/L plasma
α-HCH, β-HCH mixture	H	–	MAK: 0.5 mg/m^3
γ-HCH (lindane)	H	4	MAK: 0.1 mg/m^3 BAT: 25 µg/L plasma
PCB mixture with 42% chlorine content	H	3B	MAK: 1.1 mg/m^3
PCB mixture with 54% chlorine content	H	3B	MAK: 0.7 mg/m^3

Author: *H.-W. Hoppe, T. Weiss*
Examiners: *M. Ball, J. Lewalter*

Organochlorine compounds in whole blood and plasma

Application   Determination in whole blood and plasma

Analytical principle   Capillary gas chromatography/
mass spectrometric detection (MS)

Completed in   May 2001

Contents

Essential Biomonitoring Methods. DFG, Deutsche Forschungsgemeinschaft
Copyright © 2006 WILEY-VCH Verlag GmbH & Co. KGaA, Weinheim
ISBN: 3-527-31478-4

1 General principles

Blood and plasma are shaken with formic acid and thus homogenised. The analytes are subsequently extracted in a mixture of hexane and toluene. The extracts are purified and enriched using silica gel, and after capillary gas chromatographic separation they are quantified by means of mass selective detection in the SIM mode. Calibration standards prepared in bovine blood are processed and measured in the same manner as the samples to be analysed. 4,4′-Dibrombiphenyl, δ-hexachlorocyclohexane and Mirex serve as internal standards.

2 Equipment, chemicals and solutions

2.1 Equipment

Gas chromatograph with mass selective detector, split-splitless injection system, autosampler and data processing system for evaluation

Capillary gas chromatographic column:
Length: 30 m, inner diameter: 0.25 mm; stationary phase: 5% phenyl / 95% methylpolysiloxane; film thickness: 0.25 µm (e.g. from Hewlett-Packard)

Thermal evaporation block equipped with a fan (e.g. from Barkey)

Laboratory shaker (e.g. Multi-Tube Vortex from Baker)

Laboratory centrifuge (e.g. from Heraeus)

Vacuum centrifuge (e.g. SpeedVac from Savant)

Adjustable 100 µL and 1 mL pipettes (e.g. from Eppendorf)

It is advisable to heat all glassware for 6 hours at 480 °C before use. Alternatively, the glassware can be cleaned by rinsing it several times with acetone.

250 mL and 500 mL glass beakers

100 mL and 500 mL glass cylinders

10 mL and 25 mL volumetric flasks

10 µL to 100 µL glass syringes (e.g. from Hamilton)

Autosampler vials, 2 mL (e.g. from Agilent)

Microvial inserts, 250 µL (e.g. from Agilent)

10 mL Glass tubes with screw caps

20 mL Glass tubes with ground glass stoppers

2.2 Chemicals

Toluene p.a. (e.g. from Merck)

n-Hexane p.a. (e.g. from Merck)

Ethanol p.a. (e.g. from Sigma)

Bovine blood (e.g. from ACILA GMN mbH)

Bovine plasma (e.g. from ACILA GMN mbH)

Formic acid 98–100% (e.g. from Fluka)

Silica gel, 0.063–0.100 mm (Silica Gel 60 from Merck, No. 15101)

Silanised glass wool (e.g. from Supelco)

n-Decane (e.g. from Fluka, No. 30550)

PCB mixture (congeners 28, 52, 101, 138, 153, 180: e.g. from Dr. Ehrenstorfer, L 200301)

Hexachlorocyclohexane isomers (e.g. from Dr. Ehrenstorfer)

p,p'-DDE (e.g. from Dr. Ehrenstorfer)

p,p'-DDT (e.g. from Dr. Ehrenstorfer)

Hexachlorobenzene (HCB) (e.g. from Dr. Ehrenstorfer)

δ-HCH (e.g. from Dr. Ehrenstorfer)

Mirex (e.g. from Dr. Ehrenstorfer)

4,4'-Dibromobiphenyl (e.g. from Aldrich)

2.3 Solutions and conditioning of the clean-up material

The quantities given in this section are sufficient for 20 individual analyses including calibration. The quantities for larger analytical series must be adapted accordingly. All the solutions must be freshly prepared on the day that the analysis is carried out.

Formic acid:
60 mL formic acid are placed in a 250 mL separation funnel. 50 mL n-hexane are added. The mixture is shaken intensively for 5 minutes. After separation of the phases, the lower formic acid phase is poured into a 100 mL glass beaker and used immediately.

Extraction solution (toluene/n-hexane; 1:1, v/v):
210 mL toluene are placed in a 500 ml L glass beaker. 210 mL n-hexane are added while stirring.

Elution solution for cleaning up (toluene/n-hexane; 1:4, v/v):
80 mL toluene are placed in a 500 L glass beaker. 320 mL n-hexane are added while stirring.

Separation column for cleaning up:
A glass column (25 cm, 6 mm diameter, e.g. 5 mL enzyme test pipettes from BRAND, No. 27947) are filled with some glass wool and 1 g silica gel. This column is heated at 480 °C for 6 hours, and after cooling, it is stored in a desiccator until use.

2.4 Calibration standards

2.4.1 Internal standard

Starting solution δ-HCH (internal standard 1):
Approximately 25 mg δ-HCH are weighed exactly into a 25 mL volumetric flask. The volumetric flask is subsequently filled to its nominal volume with toluene (1 g/L).

Starting solution 4,4'-dibromobiphenyl (internal standard 2):
Approximately 25 mg 4,4'-dibromophenyl are weighed exactly into a 25 mL volumetric flask. The flask is subsequently filled to its nominal volume with toluene (1 g/L).

Starting solution Mirex (internal standard 3):
Approximately 25 mg Mirex are weighed exactly into a 25 mL volumetric flask. The flask is subsequently filled to its nominal volume with toluene (1 g/L).

Spiking solution for the internal standard (mixture):
Approx. 2 mL ethanol are placed in a 10 mL volumetric flask. Then 5 µL of the starting solution of δ-HCH in toluene, 5 µL of the starting solution of 4,4'-dibromobiphenyl and 2 µL of the starting solution of Mirex are added using a pipette. The flask is subsequently filled to its nominal volume with ethanol (δ-HCH 0.5 mg/L, 4,4'-dibromobiphenyl 0.5 mg/L; Mirex 0.2 mg/L).

2.4.2 Calibration standards

Starting solution A (PCB mixture):
A commercially available mixture of PCBs Nos. 28, 52, 101, 153 and 180 in iso-octane is used as the starting solution (e.g. Ehrenstorfer, L 200301). The concentration of each PCB congener is 10 mg/L.

Starting solution B (HCH isomers, HCB, p,p'-DDT and p,p'-DDE):
Approximately 25 mg of each substance are weighed exactly into a 25 mL volumetric flask. The flask is subsequently filled to its nominal volume with toluene (1 g/L).

Dilution solution B (HCH isomers, HCB, p,p'-DDT and p,p'-DDE):
Approx. 2 mL toluene are placed in a 10 mL volumetric flask. Then 100 µL of starting solution B are added using a pipette. The flask is subsequently filled to its nominal volume with ethanol (10 mg/L).

Stock solution 1:
Approx. 2 mL ethanol are placed in a 10 mL volumetric flask. 25 μL of dilution solution B and 25 μL of starting solution A are added using a pipette. The flask is then filled to its nominal volume with ethanol (0.025 mg/L).

Stock solution 2:
Approx. 2 mL ethanol are placed in a 10 mL volumetric flask. 250 μL of dilution solution B and 250 μL of starting solution A are added using a pipette. The flask is then filled to its nominal volume with ethanol (0.25 mg/L).

Calibration standards containing between 0.1 and 20 μg/L of the substances are prepared from the stock solutions by dilution with commercially available bovine blood. For this purpose 2.5 mL of the bovine blood or bovine plasma are first placed into a 20 mL glass tube with a ground-glass stopper, and the appropriate volumes of the stock solutions are added using a pipette in accordance with the pipetting scheme shown in Table 8. The sample (calibration standard) is then intensively shaken on a laboratory shaker (Vortex) for 10 seconds.

Table 8. Pipetting scheme for the preparation of calibration standards in blood or plasma

Volume of stock solution 1 [μL]	Volume of stock solution 2 [μL]	Final volume of the calibration standard [mL]	Concentration of the calibration standard [μg/L]
–	–	2.5	0
10	–	2.51	0.1
25	–	2.53	0.25
100	–	2.6	1.0
–	25	2.53	2.5
–	50	2.55	5
–	100	2.6	10
–	200	2.7	20

The given concentrations are based on an ideal final volume of exactly 2.5 mL, which may deviate from the real volume obtained by spiking by no more than 8%. This procedure was selected, as it is relatively difficult to pipette bovine blood. Possible dilution effects were not observed. As evaluation is carried out using internal standards, the deviation is compensated when the calibration curve is plotted and the analytical result is calculated.
As the organochlorine compounds are very stable, the starting solutions, spiking solutions, dilution solutions and stock solutions described in Section 2.4 can be kept in glass vessels in the refrigerator for a practically unlimited period. The actual calibration standards in bovine blood or bovine plasma should be prepared and processed at the same time as the samples.

3 Specimen collection and sample preparation

Approx. 5 to 8 mL blood are withdrawn slowly using a disposable syringe containing an anticoagulant (e.g. EDTA-K Monovettes). If plasma is to be analysed, then it must be prepared by centrifugation (<3500 g) as soon as possible after the blood sample has been taken.

In order to avoid substance loss due to adsorption on the wall of the plastic vessel, the blood or plasma should be transferred to a sealable 10 mL glass tube. The sample can be stored for at least a week at room temperature and at least 21 days in the refrigerator (at approx. 4 °C).

3.1 Sample preparation

2.5 mL blood or plasma, 2.5 mL purified formic acid and 50 µL spiking solution of the internal standards are placed in a 20 mL glass tube with a ground-glass stopper. The sample is homogenised by shaking for 1 minute on the laboratory shaker. Then 10 mL of a mixture of toluene/n-hexane (1:1, v/v) are added using a pipette, and the sample is extracted by shaking on the laboratory shaker. After centrifugation (5 min, 3000 g), 9 mL of the organic phase are taken up with a 10 mL pipette and transferred to a 10 mL glass tube. The samples are evaporated to approx. 100 to 500 µL in a vacuum centrifuge. The concentrated extract is transferred to the prepared cleanup columns using a Pasteur pipette. Then the analytes are eluted by adding 9 mL of a mixture of toluene/n-hexane (1:4, v/v). 50 µL n-decane are added to the eluate as a keeper, and it is evaporated to approx. 100 to 500 µL in a vacuum centrifuge. The solution is subsequently transferred to an autosampler ampoule and concentrated to approx. 50 µL in a stream of nitrogen at 35 °C. The concentrated measurement solution is then transferred into a GC microvial using a pipette.

4 Operational parameters

4.1 Operational parameters for gas chromatography and mass spectrometry

Capillary column:	Material:	Fused silica
	Stationary phase:	DB-5
	Length:	30 m
	Inner diameter:	0.25 mm
	Film thickness:	0.25 µm
Detector:	Mass selective detector (MSD)	

Temperatures:	Column:	Starting temperature 130 °C, 1 minute isothermal, then increase at a rate of 10 °C/min to 295 °C, then 7 min at the final temperature

Injector: 250 °C

Transfer line: 280 °C

Carrier gas: Helium 4.6 at a pre-pressure of 9 psi

Split: Splitless, split on after 60 s

Sample volume: 2 µL

Evaporation tube: 2 mm inner diameter

Ionisation type: Electron impact ionisation (EI)

Ionisation energy: 70 eV

Dwell time: See Table 9

Electron multiplier: 1400 V + 600 V

All other parameters must be optimised in accordance with the manufacturer's instructions.

5 Analytical determination

In each case 2 µL is injected into the gas chromatograph for the analytical determination of the blood samples processed as described in Section 3.1. A quality control sample and an aqueous blank sample is analysed with each analytical series. The temporal profiles of the ion fragments shown in Table 9 are recorded in the SIM mode.

Table 9. Retention times, masses and internal standards used for quantification

Compound	Retention time [min]	Masses	Dwell time [ms]	Internal standard
α-HCH	8.3	218.9 216.9*	150	δ-HCH (IS 1)
HCB	8.5	283.9* 285.9	150	δ-HCH (IS 1)
β-HCH	8.9	218.9 216.9*	150	δ-HCH (IS 1)
γ-HCH	9.0	218.9 216.9*	150	δ-HCH (IS 1)
δ-HCH (IS 1)	9.5	218.9 220.9	150	

Table 9 (continued)

Compound	Retention time [min]	Masses	Dwell time [ms]	Internal standard
PCB 28	10.2	255.9* 257.9	70	4,4′-Dibromophenol (IS 2)
PCB 52	11.0	289.9 291.9*	70	4,4′-Dibromophenol (IS 2)
4,4′-Dibromophenol (IS 2)	11.6	312* 310 314	70	
PCB 101	12.7	325.8* 327.8	40	4,4′-Dibromophenol (IS 2)
p,p′-DDE	13.3	316.0 318.0*	40	δ-HCH (IS 1)
PCB 153	14.4	359.8* 361.8	40	Mirex
p,p′-DDT	14.9	235.0* 237.0	40	δ-HCH (IS 1)
PCB 138	15.0	359.8* 361.8	40	Mirex
PCB 180	16.3	393.7* 395.7	80	Mirex
Mirex (IS 3)	16.9	272* 270 274	80	

The masses marked * are used for quantitative evaluation.

The retention times shown in Table 9 serve only as a guide. Users of the method must satisfy themselves of the separation power of the capillary column used and the resulting retention behaviour of the substances. Figure 5 shows an example of a chromatogram of a processed blood standard spiked with 5 µg/L. The chromatogram of a processed human blood sample is shown in Figure 6.

6 Calibration

The calibration standards (Section 2.4.2) are processed in the same manner as the blood samples (Section 3.1) and analysed by gas chromatography/mass spectrometry as described in Sections 4 and 5. Calibration graphs are obtained by plotting the quotients of the peak areas of the analytes and that of the relevant internal standard (see Table 9) as a function of the concentrations used. It is unnecessary to plot a complete calibration graph for every analytical series. It is sufficient to analyse one calibration standard for every analytical series. The ratio of the result obtained for this standard and the result for the equivalent standard in the complete calibration graph is calculated. Using this quotient, each result read off the calibration graph is adjusted for the relevant series.

New calibration graphs should be plotted if the quality control results indicate systematic deviation.

The calibration graph is linear between the detection limit and 20 µg per litre blood.

7 Calculation of the analytical result

Quotients are calculated by dividing the peak areas of the analytes by that of the relevant internal standard. These quotients are used to read off the corresponding concentration of the analytes in µg per litre blood from the relevant calibration graph. If the bovine blood used to prepare the calibration standards exhibits background interference, the resulting calibration graph must be shifted in parallel so that it passes through the zero point of the coordinates. (The concentrations of the background exposure can be read off from the point where the graph intercepts the axis before parallel shifting in each case.) If the aqueous blank solution indicates that there are reagent blank values, the source of the contamination must be identified and eliminated.

8 Standardisation and quality control

Quality control of the analytical results is carried out as stipulated in the guidelines of the Bundesärztekammer (German Medical Association) [22, 23] and in the special preliminary remarks to this series. In order to determine the precision of the method a spiked bovine blood sample containing a constant concentration of the analytes is analysed. As material for quality control is not commercially available, it must be prepared in the laboratory. For this purpose, bovine blood is spiked with a defined quantity of the analytes. A six-month supply of this control material is prepared, divided into aliquots in sealable 10 mL glass tubes and stored in the deep-freezer. The concentration of this control material should lie within the decisive concentration range. The theoretical value and the tolerance range for this quality control material are determined in the course of a pre-analytical period (one analysis of the control material on each of 20 different days) [24–26].

External quality control can be achieved by participation in round-robin experiments. The Deutsche Gesellschaft für Arbeits- und Umweltmedizin (German Association for Occupational and Environmental Medicine) offers polychlorinated biphenyls, hexachlorocyclohexane as well as DDT and DDE as parameters for toxicological occupational and environmental analyses in their round-robin programme [27].

9 Evaluation of the method

9.1 Precision

Bovine blood samples spiked to give concentrations of 0.25 µg/L and 1 µg/L were processed and analysed to check the precision in the series. Ten replicate determinations of these blood samples yielded the precision in the series documented in Table 10.

Table 10. Precision in the series

Parameter	n	Concentration [µg/L]	Standard deviation (rel.) [%]	Prognostic range [%]
p,p'-DDT	10	0.25	12.0	26.7
	6	1	6.4	15.8
p,p'-DDE	10	0.25	9.0	20.1
	6	1	6.7	16.6
HCB	10	0.25	5.3	11.8
	6	1	2.7	6.7
α-HCH	10	0.25	10.0	22.3
	6	1	2.8	6.9
β-HCH	10	0.25	11.0	24.5
	6	1	8.1	20.1
γ-HCH	10	0.25	11.0	24.5
	6	1	7.4	18.3
PCB 28	10	0.25	9.8	21.8
	6	1	4.3	10.7
PCB 52	10	0.25	9.2	20.5
	6	1	4.4	10.9
PCB 101	10	0.25	10.1	22.5
	6	1	4.7	11.6
PCB 138	10	0.25	9.7	21.6
	6	1	6.4	15.9
PCB 153	10	0.25	8.4	18.7
	6	1	4.9	12.1
PCB 180	10	0.25	9.9	22.1
	6	1	6.7	16.6

In addition, the precision from day to day was determined. Plasma samples from round-robin experiments [28] of the Deutsche Gesellschaft für Arbeits- und Umweltmedizin [German Society for Occupational and Environmental Medicine] (18th series of round-robin experiments, sample A containing DDE, HCHs, HCB, PCB) and the Arctic Monitoring and Assessment Programme (round-robin experiments 2001, Round 2, sample W-01-05 including p,p'-DDT) were used. These plasma solutions were processed and analysed on 6 different days. The precision results are also shown in Table 11.

Table 11. Precision from day to day

Parameter	n	Theoretical value [µg/L]	Actual value [µg/L]	Standard deviation (rel.) [%]	Prognostic range [%]
p,p'-DDT	6	0.55	0.4	14	35
p,p'-DDE	6	1.8	2.5	7.0	17.4
HCB	6	3.1	4.1	8.1	20.1
α-HCH	6	0.7	0.4	7.3	18.1
β-HCH	6	2.2	2.6	8.3	20.6
γ-HCH	6	1.5	1.5	9.0	22.3
PCB 28	6	3.4	4	7.2	17.9
PCB 52	6	2.4	2.2	6.3	15.6
PCB 101	6	2.1	2.1	5.4	13.4
PCB 138	6	2.7	2.5	7.0	17.4
PCB 153	6	2.6	2.6	5.9	14.6
PCB 180	6	3.2	3.5	6.4	15.8

9.2 Accuracy

The accuracy of the method was checked by means of recovery experiments using human blood samples. For this purpose, 6 different individual samples with a relatively low background level were selected. These samples were spiked with 0.5 and 2.0 µg/L, then processed and measured 6 times. The mean relative recovery rates can be found in Table 12.

Table 12. Mean relative recovery rates and losses due to processing in spiked human blood samples

Parameter	n	Relative recovery (0.5 µg/L) [%]	Relative recovery (2.0 µg/L) [%]	Losses due to processing [%]
p,p'-DDT	6	80	110	26
p,p'-DDE	6	90	95	19
HCB	6	88	90	16
α-HCH	6	86	85	20
β-HCH	6	80	79	5
γ-HCH	6	85	86	18
PCB 28	6	97	99	9
PCB 52	6	97	98	11
PCB 101	6	95	100	16
PCB 138	6	91	97	17
PCB 153	6	93	95	17
PCB 180	6	100	96	14

Furthermore, the accuracy of the method was objectively evaluated by participation in the 19th series of round-robin experiments of the Deutsche Gesellschaft für Arbeits- und Umweltmedizin [German Society for Occupational and Environmental

Medicine] [29]. As the results in Table 13 show, there is an excellent correlation between the theoretical values of the round-robin experiments and the values obtained using this method.

Table 13. Theoretical values and values obtained by this method during the 19th series of round-robin experiments of the DGAUM

Substance	19th Series of round-robin experiments Sample A (environmental range)		19th Series of round-robin experiments Sample B (environmental range)	
	Theoretical value (tolerance range) [µg/L]	Result [µg/L]	Theoretical value (tolerance range) [µg/L]	Result [µg/L]
p,p′-DDE	1.14 (0.66–1.62)	1.0	2.88 (2.0–3.76)	2.6
HCB	2.13 (1.35–2.92	2.2	4.35 (2.94–5.76)	4.5
α-HCH	0.09 (0.03–0.15)	0.08	0.27 (0.15–0.4)	0.36
β-HCH	1.1 (0.7–1.5)	1.35	2.91 (2.1–3.7)	3.1
γ-HCH	1.15 (0.73–1.58	1.0	2.9 (1.9–3.9)	3.0
PCB 28	0.46 (0.26–0.66)	0.43	1.78 (1.03–2.54)	1.4
PCB 52	0.46 (0.28–0.65)	0.40	1.45 (0.932–1.98)	1.2
PCB 101	0.71 (0.41–1.0)	0.65	2.37 (1.47–3.2)	2.2
PCB 138	1.78 (1.19–2.4)	1.4	3.88 (2.67–5.1)	4.0
PCB 153	0.45 (0.28–0.62)	0.43	4.63 (3.33–5.93)	4.7
PCB 180	1.39 (0.97–1.82)	1.4	4.94 (3.51–6.37)	4.5

9.3 Detection limits

Under the conditions described here the detection limits, calculated as three times the signal/noise ratio of the analytical background interference in the temporal environment of the analyte signals, were between 0.02 and 0.05 µg/L. The detection limits for the parameters determined using this method are shown in Table 14.

Table 14. Detection limits in µg/L

Substance	Detection limits
p,p′-DDT	0.05
p,p′-DDE	0.02
HCB	0.02
α-HCH	0.05
β-HCH	0.05
γ-HCH	0.05
PCB 28	0.02
PCB 52	0.02
PCB 101	0.02
PCB 138	0.03
PCB 153	0.03
PCB 180	0.03

9.4 Sources of error

Interference due to matrix components or exogenous substances was occasionally observed in the case of γ-HCH, and this was attributed to the deteriorating separation capability of ageing separation columns. This interference was clearly recognisable from the peak pattern and the isotope ratio. In such cases the 219 mass fraction can be used for evaluation instead. The separation of β-HCH and γ-HCH gives an indication of performance of the separation column. These isomers must be separated to the baseline. Otherwise a new column must be used.

p,p'-DDT is a critical substance. High sensitivity and acceptable precision can only be achieved with clean, deactivated evaporation tubes and good columns.

In addition, it was checked whether the following substances or substance groups caused interference to the method:

Chlordane	Heptachloroepoxide	Phthalates
Oxychlordane	Aldrin	Phosphoric acid esters
Nonachlor	Dieldrin	Hexachloronaphthalene
Chlorophenols	Quintozene	Dichlorfluanid
Chlorobenzenes	Pentachloroaniline	Tolylfluanid
Methoxychlor	Endosulfan	Chlorthalonil
Methoxyolefin	Musk xylene	Polyaromatic hydrocarbons
Heptachlor	Musk ketone	Pyrethroids
Endrin		

Toxaphene congeners (nomenclature according Parlar): Palar 26, 32, 50 and 63

No interference was observed under the conditions given for the method.

The evaporation of the blood extract and the eluate from the silica gel column can be a critical step. In order to prevent the loss of HCH isomers, PCB 28 and PCB 52, the sample may not be evaporated to dryness. n-Decane, which is added as a keeper to prevent losses, may be contaminated by phthalates, but no analytical problems result from the impurities. Decane from Fluka (No. 30550) exhibits a distinctly higher degree of purity than that from Aldrich (No. D90-1). No contamination of the measurement solution caused by the vacuum centrifuge or the evaporation block was observed.

It is essential to find the source of any reagent blank values detected as a result of the inclusion of an aqueous blank solution. The blank values of the individual reagents must be determined, and if necessary the reagents must be replaced by non-contaminated chemicals. Possible contamination due to the glassware or plastic devices used must also be taken into account.

10 Discussion of the method

The GC-MS procedure presented here permits the sensitive, specific and reliable determination of polychlorinated biphenyls (6 indicator PCBs), hexachlorocyclohexane isomers, DDT and DDE as well as hexachlorobenzene in blood.

It is based on the DFG method according to Schulte, Lewalter and Ellrich [30, 31]. There are two important differences. Firstly, extraction is performed with toluene/ hexane instead of n-heptane or isooctane, and secondly detection is carried out by means of MSD instead of ECD. Thus an improved extractability of the HCH isomers and a generally higher specificity and sensitivity of detection is achieved.

On account of its sensitivity the method is suitable for reliably detecting the background levels of the persistent organochlorine compounds in the general population. The described method is extremely specific due to the use of mass spectrometry. It allows a more reliable determination than an electron capture detector, even in the lower concentration range (<0.1 µg/L). The interference due to the analytical background at concentrations below 0.1 µg/L, which occasionally occurs in methods that are quantified by means of electron capture detector [30, 31], was not observed for this method. The blood concentrations of non-persistent organochlorine compounds (a-HCH and γ-HCH, PCB 28 and PCB 52) are seldom detected and, if so, usually only in the range near the detection limit. However, in the case of current exposure (e. g. from sources in internal rooms and due to occupational exposure) concentrations above 0.1 µg/L can also be present in blood.

Despite the relative laborious processing and evaporation steps, this method is easily usable under routine conditions. An experienced technician can certainly process 35 to 40 samples a day.

The reliability criteria of the method are regarded as good for all the analytes. As the round-robin experiments (Section 9) have shown, accurate results have been achieved for all the analytes using this method.

A mixture of n-hexane/toluene has proved very effective for the extraction of the organochlorine compounds including DDT and the HCH isomers. The only drawback is the relatively time-consuming step required to evaporate the extract. If the analysis is limited to HCB and PCB, then heptane or isooctane can be used without problems [30, 31]. In addition to the liquid/liquid extraction, a clean-up on a column with silica gel is prescribed in order to remove the matrix components, especially the blood lipids. Readers are advised to fill the columns with silica gel themselves and then to heat them to glowing. Deactivation of the silica gel before use is not necessary. This type of clean-up column is easier to keep free of contamination than commercially available columns.

Before extraction, δ-HCH, 4,4'-dibromobiphenyl and Mirex are added as internal standards. If calibration is carried out in the matrix as described, addition of these calibration substances is sufficient to obtain accurate and reliable values. Experiments were performed to carry out calibration directly with stock solutions diluted in n-decane in order to simplify the analytical procedure. However, comparison of the calibration graphs in bovine blood with those in n-decane showed that the gradients differed significantly for some substances. Therefore losses due to processing must be assumed (see Table 12). However, the accuracy of the analytical result is not influenced, as these losses are compensated arithmetically because the results are based on the relevant internal standard.

One of the principal advantages of GC-MS analysis is that stable isotopes can be used as internal standards if necessary ("isotope dilution") in order to optimise the accuracy and precision of the results. This technique is increasingly being used for routine analyses. Deuterated and ^{13}C-labelled isotopes are commercially available (e. g. from Promochem) for all the organochlorine compounds included in this method. However, their cost is relatively high and in the experience of the author general use of such isotopes is not necessary in order to meet the quality criteria.

The sensitivity of the method can be enhanced, if necessary. This can be achieved by increasing the size of the sample batch, by evaporating the extracts to approx. 10 µL in special glass flasks and by the use of negative chemical ionisation (NCI) in combination with mass spectrometric detection.

This method describes the determination of relevant organochlorine compounds in blood and plasma. At present the quality criteria for precision and recovery have only been explicitly established for blood.

The efficiency of the method for determination of organochlorine compounds (also in plasma) has been confirmed by the comparison of results with other laboratories and the successful participation of the author in round-robin experiments (Deutsche Gesellschaft für Arbeits- und Umweltmedizin [German Society for Occupational and Environmental Medicine], Arctic Monitoring and Assessment Programme) for many years.

On principle, plasma is preferable to blood as an investigation matrix, as the organochlorine compounds are only present in the erythrocytes in insignificant amounts, so that higher concentrations can be detected when plasma is used. However, when EDTA whole blood is used, one process step can be saved, and this further reduces the risk of contamination. Reference values for the German general population which provide an orientation with respect to β-HCH, HCB and PCB are discussed in the toxicological section of this chapter.

Instruments used:

Hewlett-Packard HP5890 gas chromatograph with Hewlett-Packard HP5971A mass selective detector, with split-splitless injection system, Hewlett-Packard HP7673 autosampler and Hewlett-Packard MS HPChemstation data system.

11 References

[1] *K. Ballschmiter* and *M. Zell:* Analysis of polychlorinated biphenyls (PCB) by gas capillary chromatography. Fresenius Z. Anal. Chem. 302, 20–23 (1980)

[2] *H. Beck* and *W. Mathar:* Analysenverfahren zur Bestimmung von ausgewählten PCB-Einzelkomponenten in Lebensmitteln. Bundesgesundhbl. 28, 1–12 (1985)

[3] *L. J. Fischer, R. F. Seegal, P. E. Ganey, I. N. Pessah* and *P. R. Kodavanti:* Symposium overview: toxicity of non-coplanar PCBs. Toxicol. Sci. 41(1), 49–61 (1998)

[4] *R. D. Kimbrough:* Polychlorinated biphenyls (PCBs) and human health: an update. Crit. Rev. Toxicol. 25(2), 133–163 (1995)

[5] *H. Tryphonas:* Immunotoxicity of polychlorinated biphenyls: present status and future considerations. Exp. Clin. Immunogenet. 11(2/3), 149–162 (1994)

[6] *S. Safe:* Toxicology, structure-function relationship, and human and environmental health impacts of polychlorinated biphenyls: progress and problems. Environ. Health Perspect. 100, 259–268 (1993)

[7]   *F. Iverson* and *D.L. Grant:* Toxicology of the polychlorinated biphenyls, dibenzofurans and di-benzodioxins. IARC Sci. Publ. 108, 5–29 (1991)

[8]   *U. Heudorf* and *J. Angerer:* Aktuelle PCB-Belastung einer Wohnbevölkerung in Deutschland 1998. Umweltmed. Forsch. Prax. 5(3), 137–142 (2000)

[9]   *T. Gabrio, I. Piechotowski, T. Wallenhorst, M. Klett, L. Cott, P. Friebel, B. Link* and *M. Schwenk:* PCB-blood levels in teachers, working in PCB-contaminated schools. Chemosphere 40, 1055–1062 (2000)

[10] *H. Wingfors, G. Lindstrom, B. van Bavel, M. Schuhmacher* and *L. Hardell:* Multivariate data evaluation of PCB and dioxin profiles in the general population in Sweden and Spain. Chemosphere 40, 1083–1088 (2000)

[11] *A.W. Glynn, A. Wolk, M. Aune, S. Atuma, S. Zettermark, M. Maehle-Schmid, P.O. Darnerud, W. Becker, B. Vessby* and *H.O. Adami:* Serum concentrations of organochlorines in men: a search for markers of exposure. Sci. Total Environ. 263, 197–208 (2000)

[12] *K. Schmid, P. Lederer, T. Göen, K.H. Schaller, H. Strebl, A. Weber, J. Angerer* and *G. Lehnert:* Internal exposure to hazardous substances of persons from various continents: investigations on exposure to different organochlorine compounds. Int. Arch. Occup. Environ. Health 69, 399–406 (1997)

[13] *Umweltbundesamt:* Referenzwerte für die PCB-Kongenere Nr. 138, 153, 180 und deren Summe im Humanblut. Bundesgesundhbl. 41(9), 416 (1998)

[14] *G. Koss:* Polychlorierte Biphenyle (PCB). In: *H. Marquardt* and *S.G. Schäfer (eds.):* Lehrbuch der Toxikologie. Spektrum Verlag, Heidelberg (1997)

[15] *DFG – Kommission zur Prüfung von Rückständen in Lebensmitteln:* Hexachlorcyclohexan-Kontamination – Ursachen, Situation und Bewertung, Report IX. Verlag Harald Boltd, Boppard, Deutschland (1982)

[16] *J. Angerer, R. Maass* and *R. Heinrich:* Occupational exposure to hexachlorocyclohexane. VI. Metabolism of gamma-hexachlorocyclohexane in man. Int. Arch. Occup. Environ. Health 52(1), 59–67 (1983)

[17] *H. Greim (ed.):* Lindan. Toxikologisch-arbeitsmedizinische Begründung von MAK-Werten, 27th issue. Wiley-VCH, Weinheim (1998)

[18] *D. Henschler (ed.):* alpha-Hexachlorcyclohexan, beta-Hexachlorcyclohexan. Toxikologisch-arbeitsmedizinische Begründung von MAK-Werten, 9th issue. VCH Verlagsgesellschaft, Weinheim (1983)

[19] *Umweltbundesamt:* Statusbericht zur Hintergrundbelastung mit Organochlorverbindungen in Humanblut. Bundesgesundhbl. 42 (5), 446–448 (1999)

[20] *H. Greim (ed.):* Hexachlorbenzol. Toxikologisch-arbeitsmedizinische Begründung von MAK-Werten, 26th issue. Wiley-VCH, Weinheim (1998)

[21] *Deutsche Forschungsgemeinschaft:* MAK- und BAT-Werte-Liste, Report 37. Wiley-VCH, Weinheim (2001)

[22] *Bundesärztekammer:* Qualitätssicherung der quantitativen Bestimmungen im Laboratorium. Neue Richtlinien der Bundesärztekammer. Dt. Ärztebl. 85, A699–A712 (1988)

[23] *Bundesärztekammer:* Ergänzung der „Richtlinien der Bundesärztekammer zur Qualitätssicherung in medizinischen Laboratorien". Dt. Ärztebl. 91, C159–C161 (1994)

[24] *J. Angerer, T. Göen* and *G. Lehnert:* Mindestanforderungen an die Qualität von umweltmedizinisch-toxikologischen Analysen. Umweltmed. Forsch. Prax. 3, 307–312 (1998)

[25] *G. Lehnert, J. Angerer* and *K.H. Schaller:* Statusbericht über die externe Qualitätssicherung arbeits- und umweltmedizinisch-toxikologischer Analysen in biologischen Materialien. Arbeitsmed. Sozialmed. Umweltmed. 33(1), 21–26 (1998)

[26] *J. Angerer* and *G. Lehnert:* Anforderungen an arbeitsmedizinisch-toxikologische Analysen – Stand der Technik. Dt. Ärztebl. 37, C1753–C1760 (1997)

[27] *Ringversuch Nr. 28.* Qualitätsmanagement in der Arbeits- und Umweltmedizin, Projektgruppe Qualitätssicherung. Organisation: Institut für Arbeits-, Sozial- und Umweltmedizin der Universität Erlangen-Nürnberg (2001)

[28] *Ringversuch Nr. 18.* Qualitätsmanagement in der Arbeits- und Umweltmedizin, Projektgruppe Qualitätssicherung, Organisation: Institut für Arbeits-, Sozial- und Umweltmedizin der Universität Erlangen-Nürnberg (1996)

[29] *Ringversuch Nr. 19.* Qualitätsmanagement in der Arbeits- und Umweltmedizin, Projektgruppe Qualitätssicherung. Organisation: Institut für Arbeits-, Sozial- und Umweltmedizin der Universität Erlangen-Nürnberg (1997)

[30] *E. Schulte, J. Lewalter* and *D. Ellrich:* Polychlorinated Biphenyls. In: *J. Angerer* and *K.-H. Schaller (eds.):* DFG – Analysis of Hazardous Substances in Biological Materials, Vol. 3. Wiley-VCH, Weinheim (1991)

[31] *E. Schulte, J. Lewalter* and *D. Ellrich:* Polychlorierte Biphenyle. In: *J. Angerer* and *K.-H. Schaller (eds.):* DFG – Analysen in biologischem Material. Loose-leaf collection, 10th issue. Verlag Wiley-VCH, Weinheim (1991)

Author: *H.-W. Hoppe, T. Weiss*
Examiners: *M. Ball, J. Lewalter*

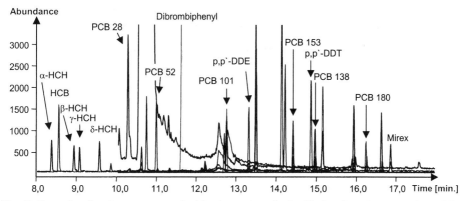

Fig. 5. Example of a chromatogram of a blood sample spiked with 5 µg/L and processed (qualifier ion traces)

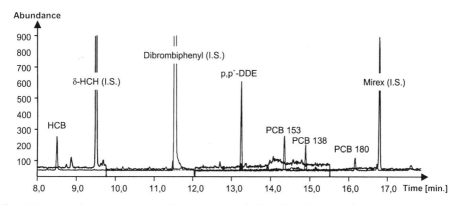

Fig. 6. Example of a chromatogram of a processed individual blood sample of a test person

PAH metabolites (1-hydroxyphenanthrene, 4-hydroxyphenanthrene, 9-hydroxyphenanthrene, 1-hydroxypyrene)

Application   Determination in urine

Analytical principle   High performance liquid chromatography

Completed in   May 1998

Summary

The high performance liquid chromatographic method described here permits the determination of 1-hydroxyphenanthrene, 4-hydroxyphenanthrene, 9-hydroxyphenanthrene and 1-hydroxypyrene in the urine of persons who have been exposed to PAH at the workplace or in the environment. After enzymatic hydrolysis of the relevant glucuronic acid and sulphate conjugates, the urine sample is centrifuged, an aliquot of the supernatant is injected into an HPLC instrument for a system-internal sample processing, then it is analyzed. The PAH metabolites are first enriched on a phthalocyanine-modified silica gel within the HPLC instrument and thus separated from the urine matrix. Then the analytes are transferred onto a reverse phase column by means of an automatic switch valve, and quantified with the aid of a fluorescence detector. The calibration is carried out using aqueous standards which are processed and analyzed in the same manner as the urine samples.

1-Hydroxyphenanthrene

Within-series imprecision: Standard deviation (rel.) $s_w = 5.5\%$
                           Prognostic range $u = 13.0\%$
                           at a concentration of 7.9 µg per litre urine
                           and where $n = 8$ determinations

Between-day imprecision: Standard deviation (rel.) $s = 5.1\%$
                          Prognostic range $u = 11.4\%$
                          at a concentration of 8.7 µg per litre urine
                          and where $n = 10$ days

Essential Biomonitoring Methods. DFG, Deutsche Forschungsgemeinschaft
Copyright © 2006 WILEY-VCH Verlag GmbH & Co. KGaA, Weinheim
ISBN: 3-527-31478-4

Accuracy: Recovery $r = 101\%$

Detection limit: 0.3 µg/L urine

4-Hydroxyphenanthrene

Within-series imprecision: Standard deviation (rel.) $s_w = 6.9\%$
Prognostic range $u = 16.3\%$
at a concentration of 1.6 µg per litre urine
and where $n = 8$ determinations

Between-day imprecision: Standard deviation (rel.) $s = 6.1\%$
Prognostic range $u = 13.8\%$
at a concentration of 1.6 µg per litre urine
and where $n = 10$ days

Accuracy: Recovery $r = 99\%$

Detection limit: 0.5 µg/L urine

9-Hydroxyphenanthrene

Within-series imprecision: Standard deviation (rel.) $s_w = 3.9\%$
Prognostic range $u = 9.2\%$
at a concentration of 1.2 µg per litre urine
and where $n = 8$ determinations

Between-day imprecision: Standard deviation (rel.) $s = 12.3\%$
Prognostic range $u = 29.1\%$
at a concentration of 5.8 µg per litre urine
and where $n = 10$ days

Accuracy: Recovery $r = 93\%$

Detection limit: 0.4 µg/L urine

1-Hydroxypyrene

Within-series imprecision: Standard deviation (rel.) $s_w = 7.6\%$
Prognostic range $u = 18.0\%$
at a concentration of 2.6 µg per litre urine
and where $n = 8$ determinations

Between-day imprecision: Standard deviation (rel.) $s = 9.2\%$
Prognostic range $u = 20.8\%$
at a concentration of 2.8 µg per litre urine
and where $n = 10$ days

Accuracy: Recovery $r=91\%$

Detection limit: 0.1 µg/L urine

Polycyclic aromatic hydrocarbons (PAH)

Polycyclic aromatic hydrocarbons (PAH) form a group of several hundred compounds. They are composed of at least two condensed ring systems and, in addition to carbon and hydrogen atoms, some of them contain heteroatoms such as sulphur and nitrogen. Besides, PAH are known in which one or more hydrogen atoms are substituted by the nitro group. PAH are formed by pyrolysis and the incomplete combustion of organic material [1]. The more completely a fuel containing carbon and hydrogen is burned with oxygen to form carbon dioxide and water, the smaller the amount of any PAH which is formed as a by-product. The relative proportions of the compounds depend on the pyrolytical process and on the starting material [2]. The most important non-anthropogenic source of PAH is coal and oil. In addition, PAH are released from forest fires and volcanic eruptions. The PAH which occur in the environment are largely of anthropogenic origin. They are formed in industrial pyrolysis processes such as high-temperature carbonization of coal or in the petrochemical cracking process. PAH are also formed in various combustion processes, e.g. industrial energy and heat generation, in the domestic heating supply or powering of automobiles. In addition, cigarette smoke contains PAH [1]. As a result of the numerous sources of PAH and their thermodynamic stability, they are distributed ubiquitously in the environment. They are found in the air, in the earth, in water, in plants and also in food. In Table 1 the concentrations of benzo[a]pyrene in various environmental compartments are summarized.
PAH with a boiling point of less than about 400 °C, e.g. phenanthrene, fluoroanthene or pyrene primarily occur as gases in the air. The higher-boiling PAH, e.g. benz[a]anthracene or chrysene, are mainly bound to particles [2].

PAH are present at numerous workplaces. Very high concentrations were found in coking plants, especially in the vicinity of the ovens. Benzo[a]pyrene concentrations

Table 1. Benzo[a]pyrene concentrations in various environmental compartments [3].

Environmental compartment	Concentration
Air	1.3–500 ng/m^3
Earth	0.8 ng/kg–100 mg/kg
Domestic water supply	2.5–9 ng/L
Surface water	130–500 ng/L
Plants	up to 150 µg/kg
Food	0.1–20 µg/kg

up to 90 µg/m^3 were measured there [4]. High concentrations also occur in the processing of pitch, e.g. the production and loading of pencil pitch. Elevated benzo[a]pyrene concentrations are also detected in foundries, and in production plants for aluminium, graphite electrodes and fire-resistant products [5]. In the Federal Republic of Germany there is a splitted TRK (Technical Exposure Limit) value for benzo[a]pyrene, which serves as an indicator of PAH contamination in the air. It is 5 µg/m^3 for production and loading of pencil pitch and near the ovens in coking plants. For all other workplaces the currently valid TRK value (1998) is 2 µg/m^3 [6].

PAH can enter the human body by means of inhalation, or by dermal and oral intake. As PAH are very non-polar substances, which can readily diffuse through the lipoprotein layers of the skin, dermal intake plays a major role in the absorption of the PAH [7–9]. In the case of workers in a coking plant, approximately 70% of the total pyrene found in the human body was absorbed through the skin [8].

Only about 10% of the benzo[a]pyrene which is taken orally with food is absorbed into the body in humans. During the first 24 hours after intake, a large proportion of the orally supplied PAH remains in the intestinal lumen and is distributed throughout the body only in the course of 3 to 4 days [10].

As soon as the PAH reach the blood stream, they are distributed in the body within a period of minutes to hours. In particular, the fatty tissues offer ready depots for PAH [1]. The substances are still detectable here months after application [2].

Benzo[a]pyrene is a typical representative of the PAH group and as such its metabolism has been investigated in many studies.

Phase 1 of the biotransformation of benzo[a]pyrene (Fig. 1) involves a metabolic conversion catalyzed by cytochrome P450-dependent monooxigenases, whereby initially various aryl oxides are formed (1). These aryl oxides can undergo spontaneous rearrangement to phenols (2) or can be catalytically hydrolyzed to the corresponding dihydrodiols (3). Dihydrodiols can be further converted to epoxides (4), whereby the corresponding dihydrodiol epoxides are formed, which can undergo spontaneous hydrolysis to form tetrols (5). Alternatively, benzo[a]pyrene can also be directly oxidized to form various phenols (6). Certain phenols, e.g. 6-hydroxybenzo[a]pyrene are oxidized either spontaneously or metabolically to quinones (7), which under redox cycling conditions can form reactive oxygen species. The epoxides are conjugated with glutathione by glutathione S-transferases, sulfotransferases and by UDP-glucuronosyltransferases, or the phenolic metabolites are converted to sulphates and glucuronides [10].

Figure 1 illustrates the metabolism of benzo[a]pyrene as an example.

Animal studies with mice and rats have shown that a very small proportion of PAH are excreted unchanged. In mice 4 to 12% of a dose of benzo[a]pyrene are recovered in the urine within 6 days of administration by subcutaneous injection. The larger part is excreted in bile via the faeces [1].

In humans experience has largely been gained with regard to the excretion of pyrene. As its main metabolite 1-hydroxypyrene is readily quantified in urine. Jongeneelen et al. [11] found that the elimination kinetics exhibited a biphasic curve with a short

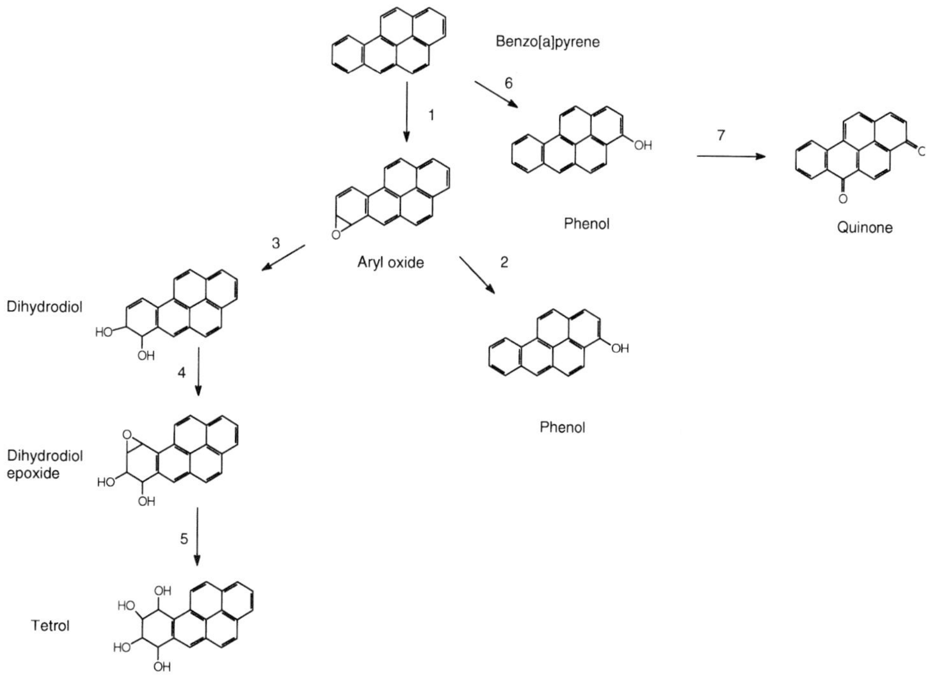

Fig. 1. Schematic illustration of the metabolism of benzo[a]pyrene.

half-life of 1 to 2 days followed by a longer elimination phase with a half-life of about 16 days. According to these findings, part of the absorbed pyrene is immediately bioavailable. Another part of the pyrene is deposited in deeper compartments, such as the fatty tissues, from which it is only released into the blood stream after a time lag. The first elimination phase was investigated in further studies on workers from various branches of industry [12–14] and on control persons [15–17]. Half-lives between 4 and 35 hours were found.

Few investigations have been carried out on the acute toxicity of the PAH. The LD_{50} of benzo[a]pyrene for mice following intraperitoneal injection was 250 mg/kg body weight. A single intraperitoneal injection of 2 mg/kg led to a growth standstill in young rats [2].

A 1% solution of benzo[a]pyrene in olive oil was applied daily up to 120 times to the skin of 26 patients in a dermatological clinic and the area of skin was subsequently excised. Pigmentation resulted in every case, occurring earlier in the case of older patients than in younger ones. Warts developed in the late stage of the test but disappeared soon after the treatment was discontinued.

On the basis of the results of many animal studies, there is good justification for the assumption that several PAH have carcinogenic effects, even in humans [6]. Many epidemiological studies have shown that workers exposed to PAH have a higher risk of contracting cancer [18–21]. Those affected are, for example, workers in the following industrial areas: coking plants, processing of pitch, the carbon industry, aluminium production, foundries, the manufacture of silicon carbide, the impregnation of wood with coal tar, and road construction [6].

Pyrolysis products, such as brown coal tar, coal tar, coal-tar pitch, coal-tar oil and coke-oven raw gases contain a large proportion of PAH. These mixtures are documented as carcinogenic working materials in the MAK and BAT value lists of the Deutsche Forschungsgemeinschaft's Commission for the Investigation of Health Hazards of Chemical Compounds in the Work Area [6]. The carcinogenic effect on humans who handle these substances at the workplace has been established by means of epidemiological methods. These PAH-containing mixtures have therefore been assigned to category 1 of the carcinogenic working materials.

Quantitative estimates of the health risks for workers in coking plants have been made in various studies. The incidence of cancer is estimated as between 8.7 and 63 cases for life-long exposure of 10^5 persons for each ng of benzo[a]pyrene per m^3 air. The Länderausschuß für Immissionsschutz (German Länder Committee for Immission Control) recommends using a risk estimation value of 7 cases of cancer per 10^5 exposed persons for each ng of benzo[a]pyrene per m^3 air as a basis for consideration in preventive medicine [2].

"Malignant neoplasms of the respiratory tract and the lungs caused by coke-oven raw gases" has been included under Number 4110 in the list of recognized occupational diseases since 1988. The lung tumours of other workers, who are also exposed to high concentrations of PAH, are not taken into account in this regulation. The Ärztliche Sachverständigen-Beirat des Bundesministeriums für Arbeits- und Sozialordnung – Sektion "Berufskrankheiten" – (Medical Expert Advisory Board of the German Ministry for Occupational and Social Order – Section "Occupational Diseases") recently recommended the inclusion as a new occupational disease under Number 4110 of the Regulations governing Occupational Diseases of lung cancer caused by polycyclic aromatic hydrocarbons when exposure to a cumulative dose of at least 100 benzo[a]pyrene-years [($\mu g/m^3$)×years] is confirmed [22].

The only determination of the PAH in the ambient air at the workplace is not sufficient to permit the estimation of the health risk to individuals. For this reason, efforts were made to devise methods for biological monitoring in addition to measurement of the contamination in the air. Jongeneelen [23] achieved a breakthrough with regard to the biological monitoring of persons exposed to PAH with the determination of 1-hydroxypyrene in urine. In recent years a series of studies has been published in which the inner stress due to PAH exposure in various industries has been described. The 1-hydroxypyrene excretion ranges from concentrations similar to those measured in the general public (e.g. in smokeries, refuse incineration plants) to peak values of 700 µg/g creatinine (manufacture of fire-resistant stones). In contrast, the concentrations of 1-hydroxypyrene in the urine of employees in coking plants do not seem to exceed 100 µg/g creatinine [12–14, 25–34].

Table 2. Excretion of the monohydroxyphenanthrenes and of 1-hydroxypyrene in the urine of occupationally exposed persons (range: μg/g creatinine).

Workplace	n	Hydroxyphenanthrenes					1-Hydroxy-pyrene	References
		1-	2-	3-	4-	9-		
Coking plants	4	1.0–18.1[1]	0.7–13.6[1]	1.6–30.6[1]	0.1–2.5[1]	0.16–1.9[1]	3.0–71.0[1]	[25]
Graphite electrode production	67	0.4–42.9	0.3–31.7[2]	0.6–77.1	0.1–3.9		0.2–325.7	[32]
Production of fire-resistant stones	9	0.6–12.3	0.5–9.9[2]	0.6–18.7	0.09–0.7		4.9–110.3	[33]

[1] Range of the mean values.
[2] Sum of 2- and 9-hydroxyphenanthrene.

In the meantime, various research teams [25, 32, 33] have also used the monohydroxylated phenanthrenes as indicators of exposure to the PAH in addition to 1-hydroxypyrene. Table 2 presents examples of the 1-hydroxypyrene, 1-, 2-, 3- and 4-hydroxyphenanthrene concentration in the urine of people who have been exposed to PAH at the workplace.

Lower concentrations of 1-hydroxypyrene as well as 1-, 2-, 3- and 4-hydroxyphenanthrene are measured in the urine of persons who have not been exposed to the PAH at the workplace . The excretion of 1-hydroxypyrene in the urine of female smokers is significantly higher than in the urine of female non-smokers. Not only the median value (0.48 μg/g creatinine) but also the 95th percentile (1.45 μg/g creatinine) is three times higher for the female smokers than the corresponding values for female non-smokers (0.15 and 0.46 μg/g creatinine). Of the hydroxyphenanthrene isomers in the urine of female non-smokers, hydroxyphenanthrene proved to be the main metabolite (median: 0.51 μg/g creatinine) followed by 2-hydroxyphenanthrene (median: 0.31 μg/g creatinine) and 3-hydroxyphenanthrene (median: 0.31 μg/g creatinine). In contrast, female smokers excreted more 3-hydroxyphenanthrene (median: 0.61 mg/g creatinine) than 1-hydroxyphenanthrene (median: 0.53 μg/g creatinine). Compared to the other hydroxyphenanthrene isomers, 4-hydroxyphenanthrene could be detected only in considerably lower concentrations in the urine of both groups: female non-smokers (median: 0.04 μg/g creatinine), female non-smokers (median: 0.1 μg/g creatinine). With the exception of 1-hydroxyphenanthrene, female smokers exhibited significantly higher total hydroxyphenanthrene concentrations in urine than non-smokers [35, 36].

Author: *J. Lintelmann*
Examiner: *J. Angerer*

PAH metabolites (1-hydroxyphenanthrene, 4-hydroxyphenanthrene, 9-hydroxyphenanthrene, 1-hydroxypyrene)

Application	Determination in urine
Analytical principle	High performance liquid chromatography
Completed in	May 1998

Contents

Essential Biomonitoring Methods. DFG, Deutsche Forschungsgemeinschaft
Copyright © 2006 WILEY-VCH Verlag GmbH & Co. KGaA, Weinheim
ISBN: 3-527-31478-4

1 General principles

In order to determine 1-hydroxyphenanthrene, 4-hydroxyphenanthrene, 9-hydroxy-phenanthrene and 1-hydroxypyrene the urine is first subjected to enzymatic hydrolysis, whereby the analytes are released from their glucuronic acid and sulphuric acid conjugates. The hydrolyzed urine is centrifuged and the supernatant is analyzed after an HPLC system-internal sample preparation. The PAH metabolites are first enriched on a phthalocyanine-modified silica gel and thus specifically separated from the urine matrix within the HPLC instrument. Then the analytes are transferred onto a reverse phase separation column by means of an automatic switch valve. There they are separated and quantified with the aid of a fluorescence detector. The calibration is carried out using aqueous standards which are processed and analyzed in the same manner as the urine samples.

2 Equipment, chemicals and solutions

2.1 Equipment

HPLC system consisting of a binary gradient pump with connections for further instruments, a device for degassing the eluents, a single-channel pump, a column thermostat, an injection valve, an autosampler, an automatic six-way valve, a fluorescence detector with a measurement range which includes the excitation wavelength of 242 nm and the emission wavelength of 388 nm; (band width each 20 nm) and an integrator.

High performance liquid chromatographic columns:
Precolumn: silica gel modified with copper phthalocyanine trisulphonic acid derivative (30 μm; length 5 mm, I.D. 4 mm; e.g. from Krannich, Göttingen)

Analytical column: Superspher 100 RP-18, 4 μm; length 250 mm, I.D. 4 mm (e.g. from Merck)

Temperature-controlled water bath

Screw-topped 20 mL centrifuge vial

5 mL Pipette

pH meter (e.g. 761 Calimatic, from Knick, Berlin)

pH electrode (e.g. InLab 422, from Mettler-Toledo, Steinbach)

Centrifuge (e.g. Minifuge 2, from Heraeus-Christ, Osterode)

1 mL Autosampler vials, sealable with crimp caps and PTFE-coated septa

Automatic microlitre pipettes, adjustable between 10 and 100 µL, and between 200 and 1000 µL (e.g. from Eppendorf)

10 mL Measuring pipette

10 and 1000 mL Volumetric flasks

500 mL Graduated cylinder

10 mL Erlenmeyer flasks

1000 mL Laboratory glass bottles with screw caps

20 mL Threaded centrifuge tube (e.g. from Laborcenter, Nürnberg)

2.2 Chemicals

1-Hydroxyphenanthrene, 4-hydroxyphenanthrene, 9-hydroxyphenanthrene (e.g. from Ehrenstorfer, Augsburg)

1-Hydroxypyrene (e.g. from Promochem)

β-Glucuronidase/arylsulphatase solution (e.g. 100000 Fishman U/mL and 800000 Roy U/mL from Boehringer)

Sodium acetate, anhydrous p.a. (e.g. from Merck)

Glacial acetic acid p.a. (e.g. from Merck)

37% Hydrochloric acid, p.a. (e.g. from Merck)

Methanol for HPLC (e.g. from Merck)

Ultrapure water (equivalent to ASTM type 1) or double distilled water

2.3 Solutions

1 M Hydrochloric acid:
About 50 mL ultrapure water are placed in a 100 mL volumetric flask. Then 8.3 mL 37% hydrochloric acid are pipetted into the water and the volumetric flask is filled to its nominal volume with ultrapure water.

0.1 M Hydrochloric acid:
10 mL 1 M HCl are pipetted into a 100 mL volumetric flask, into which about 50 mL ultrapure water have been previously placed. The volumetric flask is subsequently filled to its nominal volume with ultrapure water.

0.1 M Sodium acetate buffer (pH 5):
8.2 g Sodium acetate are placed in a 1000 mL volumetric flask and dissolved in about 300 mL ultrapure water. After swirling the solution, the volumetric flask is filled to its nominal volume with ultrapure water. The pH value is adjusted to 5.0 with glacial acetic acid.

These solutions are stable at room temperature for at least three months.

Methanol/water solution:
500 mL methanol are measured in a 500 mL graduated cylinder and subsequently filled into a 1 L glass bottle. 500 mL ultrapure water are measured in the same way and added to the methanol. The solution is thoroughly mixed.

Solvent A (methanol):
1000 mL methanol are measured using a 500 mL graduated cylinder and subsequently filled into a 1000 mL glass bottle.

Solvent B (methanol/water=6:4 (v/v)):
600 mL methanol are placed in a 1000 mL glass bottle. Then 400 mL water are added. The glass bottle is thoroughly shaken.

Solvent C (methanol/water=1:9 (v/v)):
100 mL methanol are placed in a 1000 mL glass bottle. Then 900 mL water are added. The glass bottle is thoroughly shaken.

Before the HPLC analysis, the solvents are degassed in an ultrasonic bath for 20 min to remove any dissolved oxygen.

These solutions are stable at room temperature for at least four weeks.

2.4 Calibration standards

Starting solution:
Approximately 1 mg each of 1-hydroxyphenanthrene, 4-hydroxyphenanthrene, 9-hydroxyphenanthrene and 1-hydroxypyrene are exactly weighed in a 100 mL volumetric flask. The volumetric flask is then filled to its nominal volume with methanol. This solution contains the individual compounds each at a concentration of 10 mg/L.

Stock solution A:
500 μL of the starting solution are pipetted into a 10 mL volumetric flask, into which approximately 5 mL methanol have been previously placed. The volumetric flask is then filled to its nominal volume with methanol. The individual analytes are present in this solution at a concentration of 0.5 mg/L each.

Stock solution B:
500 μL of the starting solution are pipetted into a 100 mL volumetric flask, into which approximately 50 mL methanol have been previously placed. The volumetric flask is then filled to its nominal volume with methanol. The concentration of each PAH metabolite is 0.05 mg/L.

Calibration standards containing between 0.25 and 30 µg of the appropriate PAH metabolites per litre are prepared from stock solutions A and B by means of dilution with
ultrapure water. The following pipetting scheme shows the preparation procedure:

Table 3. Pipetting scheme for the preparation of the calibration standards.

Volume of the stock solution		Final volume of the calibration standard	Concentration of the calibration standard
A (0.5 mg/L) [µL]	B (0.05 mg/L) [µL]	[mL]	[µg/L]
–	–	10	0
–	50	10	0.25
–	100	10	0.5
–	400	10	2.0
100	–	10	5.0
300	–	10	15.0
600	–	10	30.0

The starting, stock and calibration solutions can be kept at $-18\,°C$ for at least
6 months.

3 Specimen collection and sample preparation

The urine is collected in plastic bottles. If the samples are not processed immediately, they are stored in the deep-freezer at $-18\,°C$ until processing.
Before analysis, the samples are thawed in a water bath at $40\,°C$ and subsequently
brought to room temperature. Before an aliquot is taken, the samples are thoroughly
shaken. 5 mL urine are transferred to a 10 mL Erlenmeyer flask and the pH value is
adjusted with the aid of a pH meter to 5.0 using 1 M HCl. The urine sample is quantitatively transferred into a 10 mL volumetric flask. The Erlenmeyer flask is then
rinsed three times using 1 mL sodium acetate buffer in each case. The rinsing solutions are also filled into the volumetric flask. Finally, the volumetric flask is filled to
its nominal volume with sodium acetate buffer.
The dilute sample is thoroughly mixed and filled into a 20 mL centrifuge tube which
can be sealed with a screw cap. 10 µL β-glucuronidase/arylsulphatase suspension are
added, the mixture is shaken mechanically for 10 min and incubated in a water bath
at $37\,°C$ for 16 hours.
After incubation, the hydrolyzed urine is centrifuged at 2700 g. 1 mL of the supernatant
is transferred to a 1 mL autosampler vial. The vial is sealed and fed into the HPLC for
analysis. The urine samples should not be left standing in the autosampler for more than
16 hours at room temperature. If analysis by HPLC is not possible within this period,
the prepared samples can be stored at $-18\,°C$. No losses of the analytes have been ascertained, when the samples are stored up to four weeks in this way.

A reagent blank is included in each analytical series. In this case ultrapure water is subjected to the sample processing described above instead of urine.

4 Operational parameters for high performance liquid chromatography

4.1 HPLC-internal enrichment of the analytes

Precolumn: Material: Steel
  Length: 5 mm
  Inner diameter: 4 mm

Column packing: Silica gel, modified with copper phthalocyanine
                          trisulphonic acid derivative; 30 µm

Mobile phase for the
enrichment of the analytes Solvent C: 10% methanol/90% water (v/v)

Flow rate during
the enrichment   1.0 mL/min

Injection volume: 100 µL

4.2 Analytical separation

Analytical column: Material: steel
  Length: 25 cm
  Inner diameter: 4 mm

Column packing: Superspher 100 RP-18; 4 µm

Principle of the separation:Reversed phase

Detector: Fluorescence detector
  Excitation wavelength: 242 nm
  Emission wavelength: 388 nm

Column temperature: 40 °C

Mobile phase: Solvent B: 60% methanol/40% water (v/v)
  Solvent A: 100% methanol

Flow rate
during transfer and
analytical separation: 0.8 mL/min

Gradient program: cf. Table 4

4.3 Principles of the HPLC method

The HPLC system (cf. Fig. 2) permits coupling between the copper phthalocyanine column and the analytical separation column via an automatic switch valve. In the course of the HPLC program, it is switched by a timing mechanism between the load position and the inject position.

In the load position the enrichment column is washed with solvent C (10% methanol and 90% water) and the analytical separation column with solvent B (60% methanol and 40% water). After injection of the sample, the analytes are transported by solvent C onto the enrichment column where they are selectively enriched. The remaining urine constituents pass through the enrichment column largely unretarded and are discarded with the waste.

Table 4. Gradient program for high performance liquid chromatographic determination of the PAH metabolites. (A: solvent 100% methanol, B: solvent 60% methanol and 40% water).

Time [min]	A [%]	B [%]	Flow rate [mL/min]	Position of the switch valve	Procedure
0	0	100	0.8	LOAD	HPLC-integrated sample preparation, equilibration of the analytical column
10	0	100	0.8	INJECT	Transfer of the analytes Start of data recording
14	0	100	0.8	LOAD	
35	0	100	0.8	LOAD	Analytical
43	62	38	0.8	LOAD	separation
45	62	38	0.8	LOAD	
45.1	100	0	0.8	INJECT	Rinsing and
48	100	0	0.8	LOAD	reconditioning
51	100	0	0.8	LOAD	the columns

Table 5. Retention times for the PAH metabolites.

Metabolite	Retention time [min]
4-Hydroxyphenanthrene	21.0
9-Hydroxyphenanthrene	24.7
1-Hydroxyphenanthrene	25.3
4-Hydroxyphenanthrene	26.6
1-Hydroxypyrene	35.0

After 10 min enrichment, the valve is switched to the inject position. This permits solvent B to be introduced onto the enrichment column, while solvent C is directly discarded with the waste. As solvent B has a greater power of elution, the enriched analytes are eluted and transferred to the analytical separation column. This process is complete after 4 minutes.

The valve is subsequently switched to the load position again. In this case the analytes are separated by a methanol gradient (solvent A and B) and detected by fluorimetry.

After 45 min, the valve is switched to the inject position once again. This permits the enrichment column and the analytical column to be washed and purified with solvent A (100% methanol).

After 3 min, the valve is switched to the load position and the columns are reconditioned for 3 min in preparation for the next cycle.

Table 4 shows the gradient program for high performance liquid chromatographic determination of the PAH metabolites.

During the entire analytical period solvent C (10% methanol and 90% water) flows at a rate of 1 mL/min.

Under the conditions for high performance liquid chromatography given here the following retention times were found (c.f. Tab. 5) for the PAH metabolites.

The retention times serve only as an orientation. The user of the method must optimize the adjustment of the instrument in use.

Figure 3 shows the HPLC chromatogram of the urine sample of a person who had been previously exposed to PAH at the workplace.

5 Analytical determination

100 µL of the hydrolyzed and centrifuged urine sample are injected into the HPLC system with the aid of an autosampler or an injection valve with an appropriate dosing loop. If the measured values are above the linear range of the calibration curve, the urine samples are diluted and processed anew. A quality control sample is analyzed with each analytical series.

6 Calibration

The aqueous calibration standards are processed and analyzed in the same manner as the samples. The calibration graph is obtained by plotting the signal areas of the individual PAH metabolites as a function of the concentrations used. It is unnecessary to plot a complete calibration graph for every analytical series. Analysis of one aqueous calibration standard with each new series is sufficient. The ratio of the value obtained for this standard and the area unit obtained for the equivalent standard in the complete calibration graph is calculated. Using the quotients obtained for each metabolite, each of the results read off the calibration graphs is adjusted.

New calibration graphs should be plotted if the quality control results indicate systematic deviation.

The calibration graphs are each linear in the concentration range to 30 µg of the individual PAH metabolites per litre.

7 Calculation of the analytical result

The resulting signal areas of the analytes are used to read off the appropriate concentration in µg per litre from the relevant calibration graph. The results are adjusted as described in Section 6.
If a reagent blank value is detected, it must be taken into account.

8 Standardization and quality control

Quality control of the analytical results is carried out as stipulated in the guidelines of the Bundesärztekammer (German Medical Association) [37, 38] and in the special preliminary remarks of Volume 1 of this series. In order to determine the precision of the method, a urine sample containing a constant concentration of the individual PAH metabolites is analyzed. As material for quality control is not commercially available, it must be prepared in the laboratory. For this purpose, urine is spiked with a defined quantity of the PAH metabolites. A six-month supply of this control material is prepared, divided into aliquots in pierceable ampoules and stored in the deep-freezer. The concentration of this control material should lie in the middle of the most frequently occurring concentration range. The theoretical value and the tolerance range for this quality control material is determined in the course of a pre-analytical period (one analysis of the control material on 20 different days) [37, 39].
External quality control is achieved by participating in round-robin experiments. The round-robin experiments carried out to test analysis in occupational and environmental medicine in Germany offer the analysis of 1-hydroxypyrene in urine both in the concentration of interest to occupational medicine and environmental medicine in the quality control programme [40, 41].

9 Reliability of the method

9.1 Precision

The urine of a worker who had been exposed to PAH was processed and analyzed 8 times in accordance with Section 3 to determine the precision in the series. This resulted in relative standard deviations between 3.9 and 7.6%, equivalent to prognostic ranges of 9.2 and 18.0% (cf. Tab. 6).

Table 6. Precision in the series for the high performance liquid chromatographic determination of the PAH metabolites ($n=8$).

Substance	Concentration [µg/L]	Standard deviation (rel.) [%]	Prognostic range [%]
1-Hydroxyphenanthrene	7.9	5.5	13.0
4-Hydroxyphenanthrene	1.6	6.9	16.3
9-Hydroxyphenanthrene	1.2	3.9	9.2
1-Hydroxypyrene	2.6	7.6	18.0

Table 7. Precision from day to day for the high performance liquid chromatographic determination of the PAH metabolites ($n=10$).

Substance	Concentration [µg/L]	Standard deviation (rel.) [%]	Prognostic range [%]
1-Hydroxyphenanthrene	8.7	5.1	11.4
4-Hydroxyphenanthrene	1.6	6.1	13.8
9-Hydroxyphenanthrene	5.8	12.3	29.1
1-Hydroxypyrene	2.8	9.2	20.8

In addition, the precision from day to day was determined. For this purpose, urine of an occupationally exposed person was processed and analyzed on $n=10$ days in accordance with Section 3. The relative standard deviations were between 5.1 and 12.3% which are equivalent to prognostic ranges of 11.4 and 29.1% (cf. Tab. 7).

9.2 Accuracy

Recovery experiments were carried out using aqueous standard solutions to test the accuracy of the method.

In this case the aqueous calibration standards were injected directly onto the analytical column, avoiding the enrichment column. Then these standards were subjected to the system-internal sample preparation and analyzed. The recovery is calculated from the quotients of the area units which are achieved with and without the enrichment column. The results are shown in Table 8.

In order to determine the recovery of the PAH metabolites from urine samples, pooled urine from persons who were not exposed to PAH at the workplace was divided into 7 aliquots. These aliquots were spiked with different amounts of the ana-

Table 8. Recovery rate for aqueous standard solutions.

Substance	n	Concentration	Recovery rate	Standard deviation (rel.)
		[µg/L]	[%]	[%]
1-Hydroxyphenanthrene	7	0.17–87.3	99	4.0
4-Hydroxyphenanthrene	7	0.21–106.7	98	4.0
9-Hydroxyphenanthrene	7	0.10–93.1	97	3.0
1-Hydroxypyrene	7	0.07–10.9	97	6.0

Table 9. Recovery from urine samples.

Substance	n	Concentration	Recovery rate	Standard deviation (rel.)
		[µg/L]	[%]	[%]
1-Hydroxyphenanthrene	8	0.17–87.3	101	4.0
4-Hydroxyphenanthrene	8	0.21–106.7	99	6.0
9-Hydroxyphenanthrene	8	0.10–93.1	93	6.0
1-Hydroxypyrene	8	0.07–10.9	91	5.0

lytes so that concentrations between 0.07 and 106.7 µg per litre urine resulted. These samples were each processed and analyzed eight times. At the same time aqueous solutions containing the same concentrations were processed and analyzed. The recovery was calculated by comparison of the peak areas of the aqueous solutions and those of the spiked urine. The recovery was between 91 and 101% (cf. Tab. 9).

9.3 Detection limit

Under the conditions for sample preparation and high performance liquid chromatographic determination described here, the detection limit was between 0.1 and 0.5 µg of the individual PAH metabolites per litre urine (Tab. 10). As no reagent blank value was measured, the detection limit was calculated as three times the signal/background ratio.

9.4 Sources of error

The HPLC method described here for the determination of 1-, 4- and 9-hydroxyphenanthrene as well as 1-hydroxypyrene in urine permits a reliable separation of the analytes from the other constituents of the biological samples. No interference in the method has yet been observed. This means that no interfering peak at or in the vicin-

Table 10. Recovery rates obtained for the determination of the PAH metabolites.

Metabolite	Detection limit [µg/L]
1-Hydroxyphenanthrene	0.3
4-Hydroxyphenanthrene	0.5
9-Hydroxyphenanthrene	0.4
1-Hydroxypyrene	0.1

ity of the characteristic retention times for the individual PAH metabolites could be observed for the urine samples investigated until now.

2- and 3-hydroxyphenanthrene co-eluate simultaneously at a retention time of approx. 21.0 min. As the two phenanthrene isomers exhibit different fluorescence intensities, these PAH metabolites could not be quantified.

Neither the untreated urine samples nor the standard solutions should be exposed to direct sunlight, as UV radiation can cause decomposition of the analytes. Nor may urine samples after enzymatic hydrolysis and calibration standard solutions be stored or transported in plastic vessels, otherwise irreversible, unreproducible losses of the analytes occur due to non-specific absorption on the vessel walls. Thus glass vessels must be exclusively used for storage. Centrifugation of the hydrolyzed urine prior to HPLC analysis is important, as particles which would otherwise clog the HPLC system and lead to increased pressure are deposited in this way. This step should not be replaced by a filtration step, as irreversible adsorption of the analytes can occur in the conventional plastic systems. The examiner of the method additionally installed a special HPLC filter system with exchangeable precious metal filters to protect the enrichment column. His experience with this filter was very positive. However, this filter must be replaced after 40 injections or at the latest when the pressure before the filter has risen to 2 bar.

Once the conditions for analytical separation have been optimized, they must remain constant, otherwise the separation of the hydroxyphenanthrene isomers is no longer guaranteed. A constant column temperature is also of particular importance.

The solvents used for HPLC must be free of interfering substances. Contamination of the analytical column can drastically change the retention times.
Methanol can sometimes be contaminated with fluorescent substances. A new batch should be used in that event.

The eluents should be continuously degassed to remove dissolved oxygen which causes interference to the fluorescence detection.

10 Discussion of the method

This method permits a simple, reliable and accurate determination of the PAH metabolites 1-hydroxyphenanthrene, 4-hydroxyphenanthrene, 9-hydroxyphenanthrene and 1-hydroxypyrene in the urine of both occupationally and ecologically exposed persons. Its merits are that it requires little effort and is not susceptible to interference. The reliability criteria can be regarded as good and the examiner of the method was able to duplicate the quality criteria at his first attempt.

The determined PAH metabolite concentrations should be expressed in terms of the creatinine content of the urine sample to compensate for diuretic fluctuations in the analyte concentrations [42].

The method described here employs enzymatic hydrolysis to release the conjugated analytes in urine. It is also possible to carry out hydrolysis by heating the sample with concentrated acid. As this procedure is considerably more labour-intensive and leads to the same results [43], enzymatic hydrolysis was preferred. In individual cases it is not possible to confirm if the hydrolysis is complete, as the appropriate conjugates are not commercially available. Therefore, a distinct excess of enzyme (or enzymatic activity) is necessary.

As various investigations have shown [24, 44], the incubation period should not be less than 3 hours. For practical reasons, the incubation can be carried out overnight.

The high selectivity and good detection power of this method is largely based on the special HPLC-internal precolumn which performs the sample preparation. The precolumn is packed with silica gel modified with copper phthalocyanine trisulphonic acid derivative. Packing material of this type selectively absorbs from aqueous solutions those compounds which are composed of at least three condensed, aromatic rings with a planar structure. The absorption, which is the result of π,π-interactions, is reversed by the addition of organic solvents [45–49]. With careful treatment several hundred analyses can be carried out with this precolumn.

By integrating the sample preparation (which is normally performed externally) into the HPLC system the practicability and reliability are enhanced. Time-consuming, labour-intensive, error-prone manual operations are thus reduced to a minimum. These advantages become especially clear when this method is compared with gas chromatography in which several sample preparation steps and a derivatization are necessary [25, 50].

The wavelength for the fluorescence detection was selected so that all the analytes can be sensitively detected without changing the wavelength pairs during the analytical separation. When a fluorescence detector with a programmable wavelength is used, the selectivity and sensitivity can further be enhanced by choosing individual wavelengths. Moreover, a further increase in the power of detection is possible by injecting a larger volume of urine. By these means the PAH metabolites can be detected in concentrations due to environmental exposure.

Instruments used:
Double-channel pump with gradient former HPLC pump model 300 CS and gradient former model 250 B from Gynkotek
Single-channel pump model 5590 from Waters, Millipore
Automatic six-way switch valve model motor valve from H BESTA (Wilhelmsfeld)
Autosampler 655-40, Fluorescence detector F-1050 and integrator D-2000 from Merck-Hitachi.

11 References

[1] *International Agency for Research on Cancer*: Polynuclear Aromatic Compounds: Part 1, Chemical, Environmental and Experimental Data. Lyon, IARC 1983 (IARC Monograph Vol. 32).

[2] *F. Pott* and *U. Heinrich*: Polyzyklische aromatische Kohlenwasserstoffe (PAH). In: Krebsrisiko durch Luftverunreinigungen. Ministerium für Umwelt, Raumordnung und Landwirtschaft des Landes Nordrhein-Westfalen (eds.), Düsseldorf 1993.

[3] *R. Koch* (ed.): Umweltchemikalien, 2. Aufl., VCH-Verlagsgesellschaft, Weinheim 1991.

[4] *K. Hemminiki, E. Grzybowska, M. Chorazy, K. Twardowska-Saucha, J.W. Sroczynski, K.L. Putman, K. Randerath, D.H. Phillips* and *A. Hewer*: Aromatic DNA Adducts in White Blood Cells of Coke Workers. Int. Arch. Occup. Environ. Health *62*, 467–470 (1990).

[5] *J. Angerer, C. Mannschreck* and *J. Gündel*: Biological monitoring and biochemical effect monitoring of exposure to polycyclic aromatic hydrocarbons. Int. Arch. Occup. Environ. Health *70*, 365–377 (1997).

[6] *Deutsche Forschungsgemeinschaft*: MAK- und BAT-Werte-Liste 1998. Maximale Arbeitsplatzkonzentrationen und Biologische Arbeitsstofftoleranzwerte. Mitteilung 34 der Senatskommission zur Prüfung gesundheitsschädlicher Arbeitsstoffe. WILEY-VCH Verlag, Weinheim 1998.

[7] *J.G.M. van Rooij, M.M. Bodelier-Bade, A.J.A. De Looff, A.P.G. Dijkmans* and *F.J. Jongeneelen*: Dermal exposure to polycyclic aromatic hydrocarbons among primary aluminium workers. Med. Lav. *83*, 519–529 (1992).

[8] *J.G.M. van Rooij, J.H.C. de Roos, M.M. Bodelier-Bade* and *F.J. Jongeneelen*: Estimation of individual dermal and respiratory uptake of polycyclic aromatic hydrocarbons in 12 coke oven workers. Br. J. Ind. Med. *50*, 623–632 (1993).

[9] *J.G.M. van Rooij, J.H.C. de Roos, M.M. Bodelier-Bade* and *F.J. Jongeneelen*: Absorption of polycyclic aromatic hydrocarbons through human skin: differences between anatomical sites and individuals. J. Toxikol. Environ. Health *38*, 355–368 (1993).

[10] *H. Marquardt* and *S.G. Schäfer* (eds.): Lehrbuch der Toxikologie. B.I. Wissenschaftsverlag, Mannheim (1994).

[11] *F.J. Jongeneelen, R.B.M. Anzion, P.T.J. Scheepers, R.P. Bos, P.T. Henderson, E.H. Nijnhuis, E.H. Veenstra, R.M.E. Brouns* and *A. Winkes*: 1-Hydroxypyrene in urine as a biological indicator for exposure to polycyclic aromatic hydrocarbons in several work environments. Ann. Occup. Hyg. *32*, 35–43 (1988).

[12] *F.J. Jongeneelen, F.E. van Leeuwen, S. Oosterink, R.B.M. Anzion, F. van der Loop, R.P. Bos* and *H.G. van Veen*: Ambient and biological monitoring of cokeoven workers: Determinants of the internal dose of polycyclic aromatic hydrocarbons. Br. J. Ind. Med. *47*, 454–461 (1990).

[13] *J. Buchet, J.P. Gennart, F. Mercado-Calderon, J.P. Delavignette, L. Cupers* and *R. Lauwerys*: Evaluation of exposure to polycyclic aromatic hydrocarbons in a coke production and a graphite electrode manufacturing plant: assessment of urinary excretion of 1-hydroxypyrene as a biological indicator of exposure. Br. J. Ind. Med. *49*, 761–768 (1992).

[14] *P.J. Boogaard* and *M.J. van Sittert*: Exposure to polycyclic aromatic hydrocarbons in petro-chemical industries by measurement of urinary 1-hydroxypyrene. Occup. Environ. Med. *51*, 250–258 (1994).

[15] *T.J. Buckley* and *P.J. Lioy*: An examination of the time course from human dietary exposure to polycyclic aromatic hydrocarbons to urinary elimination of 1-hydroxypyrene. Br. J. Ind. Med. *49*, 113–124 (1992).

[16] *C. Viau, G. Carrier, A. Vyskocil* and *C. Dodd*: Urinary excretion of 1-hydroxypyrene in volun-teers exposed to pyrene by the oral and dermal route. Sci. Total Environ. *163*, 179–186 (1995).

[17] *C. Viau* and *A. Vyskocil*: Patterns of 1-hydroxypyrene excretion in volunteers exposed to pyrene by dermal route. Sci. Total Environ. *163*, 187–190 (1995).

[18] *R. Doll, M.P. Vessey, R.W.R. Beaseley, A.R. Buckley, E.C. Fear, R.E.W. Fisher, E.J. Gammon, W. Gunn, G.O. Hughes, K. Lee* and *G. Norman-Smith*: Mortality of gasworkers – final report of a prospective study. Br. J. Ind. Med. *29*, 394–406 (1972).

[19] *G.W. Gibbs* and *I. Horowitz*: Lung cancer mortality in aluminium reduction plant workers. J. Occup. Med. *21*, 347–353 (1979).

[20] *G.W. Gibbs* and *I. Horowitz*: Mortality of aluminium reduction plant workers, 1950 through 1977. J. Occup. Med. *27*, 761–770 (1985).

[21] *M. Kawai, H. Amamoto* and *K. Harada*: Epidemiological study of occupational lung cancer. Arch. Environ. Health *14*, 859 (1967).

[22] *Bekanntmachung einer Empfehlung des Ärztlichen Sachverständigenbeirates*, Sektion "Berufs-krankheiten". Die BG *4*, 238–245 (1998).

[23] *F.J. Jongeneelen* and *R.B.M. Anzion*: 1-Hydroxypyrene. In: *J. Angerer* and *K.H. Schaller* (eds.) Ana-lyses of hazardous substances in biological materials. Vol. 3, WILEY-VCH Verlag, Weinheim 1991.

[24] *F.J. Jongeneelen, R.B.M. Anzion* and *P.Th. Henderson*: Determination of hydroxylated metabo-lites of polycyclic aromatic hydrocarbons in urine. J. Chromatogr. *413*, 227–232 (1987).

[25] *G. Grimmer, G. Dettbarn* and *J. Jacob*: Biomonitoring of polycyclic aromatic hydrocarbons in highly exposed coke plant workers by measurement of urinary phenanthrene and pyrene meta-bolites (phenols and dihydrodiols). Int. Arch. Environ. Health *65*, 189–199 (1993).

[26] *S. Øvrebø, A. Haugen, P.B. Farmer* and *D. Anderson*: Evaluation of biomarkers in plasma, blood, and urine samples from coke oven workers: significance of exposure to polycyclic aro-matic hydrocarbons. Occup. Environ. Med. *52*, 750–756 (1995).

[27] *M. Ferreira jr., J.P. Buchet, J.B. Burrion, J. Moro, L. Cupers, J.P. Delavignette, J. Jacques* and *R. Lauwerys*: Determinants of urinary thioethers, D-glucaric acid and mutagenicity after expo-sure to polycyclic aromatic hydrocarbons assessed by air monitoring and measurement of 1-hy-droxypyrene in urine: a cross-sectional study in workers of coke and graphite-electrode-produc-ing plants. Int. Arch. Occup. Environ. Health *65*, 329–338 (1994).

[28] *G. Müller*: Personal communication 1996.

[29] *J. Angerer*: Unpublished results.

[30] *S. Tas, J.P. Buchet* and *R. Lauwerys*: Determinants of benzo[a]pyrene diol epoxide adducts to albumin in workers exposed to polycyclic aromatic hydrocarbons. Int. Arch. Occup. Environ. Health *66*, 343–348 (1994).

[31] *Th. Göen, J. Gündel, K.H. Schaller* and *J. Angerer*: The elimination of 1-hydroxypyrene in the urine of the general population and workers with different occupational exposures to PAH. Sci. Total Environ. *163*, 195–201 (1995).

[32] *J. Angerer, C. Mannschreck* and *J. Gündel*: Occupational exposure to polycyclic aromatic hy-drocarbons in a graphite electrode producing plant: biological monitoring of 1-hydroxypyrene and monohydroxylated metabolites of phenanthrene. Int. Arch. Occup. Environ. Medicine *69*, 323–331 (1997).

[33] *C. Mannschreck, J. Gündel* and *J. Angerer*: Belastung gegenüber PAH an verschiedenen Ar-beitsplätzen – Biomonitoring von monohydroxylierten Metaboliten. Vortrag auf der 36. Jahresta-gung der Deutschen Gesellschaft für Arbeitsmedizin und Umweltmedizin, Wiesbaden 1995.

[34] *Ø. Omland, D. Sherson, Å.M. Hansen, T. Sigsgaard, H. Autrup* and *E. Overgaard*: Exposure of iron foundry workers to polycyclic aromatic hydrocarbons: benzo(a)pyrene-albumin adducts and 1-hydroxypyrene as biomarkers for exposure. Occup. Environ. Med. *51*, 513–518 (1994).

[35] *J. Angerer, J. Gündel, C. Mannschreck, U. Ewers* and *K. Büttner*: Beurteilung der PAH-Bela-stung – Anwohner eines Industriegebietes in der Bundesrepublik Deutschland. Umweltmed. Forsch. Prax. *2*, 18–22 (1997).

[36] *J. Gündel, C. Mannschreck, K. Büttner, U. Ewers* and *J. Angerer*: Urinary levels of 1-hydroxy-pyrene, 1-, 2-, 3-, and 4-hydroxyphenanthrene in females living in an industrial area of Ger-many. Arch. Environ. Contam. Toxicol. *31*, 585–590 (1996).

[37] *Bundesärztekammer*: Qualitätssicherung der quantitativen Bestimmungen im Laboratorium. Neue Richtlinien der Bundesärztekammer. Dt. Ärztebl. *85*, A 699–A 712 (1988).

[38] *Bundesärztekammer*: Ergänzung der „Richtlinien der Bundesärztekammer zur Qualitätssiche-rung in medizinischen Laboratorien". Dt. Ärztebl. *91*, C 159–C 161 (1994).

[39] *J. Angerer* und *K.H. Schaller*: Erfahrungen mit der statistischen Qualitätskontrolle im arbeitsme-dizinisch-toxikologischen Laboratorium. Arbeitsmed. Sozialmed. Präventivmed. *1*, 33–35 (1977).

[40] *G. Lehnert, J. Angerer* und *K.H. Schaller*: Statusbericht über die externe Qualitätssicherung ar-beits- und umweltmedizinisch-toxikologischer Analysen in biologischen Materialien. Arbeits-med. Sozialmed. Umweltmed. *33(1)*, 21–26 (1998).

[41] *J. Angerer* and *G. Lehnert*: Anforderungen an arbeitsmedizinisch-toxikologische Analysen – Stand der Technik: Deutsches Ärztebl. *37*, C 1753–C 1760 (1997).

[42] *K. Norpoth* and *M. Heger*: Spezielle Vorbemerkung. A. Kreatinin als Bezugsgröße bei der An-gabe von Stoffkonzentrationen im Harn. In: *D. Greim* and *G. Lehnert* (eds.): Biologische Ar-beitsstoff-Toleranz-Werte (BAT-Werte) und Expositionsäquivalente für krebserzeugende Arbeits-stoffe (EKA). Arbeitsmedizinisch-toxikologische Begründungen. Deutsche Forschungsgemein-schaft, VCH-Verlagsgesellschaft, Weinheim, 2. Lieferung (1985).

[43] *S. D. Keimig, H. W. Kirby, D. P. Morgan, J. E. Keiser* and *T. D. Hubert*: Identification of 1-hy-droxypyrene as a major metabolite of pyrene in pig urine. Xenobiotica *13*, 415–420 (1983).

[44] *J. Lintelmann*: Entwicklung von Zwei-Säulen HPLC-Analysenverfahren zur Quantifizierung von planaren, oligozyklischen Verbindungen in Körperflüssigkeiten, Dissertation, Paderborn 1990.

[45] *H. Hayatsu*: Cellulose bearing covalently linked copper phthalocyanine trisulphonate as an ad-sorbent selective for polycyclic compounds and its use in studies of environmental mutagens and carcinogens. J. Chromatogr. *597*, 37–56 (1992).

[46] *H. Hayatsu, H. Kobayashi, A. Michiue* and *S. Arimoto*: Affinity of aromatic compounds having three fused rings to copper phthalocyanine trisulfonate. Chem. Pharm. Bull. *34 (2)*, 944–947 (1986).

[47] *M. Geisert, T. Rose* and *R.K. Zahn*: Extraction and trace enrichment of genotoxics from envi-ronmental samples by solid-phase adsorption on blue pearls. Fresenius J. Anal. Chem. *330*, 437–438 (1988).

[48] *K.-S. Boos, J. Lintelmann* and *A. Kettrup*: Coupled-column high-performance liquid chromato-graphic method for the determination of 1-hydroxypyrene in urine of subjects exposed to polycyclic aromatic hydrocarbons. J. Chromatogr. *600*, 189–194 (1992).

[49] *J. Lintelmann, C. Hellemann* and *A. Kettrup*: Coupled-column high-performance liquid chroma-tographic method for the determination of four metabolites of polycyclic aromatic hydrocar-bons, 1-, 4- and 9-hydroxyphenanthrene and 1-hydroxypyrene in urine. J. Chromatogr. B. *660*, 67–73 (1994).

[50] *G. Grimmer, G. Dettbarn* and *K.-W. Naujack*: Ausscheidung von Hydroxyderivaten polycycli-scher aromatischer Kohlenwasserstoffe im Harn von Kokerei- und Straßenbaubeschäftigten. In: *F. Schuckmann* and *S. Schopper-Jochum* (eds.): Berufskrankheiten, Krebserzeugende Arbeits-stoffe, Biological-Monitoring. Arbeitsmedizinisches Kolloquium der Gewerblichen Berufsgenos-senschaften, Gentner Verlag, Stuttgart 1991.

Author: *J. Lintelmann*
Examiner: *J. Angerer*

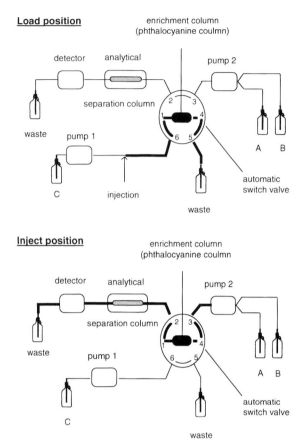

Fig. 2. Principle of the HPLC method.

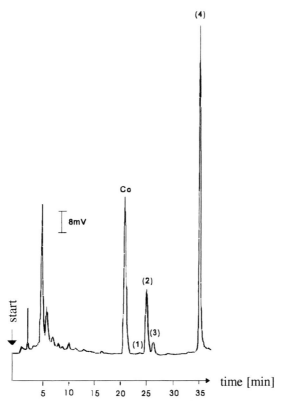

Fig. 3. HPLC chromatogram of a urine sample from a person exposed to PAH at the workplace.
Co: 2- and 3-Hydroxyphenanthrene; **(1)**: 9-Hydroxyphenanthrene (1.0 µg/L); **(2)**: 1-Hydroxyphenan-threne (16.5 µg/L); **(3)**: 4-Hydroxyphenanthrene (3.7 µg/L); **(4)**: 1-Hydroxypyrene (6.6 µg/L).

2-Thioxothiazolidine-4-carboxylic acid (TTCA)

Application   Determination in urine

Analytical principle   High pressure liquid chromatography (HPLC)

Completed in   August 1990

Summary

The procedure described here permits reliable analytical determination of 2-thioxothiazolidine-4-carboxylic acid (TTCA) excreted in the urine of persons exposed to carbon disulfide. For this reason and because of its practicability, this method is suitable for routine monitoring of the existing biological tolerance value for this working material.

After acidification of and addition of sodium chloride to the urine, TTCA is extracted with diethyl ether. The ether phase is evaporated and the residue is dissolved in methanol. TTCA is analysed by high pressure liquid chromatography.

Detection is achieved with a UV detector at 273 nm.

Calibration is carried out using standard urine solutions containing known TTCA concentrations. The standards are processed in the same way as the assay samples.

Within-series imprecision: Standard deviation (rel.) s_w = or 2.6% or 1.7%
                               Prognostic range u = 5.9% or 3.8%
                               At concentrations of 1.0 mg and 5.0 mg TTCA per litre urine and where $n = 10$ determinations

Inaccuracy: Recovery rate r = 91–95%

Detection limit: 0.2 mg TTCA per litre urine

Essential Biomonitoring Methods. DFG, Deutsche Forschungsgemeinschaft
Copyright © 2006 WILEY-VCH Verlag GmbH & Co. KGaA, Weinheim
ISBN: 3-527-31478-4

2-Thioxothiazolidine-4-carboxylic acid (TTCA)

$$H_2C - CH - COOH$$
$$S \diagdown \qquad \diagup NH$$
$$C$$
$$||$$
$$S$$

is a metabolite of carbon disulfide (CS_2). Carbon disulfide is mainly used in the artificial silk and wood pulp industries. It is still permissible (with restrictions) to use CS_2 against the vine louse (Phylloxera) in vineyards. Furthermore, it serves as a solvent for sulfur, iodine, phosphorus, rubber, guttapercha, wax, paraffin. It is used as a reagent for producing secondary amines, for the vulcanization of rubber, in the manufacture of vulcanizing agents, flotation agents and carbon tetrachloride as well as for the synthesis of sulfurous heterocylic compounds and other organic compounds containing sulfur.

The toxicology of carbon disulfide has been comprehensively described in the documentation of the MAK values [6]. Like all organic solvents it has a depressive effect on the central nervous system. Thus, typical symptoms, such as headache and fatigue, etc. could be observed as a result of the intake of CS_2 at the workplace. Characteristic for CS_2 is that depressive psychoses can occur after long-term occupational exposure. Polyneuritis is also observed. Moreover, intake of CS_2 over many years can alter the blood vessels, as in arteriosclerosis, which can be manifested in various clinical symptoms. Carbon disulfide inhibits a number of enzymes, especially hydrolases or oxidases, which explains certain effects including alcohol intolerance after CS_2 exposure. Chronic exposure causes functional disturbances in the liver, kidneys, pancreas and endocrine glands.

Berufsgenossenschaftlicher Grundsatz G6 (Principle G6 of the Professional Association) on the „Risks of exposure to carbon disulfide" [7] describes the differential diagnostic considerations and the occupational medicinal exclusion criteria of a carbon disulfide intoxication based on organic evidence and the symptoms of disease.

The metabolism and kinetics of carbon disulfide are presented in detail in the documentation of the BAT values (cf. Fig. 1) [1]. According to this publication 2/3 of the inhaled CS_2 is exhaled unchanged. About 30 % of the carbon disulfide is metabolized and reacts mainly with groups bearing lone electron pairs, such as aminosulfhydryl or hydroxy groups. The urine of exposed persons contains metabolic products which are formed when CS_2 reacts with glycine (2-mercapto-2-thiazoline-5-one) as well as cysteine (2-thioxothiazolidine-4-carboxylic acid, TTCA).

Van Doorn et al. [2] found the metabolite TTCA in the urine of people who were exposed to carbon disulfide. Less than 6 % of the CS_2 taken up by the body is metabolized to TTCA [3, 4].

Field studies have shown that TTCA excretion is dependent on exposure to carbon disulfide [1, 5]. Taking these field studies into account the BAT value was set at 8 mg TTCA per litre urine.

No TTCA is excreted by persons who are not occupationally exposed to carbon disulfide.

Authors: *A. Eben, F.P. Freudlsperger*
Examiner: *J. Angerer*

2-Thioxothiazolidine-4-carboxylic acid (TTCA)

Application   Determination in urine

Analytical principle   High pressure liquid chromatography (HPLC)

Completed in   August 1990

Contents

Essential Biomonitoring Methods. DFG, Deutsche Forschungsgemeinschaft
Copyright © 2006 WILEY-VCH Verlag GmbH & Co. KGaA, Weinheim
ISBN: 3-527-31478-4

1 General principles

After acidification and addition of sodium chloride to the urine, 2-thioxothiazolidine-4-carboxylic acid is extracted with diethyl ether. The ether phase is evaporated and the residue is dissolved in methanol. TTCA is analysed by high pressure liquid chromatography.

Detection is achieved with a UV detector at 273 nm.

Calibration is carried out using standard urine solutions containing known TTCA concentrations. The standards are processed in the same way as the assay samples.

2 Equipment, chemicals and solutions

2.1 Equipment

High pressure liquid chromatograph with a UV detector capable of measuring at 273 nm, preferably with autosampler and column thermostat

Integrator

Steel column: length: 25 cm; inner diameter: 4.6 mm

Column packing: Hypersil ODS 5 μm (supplied e.g. by Bischoff, Böblingerstr. 23, D-71229 Leonberg)

LiChrospher 5 μm (e.g. from Merck)

20 μL Syringes for HPLC

15 mL Test tubes (graduated) with ground glass stoppers

Vortex mixer (e.g. Vortex from Cenco, the Netherlands)

10, 20, 100 and 1000 mL Volumetric flasks

10 mL Flasks with tapered necks

Rotary evaporator

Centrifuge

Eppendorf or Pasteur pipettes to remove the organic phase

0.1, 0.2, 0.3, 0.5, 0.8, and 1 mL Pipettes

Universal indicator (e.g. from Merck)

2.2 Chemicals

2-Thioxothiazolidine-4-carboxylic acid (TTCA) (e.g. from Aldrich)

Sodium chloride, p.a.

Hydrochloric acid, 30%, p.a.

Diethyl ether p.a.

HPLC-grade methanol (e.g. from Baker or LiChrosolv from Merck)

Acetic acid 100%, p.a.

Acetonitrile (e.g. from Baker or LiChrosolv from Merck)

Ultrapure water (ASTM type 1) or double-distilled water

Urine from persons who have not been exposed to carbon disulfide

2.3 Solutions

Mobile phase A:
50% Acetonitrile/50% ultrapure water

Mobile phase B:
0.1 M acetic acid, pH 3.0
About 500 mL ultrapure water are placed in a 1000 mL volumetric flask, 6.0 mL glacial acetic acid are added and the flask is filled to the mark with ultrapure water. The
pH value is checked with universal indicator.

2.4 Calibration standards

Starting solution:
The calibration standards are prepared using pooled urine of persons who have not
been exposed to carbon disulfide. 20 mg TTCA are weighed exactly and placed in a
100 mL volumetric flask. After filling the flask to the mark with urine it is shaken to
ensure that the TTCA dissolves and is homogeneously dispersed (200 mg/L). The
solution can be stored in the refrigerator at 4 °C for about 2 months.

Calibration standards with TTCA concentrations up to 20 mg/L urine (cf. Table 1) are
prepared by diluting the starting solution with urine from unexposed persons.

Table 1: Pipetting scheme for the preparation of the calibration standards.

Volume of the starting solution mL	Final volume of calibration standard mL	Concentration of calibration standard mg/L
0	10	0 (blank value)
0.1	20	1.0
0.1	10	2.0
0.2	10	4.0
0.3	10	6.0
0.5	10	10.0
0.8	10	16.0
1.0	10	20.0

The calibration standards are freshly prepared every day and processed immediately.

3 Specimen collection and sample preparation

Spontaneous urine specimens are collected in plastic bottles at the end of a working shift. They are acidified with glacial acetic acid (1 mL glacial acetic acid to 100 mL urine). If the specimens are not processed immediately they should be deep frozen. They should not be stored for more than 4 weeks in this state.

5 mL Urine are pipetted into a 15 mL test tube with a ground glass stopper. After saturation with sodium chloride, 0.1 mL of 30 % hydrochloric acid are added. When the sample has been mixed for 1 min on the shaker, 3.5 mL diethyl ether are added and it is mixed again for 1 min. The mixture is subsequently centrifuged at 3000 g for 5 min. The ether phase is taken up as completely as possible using an automatic pipette or a Pasteur pipette and transferred to a 10 mL flask with a tapered neck. The solution is evaporated to dryness in a rotary evaporator (at 30 °C). The residue is dissolved in 0.5 mL methanol. This solution is used for the high pressure liquid chromatographic analysis.

4 Operational parameters for HPLC

Column:	Material:	Steel
	Length:	25 cm
	Inner diameter:	4.6 mm

Column packing: Hypersil ODS; 5 μm
Separation mode: Reversed phase
Detector: UV detector which is capable of measurement
                     at 273 nm
Column temperature: column thermostat 38 °C
Mobile phase: A: 50% acetonitrile/50% ultrapure water
                     B: 0.1 M acetic acid, pH 3
                     The eluents must be filtered
Flow rate: 1.0 mL/min
Sample volume: 20 μL

Gradient program:

Time (min)	A (%)	B (%)
0	2	98
8.0	70	30
18.0	70	30
24.0	2	98
37.0	2	98

Figure 2 shows the high pressure liquid chromatograms for a urine sample of a person exposed to carbon disulfide, a spiked urine sample as well as two calibration standards. The retention time for TTCA is approx. 5.3 min.

5 Analytical determination

The operational parameters are set and 20 μL of the methanolic extracts are each injected into the high pressure liquid chromatograph. If the analytical results do not lie within the range of the calibration curve the urine samples should be diluted and reprocessed.

6 Calibration

The calibration standards as well as the unspiked urine used for their preparation (cf. Section 2.4) are processed and analysed as described (cf. Section 2.3). A calibration curve is obtained by plotting the resulting peak areas as a function of the TTCA concentrations added. Under the working conditions described here the calibration curve is linear up to 20 mg/L.

The urine mixture used to prepare the calibration standards does not normally cause a peak in the chromatogram at the retention time of TTCA (cf. Fig. 3). If, however, a peak occurs another pooled urine sample must be used for calibration.

It is not necessary to plot a calibration curve for every analytical series. The validity of the calibration curve can be checked by analysing an aqueous standard solution of known TTCA concentration. This solution functions as an external standard.

7 Calculation of the analytical result

When the peak areas are obtained for the assay samples their TTCA concentration in mg per litre can be read directly from the calibration curve.

8 Standardization and quality control

As described in Section 6, an aqueous solution of known TTCA concentration, e.g. 10 mg/L, is included in each analytical series. The concentrations obtained from the calibration curve can be arithmetically related to this external standard. Control urine containing a constant TTCA concentration is included in each series to test the imprecision for the purpose of internal quality control. This control material is prepared in the laboratory, divided into aliquots and stored in a deep freezer. It is advisable to adjust the TTCA concentration to correspond to the BAT value (8 mg/L; 1993).

9 Reliability of the method

9.1 Precision

To determine the within-series imprecision, pooled urine was divided into two aliquots and spiked with different amounts of TTCA. The urine samples, which contained 1.0 or 5.0 mg/L TTCA per litre urine, were each analysed ten times. The resulting relative standard deviations were 2.6 or 1.7% which correspond to prognostic ranges of 5.9 or 3.8% respectively.

9.2 Accuracy

The accuracy of the method was determined with recovery experiments and checked by means of intercomparison programmes. To find the recovery rate the urine samples were processed and analysed according to the instructions given in Section 9.1. The results were evaluated using calibration standards added to a different pool of urine than the spiked urine samples used in the recovery experiments. The recovery rate was 95% for a concentration of 1 mg/L and 91% for a concentration of 5 mg/L. In an intercomparison programm 10 urine samples from exposed persons were analysed in two different laboratories. The results are presented in Table 2.

Table 2: Comparative investigations of urine samples from exposed persons.

Sample	TTCA concentration Laboratory 1 mg/L	Laboratory 2 mg/L
1	2.90	3.2
2	4.33	7.9
3	3.50	4.2
4	0.98	2.0
5	3.11	3.8
6	3.95	3.3
7	2.54	3.8
8	1.82	3.3
9	< DL	< DL
10	8.47	11.3
Mean value (n = 9)	3.51	4.76
Standard deviation (n = 9)	2.13	2.94

DL = detection limit

The values obtained by both laboratories using this method show significant linear correlation (y = 1.27 x + 0.28, r = 0.932).

To determine the losses due to sample preparation TTCA solutions prepared in methanol are analysed without further processing. Comparison of the results with those obtained for the processed standard solutions in urine show that only 50 % of the added TTCA is detected after a single extraction. When extraction is carried out three times 85 % of the added TTCA is detected. However, a single extraction is preferable for practical reasons. This is justifiable because the calibration standards are processed and analysed in the same manner as the investigated urine samples.

9.3 Detection limit

Under the analytical conditions described here the detection limit is 0.2 mg TTCA per litre urine.

9.4 Sources of error

As shown in the chromatogram in Figure 3, peaks do not generally occur at the characteristic retention time for TTCA when extracts of urine samples from unexposed persons are analysed.

Using acetic acid as an eluent in HPLC can present problems. Long-term use can cause corrosion at the pumps. Therefore, capillaries and pumps must be thoroughly rinsed with ultrapure water after the analysis has been completed. Exhaustive in-vestigations have shown that it is essential to use acetic acid if good separation is to be achieved.

Contamination caused by reagents or glass vessels has not been observed.

10 Discussion of the method

The procedure described here is essentially based on a method developed by *Van Doorn* [2, 8]. At present no other methods for the determination of TTCA are avail-able.

The method is diagnostically specific. No TTCA is found in the urine of persons who have not been exposed to carbon disulfide. Under the conditions for HPLC recom-mended here TTCA is separated from all the other urinary components, i.e. deter-mination of this metabolite is also analytically specific. Separation is achieved by the eluent program selected in this case; the TTCA retention time is approximately

5 minutes, while most of the components in urine are eluted in 10 to 25 minutes. The sensitivity of this method as well as its detection limit permit the surveillance of the BAT value for TTCA which is 8 mg/L at present (1993). Furthermore, the method is so practicable that it is suitable for routine biological monitoring investigations. The within-series precision is excellent (1.7–2.6%). Unfortunately no suitable internal standard has been found which can guarantee precision over a longer period. Therefore, it is advisable to include an aqueous control solution of known TTCA concentration in every analytical series instead of an internal standard. Then the results can be appropriately corrected if necessary. The accuracy of the method was checked by means of intercomparison programmes.

Instruments used:
High pressure liquid chromatograph 1090 with diodenarray detector and DPU from Hewlett-Packard
Steel column Hypersil ODS 5 µm
Steel column Nucleosil 120 120–5 C18 from Macherey and Nagel
Precolumn Lichrosorb RP-C18, 5µm from Merck

11 References

[1] *H.M. Bolt*: Kohlendisulfid. In: *D. Henschler and G. Lehnert* (eds.): Biologische Arbeitsstoff-Toleranz-Werte (BAT-Werte) und Expositionsäquivalente für krebserzeugende Arbeitsstoffe (EKA). Arbeitsmedizinisch-toxikologische Begründungen. Deutsche Forschungsgemeinschaft, VCH Verlagsgesellschaft, Weinheim 1989, 4th issue.

[2] *R. Van Doorn, Ch.P.J.M. Leijdeekkers, P.Th. Henderson, M. Vanhoorne and P.G. Verting*: Determination of thio compounds in urine of workers exposed to carbon disulfide. Arch. Environ. Health *36*, 289–297 (1981).

[3] *L. Campbell, A.H. Jones and H.K. Wilson*: Evaluation of occupational exposure to carbon disulfide by blood, exhaled air, and urine analysis. Am. J. Ind. Med. *8*, 143–153 (1985).

[4] *J. Rosier, R. Van Doorn, R. Grosjean, A. van de Walle, G. Billemont and C. van Petegham*: Preliminary evaluation of urinary 2-thio-thiazolidine-4-carboxylic acid levels as a test for exposure to carbon disulfide. Int. Arch. Occup. Environ. Health *51*, 159–167 (1982).

[5] *F.P. Freudlsperger and W.-P. Madaus*: Erfahrungen mit dem BAT-Wert für Schwefelkohlenstoff. Arbeitsmed. Sozialmed. Präventivmed. *24*, 71–74 (1989).

[6] *D. Henschler* (ed.): Gesundheitsschädliche Arbeitsstoffe. Toxikologisch-arbeitsmedizinische Begründung von MAK-Werten: Schwefelkohlenstoff. Deutsche Forschungsgemeinschaft, Verlag Chemie, Weinheim 1975, 4th issue.

[7] *Hauptverband der gewerblichen Berufsgenossenschaften* (ed.): G6: Gefährdung durch Schwefelkohlenstoff. Fassung 5, 1981. In: Berufsgenossenschaftliche Grundsätze für arbeitsmedizinische Vorsorgeuntersuchungen. Gentner Verlag, Stuttgart 1981.

[8] *R. Van Doorn, L.P.C. Delbressine, C.M. Leijdeekkers, P.G. Vertin and P.Th. Henderson*: Identification and determination of 2-Thiazolidine-4-carboxylic acid. Arch. Toxikol. *47*, 51–58 (1981).

Authors: *A. Eben, F.P. Freudlsperger*
Examiner: *J. Angerer*

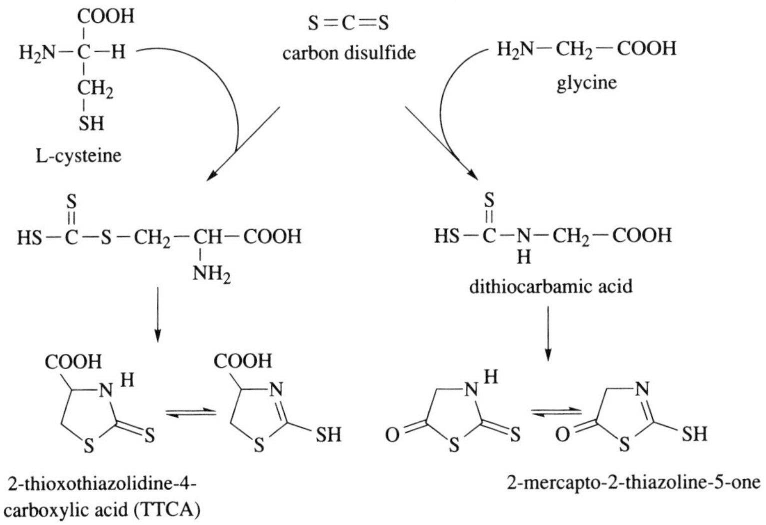

Fig. 1: Metabolism of carbon disulfide.

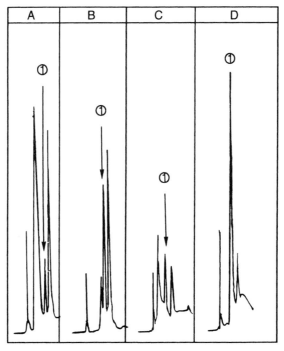

Fig. 2: High pressure liquid chromatograms of:
A Urine extract of a person exposed to CS_2. TTCA 6.1 mg/L urine
B Control urine extract – spiked with 3 mg TTCA
C Calibration standard. 3.4 mg/L urine
D Calibration standard. 14.0 mg/L urine

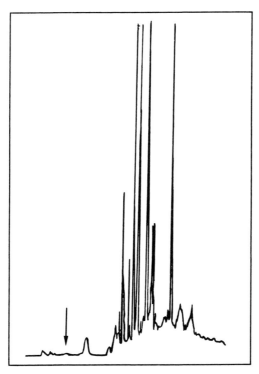

Fig. 3: High pressure liquid chromatograms of urine extract of a person who was not exposed to CS$_2$.

Trichloroacetic Acid (TCA)

Application	Determination in urine
Analytical principle	Photometry
Completed in	1976
Revised in	May 1983

Summary

Trichloroacetic acid is excreted with the urine as the metabolic product of several chlorinated hydrocarbons. Hence, the method described here is diagnostically unspecific. However, due to the ease with which it can be carried out and its good analytical reliability, this method is still quite useful for screening purposes.

The TCA is reacted with pyridine in an alkaline medium at 65 °C to give a red reaction product whose extinction is measured photometrically at 530 nm and in a cuvette with a path length of 10 mm. The calibration curve is linear for concentrations between 3.0 and 55.0 mg/L.

Sensitivity:	Reciprocal calibration factor $k' = 83.4$ mg/L for the given experimental conditions
Within-series imprecision:	Standard deviation (rel.) $s = 1.0\%$
	Prognostic range $u = 2.0\%$
	At a concentration of 30.4 mg/L TCA in urine and where $n = 25$ determinations
Between-day imprecision:	Standard deviation (rel.) $s = 4.1\%$
	Prognostic range $u = 8.6\%$
	At a concentration of 25.0 mg/L TCA in urine and where $n = 20$ days
Inaccuracy:	Recovery rate $r = 92\text{–}96\%$
Detection limit:	3.0 mg/L TCA in urine

Essential Biomonitoring Methods. DFG, Deutsche Forschungsgemeinschaft
Copyright © 2006 WILEY-VCH Verlag GmbH & Co. KGaA, Weinheim
ISBN: 3-527-31478-4

Trichloroacetic acid (TCA)

$$\text{Cl}-\overset{\displaystyle \text{Cl}}{\underset{\displaystyle \text{Cl}}{\text{C}}}-\text{C}\overset{\displaystyle \text{O}}{\underset{\displaystyle \text{OH}}{}}$$

At room temperature trichloroacetic acid is a white, very hygroscopic and very caustic crystalline substance (mp 57.5 °C).

In humans TCA is formed as a metabolic product of chlorinated hydrocarbons such as 1,1,2-trichloroethylene (TRI); 1,1,1-trichloroethane; 1,1,1,2-tetrachloroethane; pentachloroethane; and tetrachloroethylene.

Since most of the TCA in the body is bound to plasma proteins [13, 14], its excretion via the kidneys is very slow, having a half-life of about 100 h [15]. TCA thus tends to accumulate in the body [16–18].

Author: *A. Eben*
Examiners: *D. Henschler, W. Pilz*

Trichloroacetic Acid (TCA)

Application	Determination in urine
Analytical principle	Photometry
Completed in	1976
Revised in	May 1983

Contents

1 General principles

The trichloroacetic acid excreted in the urine is reacted with pyridine in an alkaline medium, according to the method of *Tanaka* and *Ikeda* [3]. The extinction of the red reaction product is determined photometrically at 530 nm.

Aqueous solutions with known concentrations of trichloroacetic acid are used to set up a calibration curve for quantitative analysis of the urine samples.

Essential Biomonitoring Methods. DFG, Deutsche Forschungsgemeinschaft
Copyright © 2006 WILEY-VCH Verlag GmbH & Co. KGaA, Weinheim
ISBN: 3-527-31478-4

2 Equipment, chemicals and solutions

2.1 Equipment

UV-vis photometer with filter Hg 546 nm or spectrophotometer
Waterbath
10 mm Cuvettes (no plastic material!)
20 mL Reagent flasks (graduated) with ground-glass rim (14.5) and fitted glass tube (inner
diameter, 3 mm; length, 100 mm)
0.5, 1, 2, 4 and 5 mL Volumetric pipettes
50 and 100 mL Volumetric flasks
10 mL Graduated test tubes with ground-glass stoppers
Desiccator
Vacuum pump

2.2 Chemicals

Trichloroacetic acid (TCA) p.a.
Phosphorus pentoxide for drying (e.g., Sikapent from Merck)
Pyridine p.a.
Toluene p.a.
Potassium hydroxide p.a.
The commercial trichloroacetic acid is dried under vacuum (10^{-2} mbar) over phosphorus
pentoxide until its weight is constant.
Double-distilled water

2.3 Solutions

7.8 M Potassium hydroxide solution:
43.76 g Potassium hydroxide is made up to 100 mL with double-distilled water.

2.4 Calibration standards

Stock solution:
About 50 mg trichloroacetic acid (very hygroscopic!) is weighed out as quickly and accurately as possible and made up to 100 mL with double-distilled water (0.5 g/L).
Calibration standards, ranging in TCA concentration from 5–50 mg/L, are made by diluting
the stock solution with double-distilled water according to the following pipetting schedule:

| Stock solution | Calibration standards | |
| Volume | Final volume | Concentration |
mL	mL	mg/L
0.5	50	5
1.0	50	10
2.0	50	20
4.0	50	40
5.0	50	50

3 Specimen collection and sample treatment

A 1 mL urine sample is pipetted into a 20 mL reagent flask to which 2.5 mL 7.8 M potassium hydroxide solution, 5.0 mL pyridine and 0.5 mL toluene are added consecutively. After the contents of the flask have been thoroughly mixed, the glass tube is inserted into the flask and the flask is placed in a waterbath for 50 min at 65 °C. The sample solution is then cooled to room temperature and 3.0 mL of the pyridine layer is pipetted into 10 mL test tubes and mixed well with 0.6 mL water. A reagent blank is prepared in the same way, using double-distilled water instead of urine.

4 Analytical determination

Within 20 min after the pyridine layer and water have been mixed thoroughly, the extinction of the sample is measured against the reagent blank at 530 nm and in a cuvette with a path length of 10 mm.

5 Calibration

The calibration standards (compare Sect. 2.4) are subjected to the same treatment as the urine samples and determined photometrically (compare Sects. 3 and 4) to set up a calibration curve. The measured extinction values are plotted as a function of TCA concentration (Fig. 2). In the range of 3–55 mg/L the calibration curve is linear. The reciprocal calibration factor k' for a cuvette with a path length of 10 mm is:

$$k' = \frac{\varrho}{\Delta E_{530\,nm/10\,mm}} = 83.4 \text{ mg/L}$$

where

ϱ = mass concentration of TCA in the urine in mg/L

$E_{530\ nm/10\ mm}$ = extinction measured at 530 nm and in a cuvette with a path length of 10 mm

6 Calculation of the analytical result

The TCA concentration in the urine in mg/L that corresponds to the measured extinction value of the urine sample is read off the calibration curve.

7 Reliability of the method

7.1 Precision

To determine the imprecision within a series, 25 1 mL samples of urine collected over 24 h from a person who had been exposed to trichloroethene were analyzed. The relative standard deviation (s) was 1% and the prognostic range (u) was 2% for a TCA concentration of 30.4 mg/L.

Between-day imprecision was determined using commercially available synthetic urine with an average TCA concentration of 25.0 mg/L. Analyses over 20 days had a relative standard deviation (s) of 4.1% and a prognostic range (u) of 8.6%.

7.2 Accuracy

To determine the recovery rate, various concentrations of pure TCA were added to urine. As shown in Table 1, 92–96% of the TCA was recovered.

7.3 Detection limit

Under the given analytical conditions the detection limit was 3.0 mg/L TCA in urine.

7.4 Sources of error

There was no interference from trichloroethanol under the given conditions. Normally human urine contains no TCA or only very low concentrations (< 1 mg/L urine) [1, 2].

8 Discussion of the method

The method of *Tanaka* and co-workers [3] is based on the *Fujiwara* reaction [11], which was first used for the determination of chloroform. The mechanism of this reaction is still not completely understood. According to *Moss* and *Rylance* [4] glutaconic aldehyde is one of the end products:

glutaconic aldehyde

Further details on the determination of TCA according to the pyridine-alkali method [5–7] can be found in the literature, as well as descriptions of a few gas chromatographic methods [8–10, 12].

Figure 1 shows the absorption spectrum between 430 and 610 nm of the reaction product of TCA and pyridine in alkaline solution under the conditions described here. The maximum is at 530 nm. The calibration curve for TCA at 530 nm and in a cuvette with a path length of 10 mm is linear for the concentration range 3.0–55.0 mg/L (Fig. 2). Thus, the conditions for the Lambert-Beer law are fulfilled in this concentration range and the analytical result can be determined directly from the measured extinction values using the reciprocal calibration factor k'.

9 References

[1] *P. Hassman* and *V. Hassmanova:* Werte von Trichloressigsäure und Trichloräthanol im Harn bei den unter einem Tetrachlormethan- und Methylchloridrisiko arbeitenden Personen. Int. Arch. Gewerbepath. Gewerbehyg. *25,* 299–306 (1969).

[2] *M. Ikeda* and *H. Ohtsuji:* Hippuric acid, phenol, and trichloroacetic acid levels in the urine of Japanese subjects with no known exposure to organic solvents. Brit. J. Industr. Med. *26,* 162–164 (1969).

[3] *S. Tanaka* and *M. Ikeda:* Determination of trichloroethanol and trichloroacetic acid in the urine. Brit. J. Industr. Med. *25,* 214–219 (1968).

[4] *M. S. Moss* and *H. J. Rylance:* The Fujiwara reaction: some observations on the mechanism. Nature *210,* 945–946 (1966).

[5] *G. A. Hunold:* Die Fujiwara-Reaktion und ihre Anwendung zur Bestimmung von Trichloräthylen in der Raumluft und von Trichloressigsäure im Harn. Arch. Gewerbepath. Gewerbehyg. *14,* 77–90 (1955).

[6] *B. Souček* and *E. Franková:* Bestimmung geringer Mengen von Trichloräthylen und Trichloressig-säure. Pracov. Lék. *4,* 264 (1952).

[7] *M. Ogata, Y. Takastuka,* and *K. Tomokuni:* A simple method for the quantitative analysis of urinary trichloroethanol and trichloroacetic acid as an index of trichloroethylene exposure. Brit. J. Industr. Med. *27,* 378–381 (1970).

[8] *E. R. Garret* and *H. J. Lambert:* Gas chromatographic analysis of trichloroethanol, chloral hydrate, trichloroacetic acid, and trichloroethanol glucuronide. J. Pharm. Sci. *55*, 812–817 (1966).

[9] *M. Ogata* and *T. Saeki:* Measurement of chloral hydrate, trichloroethanol, trichloroacetic acid and monochloroacetic acid in the serum and the urine by gas chromatography. Int. Arch. Arbeitsmed. *33*, 49–58 (1974)

[10] *H. Ehrner-Samuel, K. Balmer,* and *W. Thorsell:* Determination of trichloroacetic acid in the urine by a gas chromatographic method. Am. Ind. Hyg. Ass. J. *34*, 93–96 (1973).

[11] *K. Fujiwara:* Über eine neue sehr empfindliche Reaktion zum Chloroformnachweis. Sitzber. Abhandl. Naturforsch. Ges. Rostock *6*, 33–43 (1914).

[12] *H. C. J. Ketelaars* and *J. M. Rossum:* Gas chromatographic determination of chloral hydrate, trichloroethanol and trichloroacetic acid in blood and urine employing head-space analysis. J. Chromatogr. *88*, 55–63 (1974).

[13] *O. Vesterberg, J. Gorczak,* and *M. Krasts:* Exposure to trichloroethylene. II. Metabolites in blood and urine. Scand. Work Environ. Health *4*, 212–219 (1976).

[14] *G. Müller, M. Spassowski,* and *D. Henschler:* Metabolism of trichloroethylene in man. III. Inter-action of trichloroethylene and ethanol. Arch. Toxicol. *33*, 173–189 (1975).

[15] *G. Müller, M. Spassowski,* and *D. Henschler:* Metabolism of trichloroethylene in man. II. Pharmacokinetics of metabolites. Arch. Toxicol. *32*, 283–295 (1974).

[16] *G. Kimmerle* and *A. Eben:* Metabolism, excretion and toxicology of trichloroethylene after inhalation. 2. Experimental human exposure. Arch. Toxicol. *30*, 127–138 (1973).

[17] *T. Ertl, D. Henschler, G. Müller,* and *M. Spassowski:* Metabolism of trichloroethylene in man. I. The significance of trichloroethanol in long-terme exposure conditions. Arch. Toxicol. *29*, 171–188 (1972).

[18] *D. Henschler* (ed.): Gesundheitsschädliche Arbeitsstoffe, toxikologisch-arbeitsmedizinische Begründung von MAK-Werten. Deutsche Forschungsgemeinschaft, Verlag Chemie, Weinheim, 8th supplement 1981.

Author: *A. Eben*
Examiners: *D. Henschler, W. Pilz*

Table 1. Inaccuracy of the determination of TCA in urine.

Calculated concn. mg/L	Measured concn. mg/L	Recovery rate %
3.0	2.8	93
5.4	5.2	96
14.2	13.0	92
19.4	17.8	92
27.0	25.8	96
30.4	28.5	94
48.4	45.0	93
58.1	54.0	93

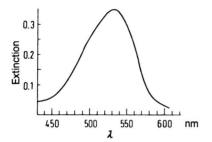

Fig. 1. Absorption spectrum of the reaction product of TCA and pyridine in alkaline solution (30.4 mg/L, 10 mm cuvette).

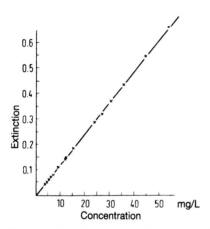

Fig. 2. Calibration curve for the determination of TCA in urine at 530 nm and a 10 mm cuvette.

Contents of Volumes 1–10

Essential Biomonitoring Methods. DFG, Deutsche Forschungsgemeinschaft
Copyright © 2006 WILEY-VCH Verlag GmbH & Co. KGaA, Weinheim
ISBN: 3-527-31478-4

The MAK-Collection for Occupational Health and Safety

MAK values (Maximum Concentrations at the Workplace) and **BAT values (Biological Tolerance Values)** promote the protection of health at the workplace. They are an efficient indicator for the toxic potential of chemical compounds.

The **MAK**-Collection

List of MAK and BAT Values

Maximum Concentrations and Biological Tolerance Values at the Workplace
German DFG-Senate-Commission, Editor

New feature: the complete list of MAK and BAT values now on a CD–ROM included in the book!

Now you can reference the largest stock of carefully revised toxicological data currently available anywhere. This book contains a list of scientifically recommended threshold limit values for about 900 chemical compounds.

MAK Values set the standards for legal regulations in many countries of the world, e.g. they are the basis for at least 30% of the threshold limits valid in the European Union.

Report 41 Paperback 260 pages August 2005
ISBN 3527-31357-5 approx € 75.00 /£ 55.00 /US$ 95.00

Part I: MAK Value Documentations

HELMUT GREIM, Editor

The volumes of this series present about 400 indispensable toxicological evaluation documents on important occupational toxicants and carcinogens. They describe the toxicological database which determines the level of a **MAK value.**

Vol. 21 Hardcover 333 pages July 2005
ISBN 3527-31134-3
Series price € 99.00 /£ 70.00 /US$ 135.00

■ Please contact the Customer Service for single volume prices

John Wiley & Sons, Ltd. • Customer Services Department
1 Oldlands Way • Bognor Regis • West Sussex
PO22 9SA England • Tel.: +44 (0) 1243-843-294
Fax: +44 (0) 1243-843-296 • www.wileyeurope.com

 WILEY

Part II: BAT Value Documentations

HANS DREXLER, Editor

You will find detailed information on dozens of pharmacokinetics, critical toxicity, exposures and effects, selections of the indicators, methodologies, background exposures, interpretation of the data, and manifesto – the toxicological database that determins the **BAT value** of threshold limits for hazardous occupational toxicants in body fluids.

Vol. 4 Hardcover approx 248 pages August 2005
ISBN 3527-27049-3
Series price approx € 109.00 /£ 80.00 /US$ 140.00

Part III: Air Monitoring Methods

Harun Palar, Editor

Vol. 9 Hardcover 216 pages July 2005
ISBN 3527-31138-6 Series price € 109.00 /£ 55.00 /US$ 90.00

Part IV: Biomonitoring Methods

Jürgen Angerer et al., Editors

Vol. 10 Hardcover approx. 354 pages January 2006
ISBN 3527-31137-8

Vol. 9 Hardcover 359 pages February 2004
ISBN 3527-27799-4 Series price € 109.00 /£ 55.00 /US$ 90.00

Detailed procedures of the determination of occupational toxicants in body fluids and in air are provided in each volume of these two series.

Throughout, considerable emphasis is placed on sample collection methods and on analytical quality control: every method is checked by at least one examine.

MAK Online Database –
convenient, comprehensive, detailed
Occupational Toxicants and MAK Values
www.mrw.interscience.wiley.com/makbat

 www.dfg.wiley-vch.de

Wiley-VCH • Customer Service Department
P.O. Box 101161 • D-69451 Weinheim • Germany
Tel.: (49) 6201 606-400 • Fax: (49) 6201 606-184
e-Mail: service@wiley-vch.de • www.wiley-vch.de

 **WILEY-VCH**